KW-483-122

LIVERPOOL JMU LIBRARY

3 1111 01524 5440

ORGANIC SYNTHESES

ADVISORY BOARD

Kay Brummond	David J. Hart	Leo A. Paquette
Elias J. Corey	Louis S. Hegedus	Andreas Pfaltz
Dennis P. Curran	Andrew B. Holmes	John A. Ragan
Huw M. L. Davies	David L. Hughes	Viresh H. Rawal
Scott E. Denmark	Sarah E. Kelly	Dieter Seebach
Jonathan Ellman	Mark Lautens	Martin F. Semmelhack
Albert Eschenmoser	Andrew S. Kende	Masakatsu Shibasaki
Margaret M. Faul	Dawei Ma	Bruce E. Smart
Ian Fleming	Steven F. Martin	James D. White
Tohru Fukuyama	David Mathre	Kenneth B. Wiberg
Alois Fürstner	Marvin J. Miller	Steven Wolff
Leon Ghosez	Koichi Narasaka	Hasashi Yamamoto
Edward J. J. Grabowski	Ryoji Noyori	

FORMER MEMBERS OF THE BOARD NOW DECEASED

Roger Adams	Nathan L. Drake	Wataru Nagata
Homer Adkins	William D. Emmons	Melvin S. Newman
C. F. H. Allen	L. F. Fieser	C. R. Noller
Richard T. Arnold	Jeremiah P. Freeman	W. E. Parham
Werner E. Bachmann	R. C. Fuson	Charles C. Price
Henry E. Baumgarten	Henry Gilman	Norman Rabjohn
Richard E. Benson	Cliff S. Hamilton	John D. Roberts
A. H. Blatt	W. W. Hartman	Gabriel Saucy
Virgil Boekelheide	E. C. Horning	R. S. Schreiber
Ronald Breslow	Herbert O. House	John C. Sheehan
Arnold R. Brossi	Robert E. Ireland	William A. Sheppard
George H. Büchi	John R. Johnson	Ichiro Shinkai
T. L. Cairns	William S. Johnson	Ralph L. Shriner
Wallace H. Carothers	Oliver Kamm	Lee Irvin Smith
James Cason	Andrew S. Kende	H. R. Snyder
Orville L. Chapman	Kenji Koga	Robert V. Stevens
H. T. Clarke	Nelson J. Leonard	Max Tishler
David E. Coffen	C. S. Marvel	Edwin Vedejs
J. B. Conant	Satoru Masamune	Frank C. Whitmore
Arthur C. Cope	B. C. McKusick	Ekkehard Winterfeltd
William G. Dauben	Albert I. Meyers	Peter Yates

ORGANIC SYNTHESES

AN ANNUAL PUBLICATION OF SATISFACTORY
METHODS FOR THE PREPARATION OF
ORGANIC CHEMICALS
VOLUME 95
2018

BRIAN M. STOLTZ
VOLUME EDITOR

BOARD OF EDITORS

RICK L. DANHEISER, *Editor-in-Chief*

KEVIN R. CAMPOS   SARAH E. REISMAN
ERICK M. CARREIRA   RICHMOND SARPONG
KUILING DING   CHRIS H. SENANAYAKE
NEIL K. GARG   BRIAN M. STOLTZ
MOHAMMAD MOVASSAGHI   KEISUKE SUZUKI
CRISTINA NEVADO   JOHN L. WOOD

CHARLES K. ZERCHER, *Associate Editor*
DEPARTMENT OF CHEMISTRY
UNIVERSITY OF NEW HAMPSHIRE
DURHAM, NEW HAMPSHIRE 03824

The procedures in this article are intended for use only by persons with prior training in experimental organic chemistry. These procedures must be conducted at one's own risk. *Organic Syntheses, Inc.*, its Editors, and its Board of Directors do not warrant or guarantee the safety of individuals using these procedures and hereby disclaim any liability for any injuries or damages claimed to have resulted from or related in any way to the procedures herein.

Copyright © 2018 by Organic Syntheses, Inc. All rights reserved.

Published by John Wiley & Sons, Inc., Hoboken, New Jersey.

Published simultaneously in Canada.

No part of this publication may be reproduced, stored in a retrieval system or transmitted in any form or by any means, electronic, mechanical, photocopying, recording, scanning, or otherwise, except as permitted under Section 107 or 108 of the 1976 United States Copyright Act, without either the prior written permission of the Publisher, or authorization through payment of the appropriate per-copy fee to the Copyright Clearance Center, Inc., 222 Rosewood Drive, Danvers, MA 01923, (978) 750-8400, fax (978) 750-4470, or on the web at www.copyright.com. Requests to the Publisher for permission should be addressed to the Permissions Department, John Wiley & Sons, Inc., 111 River Street, Hoboken, NJ 07030, (201) 748-6011, fax (201) 748-6008, or online at http://www.wiley.com/go/permission.

Limit of Liability/Disclaimer of Warranty: While the publisher and author have used their best efforts in preparing this book, they make no representations or warranties with respect to the accuracy or completeness of the contents of this book and specifically disclaim any implied warranties of merchantability or fitness for a particular purpose. No warranty may be created or extended by sales representatives or written sales materials. The advice and strategies contained herein may not be suitable for your situation. You should consult with a professional where appropriate. Neither the publisher nor author shall be liable for any loss of profit or any other commercial damages, including but not limited to special, incidental, consequential, or other damages.

For general information on our other products and services or for technical support, please contact our Customer Care Department within the United States at (800) 762-2974, outside the United States at (317) 572-3993 or fax (317) 572-4002.

Wiley also publishes its books in a variety of electronic formats. Some content that appears in print may not be available in electronic formats. For more information about Wiley products, visit our web site at www.wiley.com.

"John Wiley & Sons, Inc. is pleased to publish this volume of Organic Syntheses on behalf of Organic Syntheses, Inc. Although Organic Syntheses, Inc. has assured us that each preparation contained in this volume has been checked in an independent laboratory and that any hazards that were uncovered are clearly set forth in the write-up of each preparation, John Wiley & Sons, Inc. does not warrant the preparations against any safety hazards and assumes no liability with respect to the use of the preparations."

Library of Congress Catalog Card Number: 21-17747

ISBN: 978-1-119-59539-7

Printed in the United States of America

V10010954_061219

ORGANIC SYNTHESES

*Out of print.
†Deceased.

Out of print.
†*Deceased.*

Out of print.
†*Deceased.*

NOTICE

Beginning with Volume 84, the Editors of *Organic Syntheses* initiated a new publication protocol, which is intended to shorten the time between submission of a procedure and its appearance as a publication. Immediately upon completion of the successful checking process, procedures are assigned volume and page numbers and are then posted on the Organic Syntheses website (www.orgsyn.org). The accumulated procedures from a single volume are assembled once a year and submitted for publication. The annual volume is published by John Wiley and Sons, Inc., and includes an index. The hard cover edition is available for purchase through the publisher. Incorporation of graphical abstracts into the Table of Contents began with Volume 77. Annual volumes 70–74, 75–, 80–84 and 85–89 have been incorporated into five-year versions of the collective volumes of *Organic Syntheses*. Collective Volumes IX, X, XI and XII are available for purchase in the traditional hard cover format from the publishers.

Beginning with Volume 88, a new type of article, referred to as Discussion Addenda, appeared. In these articles submitters are provided the opportunity to include updated discussion sections in which new understanding, further development, and additional application of the original method are described.

Organic Syntheses, Inc., joined the age of electronic publication in 2001 with the release of its free web site (www.orgsyn.org). The site is accessible through internet browsers using Macintosh and Windows operating systems, and the database can be searched by key words and sub-structure. John Wiley & Sons, Inc., and Accelrys, Inc., partnered with Organic Syntheses, Inc., to develop a database (onlinelibrary.wiley.com/doi/book/10.1002/0471264229) that is available for license with internet solutions from John Wiley & Sons, Inc. and intranet solutions from Accelrys, Inc.

Both the commercial database and the free website contain all annual and collective volumes and indices of *Organic Syntheses*. Chemists can draw structural queries and combine structural or reaction transformation queries with full-text and bibliographic search terms, such as chemical name, reagents, molecular formula, apparatus, or even hazard warnings or phrases. The contents of individual or collective volumes can be browsed by lists of titles, submitters' names, and volume and page references, with or without structures.

The commercial database at onlinelibrary.wiley.com/doi/book/10.1002/0471264229 also enables the user to choose his/her preferred chemical drawing package, or to utilize several freely available plug-ins for entering queries. The user is also able to cut and paste existing structures and reactions directly into the structure search query or their preferred chemistry editor, streamlining workflow. Additionally, this database contains links to the full text of primary literature references via CrossRef, ChemPort, Medline, and ISI Web of Science. Links to local holdings for institutions using open url technology can also be enabled. The database user can limit his/her search to, or order the search results by, such factors as reaction type, percentage yield, temperature, and publication date, and can create a customized table of reactions for comparison. Connections to other Wiley references are currently made via text search, with cross-product structure and reaction searching to be added in the near future. Incorporations of new preparations will occur as new material becomes available.

INFORMATION FOR AUTHORS OF PROCEDURES

Organic Syntheses welcomes and encourages submissions of experimental procedures that lead to compounds of wide interest or that illustrate important new developments in methodology. Proposals for *Organic Syntheses* procedures will be considered by the Editorial Board upon receipt of an outline proposal as described below. A full procedure will then be invited for those proposals determined to be of sufficient interest. These full procedures will be evaluated by the Editorial Board, and if approved, assigned to a member of the Board for checking. In order for a procedure to be accepted for publication, each reaction must be successfully repeated in the laboratory of a member of the Editorial Board at least twice, with similar yields (generally ±5%) and selectivity to that reported by the submitters.

Organic Syntheses Proposals

A cover sheet should be included providing full contact information for the principal author and including a scheme outlining the proposed reactions (an *Organic Syntheses* Proposal Cover Sheet can be downloaded at orgsyn.org). Attach an outline proposal describing the utility of the methodology and/or the usefulness of the product. Identify and reference the best current alternatives. For each step, indicate the proposed scale, yield, method of isolation and purification, and how the purity of the product is determined. Describe any unusual apparatus or techniques required, and any special hazards associated with the procedure. Identify the source of starting materials. Enclose copies of relevant publications (attach pdf files if an electronic submission is used).

Submit proposals by mail or as e-mail attachments to:

Professor Charles K. Zercher
Associate Editor, *Organic Syntheses*
Department of Chemistry
University of New Hampshire
23 Academic Way, Parsons Hall
Durham, NH 03824

Electronic submission through the website (www.orgsyn.org) is strongly encouraged.

Submission of Procedures

Authors invited by the Editorial Board to submit full procedures should prepare their manuscripts in accord with the Instructions to Authors which are described below or may be downloaded at orgsyn.org. Submitters are also encouraged to consult this volume of *Organic Syntheses* for models with regard to style, format, and the level of experimental detail expected in *Organic Syntheses* procedures. Manuscripts should be submitted to the Associate Editor. Electronic submissions are encouraged; procedures will be accepted as e-mail attachments in the form of Microsoft Word files with all schemes and graphics also sent separately as ChemDraw files.

Procedures that do not conform to the Instructions to Authors with regard to experimental style and detail will be returned to authors for correction. Authors will be notified when their manuscript is approved for checking by the Editorial Board, and it is the goal of the Board to complete the checking of procedures within a period of no more than six months.

Additions, corrections, and improvements to the preparations previously published are welcomed; these should be directed to the Associate Editor. However, checking of such improvements will only be undertaken when new methodology is involved.

NOMENCLATURE

Both common and systematic names of compounds are used throughout this volume, depending on which the Volume Editor felt was more appropriate. The Chemical Abstracts indexing name for each title compound, if it differs from the title name, is given as a subtitle. Systematic Chemical Abstracts nomenclature, used in the Collective Indexes for the title compound and a selection of other compounds mentioned in the procedure, is provided in an appendix at the end of each preparation. Chemical Abstracts Registry numbers, which are useful in computer searching and identification, are also provided in these appendices.

ACKNOWLEDGMENT

Organic Syntheses wishes to acknowledge the contributions of Merck and Boehringer Ingelheim to the success of this enterprise through their support, in the form of time and expenses, of members of the Board of Editors.

INSTRUCTIONS FOR AUTHORS

All organic chemists have experienced frustration at one time or another when attempting to repeat reactions based on experimental procedures found in journal articles. To ensure reproducibility, *Organic Syntheses* requires experimental procedures written with considerably more detail as compared to the typical procedures found in other journals and in the "Supporting Information" sections of papers. In addition, each *Organic Syntheses* procedure is carefully "checked" for reproducibility in the laboratory of a member of the Board of Editors.

Even with these more detailed procedures, the experience of *Organic Syntheses* editors is that difficulties often arise in obtaining the results and yields reported by the submitters of procedures. To expedite the checking process and ensure success, we have prepared the following "Instructions for Authors" as well as a **Checklist for Authors** and **Characterization Checklist** to assist you in confirming that your procedure conforms to these requirements. Please include a completed Checklist together with your procedure at the time of submission. Procedures submitted to *Organic Syntheses* will be carefully reviewed upon receipt and procedures lacking any of the required information will be returned to the submitters for revision.

Scale and Optimization

The appropriate scale for procedures will vary widely depending on the nature of the chemistry and the compounds synthesized in the procedure. However, some general guidelines are possible. For procedures in which the principal goal is to illustrate a synthetic method or strategy, it is expected, in general, that the procedure should result in at least 5 g and no more than 50 g of the final product. In cases where the point of the procedure is to provide an efficient method for the preparation of a useful reagent or synthetic building block, the appropriate scale also should not exceed 50 g of final product. Exceptions to these guidelines may be granted in special circumstances. For example, procedures describing the preparation of reagents employed as catalysts will often be acceptable on a scale of less than 5 g.

In considering the scale for an *Organic Syntheses* procedure, authors should also take into account the cost of reagents and starting materials. In general, the Editors will not accept procedures for checking in which the cost of any one of the reactants exceeds **$500** for a single full-scale run. Authors are requested to identify the most expensive reagent or starting material on the procedure submission checklist and to estimate its cost per run of the procedure.

It is expected that all aspects of the procedure will have been optimized by the authors prior to submission, and it is required that each reaction will have been carried out at least twice on exactly the scale described in the procedure, and with the results reported in the manuscript.

It is appropriate to report the weight, yield, and purity of the product of each step in the procedure as a range. In any case where a reagent is employed in significant excess, a Note should be included explaining why an excess of that reagent is necessary. If possible, the Note should indicate the effect of using amounts of reagent less than that specified in the procedure.

The Checking Process

A unique feature of papers published in *Organic Syntheses* is that each procedure and all characterization data is carefully checked for reproducibility in the laboratory of a member of the Board of Editors. In the event that an editor finds it necessary to make any modifications in an experimental procedure, then the published article incorporates the modified procedure, with an explanation and mention of the original protocol often included in a Note. The yields reported in the published article are always those obtained by the checkers. In general, the characterization data in the published article also is that of the checkers, unless there are significant differences with the data obtained by the authors, in which case the author's data will also be reported in a Note.

Reaction Apparatus

Describe the size and type of flask (number of necks) and indicate how *every* neck is equipped.

"A 500-mL, three-necked, round-bottomed flask equipped with an 3-cm Teflon-coated magnetic stirbar, a 250-mL pressure-equalizing addition funnel fitted with an argon inlet, and a rubber septum is charged with ... "

Indicate how the reaction apparatus is dried and whether the reaction is conducted under an inert atmosphere. Note that in general balloons are not acceptable as a means of maintaining an inert atmosphere unless warranted by special circumstances. The description of the reaction apparatus can be incorporated in the text of the procedure or included in a Note.

"The apparatus is flame-dried and maintained under an atmosphere of argon during the course of the reaction."

In the case of procedures involving unusual glassware or especially complicated reaction setups, authors are encouraged to include a photograph or drawing of the apparatus in the text or in a Note (for examples, see *Org. Syn.,* Vol. 82, 99 and Coll. Vol. X, pp 2, 3, 136, 201, 208, and 669).

Use of Gloveboxes

When a glovebox is employed in a procedure, justification must be provided in a Note and the consequences of carrying out the operation without using a glovebox should be discussed.

Reagents and Starting Materials

All chemicals employed in the procedure must be commercially available or described in an earlier *Organic Syntheses* or *Inorganic Syntheses* procedure. For other compounds, a procedure should be included either as one or more steps in the text or, in the case of relatively straightforward preparations of reagents, as a Note. In the latter case, all requirements with regard to characterization, style, and detail also apply. Authors are encouraged to consult with the Associate Editor if they have any question as to whether to include such steps as part of the text or as a Note.

Authors are encouraged to consider the use of "substitute solvents" in place of more hazardous alternatives. For example, the use of *t*-butyl methyl ether (MTBE) should be considered as a substitute for diethyl ether, particularly in large scale work. Authors are referred to the articles "Sanofi's Solvent Selection Guide: A Step Toward More Sustainable Processes" (Prat, D.; Pardigon, O.; Flemming, H.-W.; Letestu, S.; Ducandas, V.; Isnard, P.; Guntrum, E.; Senac, T.; Ruisseau, S.; Cruciani, P. Hosek, P. O*rg. Process Res. Dev.* **2013**, *17*, 1517-1525) and "Solvent Replacement for Green Processing" (Sherman, J.; Chin, B.; Huibers, P. D. T.; Garcia-Valis, R.; Hatton, T. A. *Environ. Health Perspect.* **1998**, *106* (Supplement I, 253-271) as well as the references cited therein for discussions of this subject. In addition, a link to a "solvent selection guide" can be accessed via the American Chemical Society Green Chemistry website at http://www.acs.org/content/acs/en/greenchemistry/research-innovation/tools-for-green-chemistry.html.

In one or more Notes, indicate the purity or grade of each reagent, solvent, etc. It is highly desirable to also indicate the source (company the chemical was purchased from), particularly in the case of chemicals where it is suspected that the composition (trace impurities, etc.) may vary from one supplier to another. In cases where reagents are purified, dried, "activated" (e.g., Zn dust), etc., a detailed description of the procedure used should be included in a Note. In other cases, indicate that the chemical was "used as received".

"Diisopropylamine (99.5%) was obtained from Aldrich Chemical Co., Inc. and distilled under argon from calcium hydride before use. THF (99+%) was obtained from Mallinckrodt, Inc. and distilled from sodium benzophenone ketyl. Diethyl ether (99.9%) was purchased from Aldrich Chemical Co., Inc. and purified by pressure filtration under argon through activated alumina. Methyl iodide (99%) was obtained from Aldrich Chemical Co., Inc. and used as received."

The amount of each reactant must be provided in parentheses in the order mL, g, mmol, and equivalents with careful consideration to the correct number of **significant figures**. Avoid indicating amounts of reactants with more significant figures than makes sense. For example, "437 mL of THF" implies that the amount of solvent must be measured with a level of precision that is unlikely to affect the outcome of the reaction. Likewise, "5.00 equiv" implies that an amount of excess reagent must be controlled to a precision of 0.01 equiv.

The concentration of solutions should be expressed in terms of molarity or normality, and not percent (e.g., 1 N HCl, 6 M NaOH, not "10% HCl").

Reaction Procedure

Describe every aspect of the procedure clearly and explicitly. Indicate the order of addition and time for addition of all reagents and how each is added (via syringe, addition funnel, etc.).

Indicate the temperature of the reaction mixture (preferably internal temperature). Describe the type of cooling (e.g., "dry ice-acetone bath") and heating (e.g., oil bath, heating mantle) methods employed. Be careful to describe clearly all cooling and warming cycles, including initial and final temperatures and the time interval involved.

Describe the appearance of the reaction mixture (color, homogeneous or not, etc.) and describe all significant changes in appearance during the course of the reaction (color changes, gas evolution, appearance of solids, exotherms, etc.).

Indicate how the reaction can be monitored to determine the extent of conversion of reactants to products. In the case of reactions monitored by TLC, provide details in a Note, including eluent, R_f values, and method of visualization. For reactions followed by GC, HPLC, or NMR analysis, provide details on analysis conditions and relevant diagnostic peaks.

"The progress of the reaction was followed by TLC analysis on silica gel with 20% EtOAc-hexane as eluent and visualization with p-anisaldehyde. The ketone starting material has $R_f = 0.40$ (green) and the alcohol product has $R_f = 0.25$ (blue)."

Reaction Workup

Details should be provided for reactions in which a "quenching" process is involved. Describe the composition and volume of quenching agent, and time and temperature for addition. In cases where reaction mixtures are added to a quenching solution, be sure to also describe the setup employed.

"The resulting mixture was stirred at room temperature for 15 h, and then carefully poured over 10 min into a rapidly stirred, ice-cold aqueous solution of 1 N HCl in a 500-mL Erlenmeyer flask equipped with a magnetic stirbar."

For extractions, the number of washes and the volume of each should be indicated as well as the size of the separatory funnel.

For concentration of solutions after workup, indicate the method and pressure and temperature used.

"The reaction mixture is diluted with 200 mL of water and transferred to a 500-mL separatory funnel, and the aqueous phase is separated and extracted with three 100-mL portions of ether. The combined organic layers are washed with 75 mL of water and 75 mL of saturated NaCl solution, dried over 25 g of $MgSO_4$, filtered through a 250-mL medium porosity sintered glass funnel, and concentrated by rotary evaporation (25 °C, 20 mmHg) to afford 3.25 g of a yellow oil."

"The solution is transferred to a 250-mL, round-bottomed flask equipped with a magnetic stirbar and a 15-cm Vigreux column fitted with a short path distillation head, and then concentrated by careful distillation at 50 mmHg (bath temperature gradually increased from 25 to 75 °C)."

In cases where solid products are filtered, describe the type of filter funnel used and the amount and composition of solvents used for washes.

" ... and the resulting pale yellow solid is collected by filtration on a Büchner funnel and washed with 100 mL of cold (0 °C) hexane."

When solid or liquid compounds are dried under vacuum, indicate the pressure employed (rather than stating "reduced pressure" or "dried *in vacuo*").

" .... and concentrated at room temperature by rotary evaporation (20 mmHg) and then at 0.01 mmHg to provide .... "

"The resulting colorless crystals are transferred to a 50-mL, round-bottomed flask and dried overnight in a 100 °C oil bath at 0.01 mmHg."

Purification: Distillation

Describe distillation apparatus including the size and type of distillation column. Indicate temperature (and pressure) at which all significant fractions are collected.

" .... and transferred to a 100-mL, round-bottomed flask equipped with a magnetic stirbar. The product is distilled under vacuum through a 12-cm, vacuum-jacketed column of glass helices (Note 16) topped with a Perkin triangle. A forerun (ca. 2 mL) is collected and discarded, and the desired product is then obtained, distilling at 50-55 °C (0.04-0.07 mmHg) .... "

Purification: Column Chromatography

Provide information on TLC analysis in a Note, including eluent, R_f values, and method of visualization.

Provide dimensions of column and amount of silica gel used; in a Note indicate source and type of silica gel.

Provide details on eluents used, and number and size of fractions.

"The product is charged on a column (5 x 10 cm) of 200 g of silica gel (Note 15) and eluted with 250 mL of hexane. At that point, fraction collection (25-mL fractions) is begun, and elution is continued with 300 mL of 2% EtOAc-hexane (49:1 hexanes:EtOAc) and then 500 mL of 5% EtOAc-hexane (19:1 hexanes:EtOAc). The desired product is obtained in fractions 24-30, which are concentrated by rotary evaporation (25 °C, 15 mmHg) .... "

Use of Automated Column Chromatography

Automated column chromatography should not be used for purification of products unless the use of such systems is essential to the success of the procedure. When automated column chromatography equipment is employed in a procedure, justification must be provided in a Note and the consequences of carrying out the purification using conventional column chromatography must be discussed.

Purification: Recrystallization

Describe procedure in detail. Indicate solvents used (and ratio of mixed solvent systems), amount of recrystallization solvents, and temperature protocol. Describe how crystals are isolated and what they are washed with. A photograph of the crystalline product is often valuable to indicate the form and color of the crystals.

"The solid is dissolved in 100 mL of hot diethyl ether (30 °C) and filtered through a Buchner funnel. The filtrate is allowed to cool to room temperature, and 20 mL of hexane is added. The solution is cooled at -20 °C overnight and the resulting crystals are collected by suction filtration on a Buchner funnel, washed with 50 mL of ice-cold hexane, and then transferred to a 50-mL, round-bottomed flask and dried overnight at 0.01 mmHg to provide ... "

Characterization

Physical properties of the product such as color, appearance, crystal forms, melting point, etc. should be included in the text of the procedure. Comments on the stability of the product to storage, etc. should be provided in a Note.

In a Note, provide data establishing the identity of the product. This will generally include IR, MS, ^{1}H-NMR, and ^{13}C-NMR data, and in some cases UV data. Copies of the proton and carbon NMR spectra for the products of each step in the procedure should be submitted showing integration for all resonances. Submission of copies of the NMR spectra for other nuclei are encouraged as appropriate.

In the same Note, provide analytical data establishing that the purity of the **isolated** product is at least 97%. **Note that this data should be obtained for the material on which the yield of the reaction is based**, not for a sample that has been subjected to additional purification by chromatography, distillation, or crystallization. Elemental analysis for carbon and hydrogen (and nitrogen if present) agreeing with calculated values within 0.4% is preferred. However, **quantitative** NMR, GC, or HPLC analyses involving measurements versus an internal standard will also be accepted. See *Instructions for Authors* at orgsyn.org for procedures for quantitative analysis of purity by NMR and chromatographic methods. Provide details on equipment and conditions for GC and HPLC analyses.

In procedures involving non-racemic, enantiomerically enriched products, optical rotations should generally be provided, but **enantiomeric purity must be determined by another method** such as chiral HPLC or GC analysis.

In cases where the product of one step is used without purification in the next step, a Note should be included describing how a sample of the product can be purified and providing characterization data for the pure material. Copies of the proton NMR spectra of both the product both *before* and *after* purification should be submitted.

Safety Note and Hazard Warnings

Effective in August 2017, the first Note in every article is devoted to addressing the safety aspects of the procedures described in the article. The Article Template provides the required wording and format for Note 1, which reminds readers of the importance of carrying out risk assessments and hazard analyses prior to performing all experiments:

> Prior to performing each reaction, a thorough hazard analysis and risk assessment should be carried out with regard to each chemical substance and experimental operation on the scale planned and in the context of the laboratory where the procedures will be carried out. Guidelines for carrying out risk assessments and for analyzing the hazards associated

with chemicals can be found in references such as Chapter 4 of "Prudent Practices in the Laboratory" (The National Academies Press, Washington, D.C., 2011; the full text can be accessed free of charge at http://www.nap.edu/catalog.php?record_id=12654). See also "Identifying and Evaluating Hazards in Research Laboratories" (American Chemical Society, 2015) which is available via the associated website "Hazard Assessment in Research Laboratories" at https://www.acs.org/content/acs/en/about/governance/committees/chemicalsafety/hazard-assessment.html. In the case of this procedure, the risk assessment should include (but not necessarily be limited to) an evaluation of the potential hazards associated with (*enter list of chemicals here*), as well as the proper procedures for (*list any unusual experimental operations here*). *(Provide additional cautions with regard to exceptional hazards here).*

For the required list of chemicals, authors should include all reactants, solvents, and other chemicals involved in the reactions described in the article.

With regard to the list of experimental operations, this list should be limited to those operations that potentially pose significant hazards. Examples may include

- Vacuum distillations
- Reactions run at elevated pressure or in sealed reaction vessels
- Photochemical reactions

In the case of experiments that involve exceptional hazards such as the use of pyrophoric or explosive substances, and substances with a high degree of acute or chronic toxicity, authors should provide additional guidelines for how to carry out the experiment so as to minimize risk. These instructions formerly would have appeared as red "Caution Notes" in *Organic Syntheses* articles. Note that it is not essential to describe general safety procedures such as working in a hood, avoiding skin contact, using eye protection, etc., since these are discussed in the Prudent Practices reference mentioned in the "Working with Hazardous Chemicals" statement within each article. Efforts should be made to avoid the use of toxic and hazardous solvents and reagents when less hazardous alternatives are available.

Discussion Section

The style and content of the discussion section will depend on the nature of the procedure.

For procedures that provide an improved method for the preparation of an important reagent or synthetic building block, the discussion should focus on the advantages of the new approach and should describe and reference all of the earlier methods used to prepare the title compound.

In the case of procedures that illustrate an important synthetic method or strategy, the discussion section should provide a mini-review on the new methodology. The scope and limitations of the method should be discussed, and it is generally desirable to include a table of examples. Please be sure each table is numbered and has a title. Competing methods for accomplishing the same overall transformation should be described and referenced. A brief discussion of mechanism may be included if this is useful for understanding the scope and limitations of the method.

Titles of Articles

In cases where the main thrust of the article is the illustration of a synthetic method of general utility, the title of the article should incorporate reference to that method. Inclusion of the name of the final product is acceptable but not required. In the case of articles where the objective is the preparation of a specific compound of importance (such as a chiral ligand), then the name of that compound should be part of the title.

Examples

Title without name of product:

"Stereoselective Synthesis of 3-Arylacrylates by Copper-Catalyzed Syn Hydroarylation" (*Org. Synth.* **2010**, *87*, 53).

Title including name of final product (note name of product is not required):

"Catalytic Enantioselective Borane Reduction of Benzyl Oximes: Preparation of (S)-1-Pyridin-3-yl-ethylamine Bis Hydrochloride" (*Org. Synth.* **2010**, *87*, 36).

Title where preparation of specific compound is the subject:

"Preparation of (S)-3,3'-Bis-Morpholinomethyl-5,5',6,6',7,7',8,8'-octahydro-1,1'-bi-2-naphthol" (*Org. Synth.* **2010**, *87*, 59).

Heading Scheme

The title of the article should be followed by a "Heading Scheme" comprising separate equations for each step in the article. Authors should consult the article template for instructions concerning ChemDraw settings and format. In general, reaction equations should not include details such as reaction time and the number of equivalents of reagents, with the exception of reactants employed in catalytic amounts which can be labeled as "cat." or by specifying mol%.

Style and Format for Text

Articles should follow the style guidelines used for organic chemistry articles published in the ACS journals such as *J. Am. Chem. Soc.*, *J. Org.*

Chem., *Org. Lett.*, etc. as described in the ACS Style Guide (3rd Ed.). The text of the procedure should be created using the Word template available on the *Organic Syntheses* website. Specific instructions with regard to the manuscript format (font, spacing, margins) is available on the website in the "Instructions for Article Template" and embedded within the Article Template itself.

Style and Format for Tables and Schemes

Chemical structures and schemes should be drawn using the standard ACS drawing parameters (in ChemDraw, the parameters are found in the "ACS Document 1996" option) with a maximum full size width of 15 cm (5.9 inches). The graphics files should then be pasted into the Word document at the correct location and the size reduced to 75% using "Format Picture" (Mac) or "Size and Position" (Windows). Graphics files must also be submitted separately. All Tables that include structures should be entirely prepared in the graphics (ChemDraw) program and inserted into the word processing file at the appropriate location. Tables that include multiple, separate graphics files prepared in the word processing program will require modification.

Tables and schemes should be numbered and should have titles. The title for a Table should be included within the ChemDraw graphic and placed immediately above the table. The title for a scheme should be included within the ChemDraw graphic and placed immediately below the scheme. Use 12 point Palatino Bold font in the ChemDraw file for all titles. For footnotes in Tables use Helvetica (or Arial) 9 point font and place these immediately below the Table.

Photographs

Photographs illustrating key elements of procedures are required in every article published in Organic Syntheses. Authors are expected to furnish photos with their original submissions and photos may also be provided by the Checkers of procedures. Photographs should be inserted into articles at the place in the text and Notes where they are first referred to and should be numbered and labeled as Figures with descriptive titles. Particularly useful subjects for photographs include:

- Photos of reaction flasks depicting how each neck is equipped
- Photos of reaction mixtures illustrating color changes, heterogeneity, etc.
- Photos of TLC plates showing degree of resolution and the color of spots
- Photos of crystalline reaction products illustrating color and crystal type

Acknowledgments and Author's Contact Information

Contact information (institution where the work was carried out and mailing address for the principal author) should be included as footnote 1. This footnote should also include the email address for the principal author. Acknowledgment of financial support should be included in footnote 1.

Biographies and Photographs of Authors

Photographs and 100-word biographies of all authors should be submitted as separate files at the time of the submission of the procedure. The format of the biographies should be similar to those in the Volume 84 procedures found at the orgsyn.org website. Photographs can be accepted in a number of electronic formats, including tiff and jpeg formats.

DISPOSAL OF CHEMICAL WASTE

General Reference: *Prudent Practices in the Laboratory* National Academy Press, Washington, D.C. 2011.

Effluents from synthetic organic chemistry fall into the following categories:

1. **Gases**
    1a. Gaseous materials either used or generated in an organic reaction.
    1b. Solvent vapors generated in reactions swept with an inert gas and during solvent stripping operations.
    1c. Vapors from volatile reagents, intermediates and products.

2. **Liquids**
    2a. Waste solvents and solvent solutions of organic solids (see item 3b).
    2b. Aqueous layers from reaction work-up containing volatile organic solvents.
    2c. Aqueous waste containing non-volatile organic materials.
    2d. Aqueous waste containing inorganic materials.

3. **Solids**
    3a. Metal salts and other inorganic materials.
    3b. Organic residues (tars) and other unwanted organic materials.
    3c. Used silica gel, charcoal, filter aids, spent catalysts and the like.

The operation of industrial scale synthetic organic chemistry in an environmentally acceptable manner* requires that all these effluent categories be dealt with properly. In small scale operations in a research or academic setting, provision should be made for dealing with the more environmentally offensive categories.

1a. Gaseous materials that are toxic or noxious, e.g., halogens, hydrogen halides, hydrogen sulfide, ammonia, hydrogen cyanide, phosphine, nitrogen oxides, metal carbonyls, and the like.
1c. Vapors from noxious volatile organic compounds, e.g., mercaptans, sulfides, volatile amines, acrolein, acrylates, and the like.

---

*An environmentally acceptable manner may be defined as being both in compliance with all relevant state and federal environmental regulations *and* in accord with the common sense and good judgment of an environmentally aware professional.

2a. All waste solvents and solvent solutions of organic waste.
2c. Aqueous waste containing dissolved organic material known to be toxic.
2d. Aqueous waste containing dissolved inorganic material known to be toxic, particularly compounds of metals such as arsenic, beryllium, chromium, lead, manganese, mercury, nickel, and selenium.
3. All types of solid chemical waste.

Statutory procedures for waste and effluent management take precedence over any other methods. However, for operations in which compliance with statutory regulations is exempt or inapplicable because of scale or other circumstances, the following suggestions may be helpful.

Gases

Noxious gases and vapors from volatile compounds are best dealt with at the point of generation by "scrubbing" the effluent gas. The gas being swept from a reaction set-up is led through tubing to a large trap to prevent suck-back and into a sintered glass gas dispersion tube immersed in the scrubbing fluid. A bleach container can be conveniently used as a vessel for the scrubbing fluid. The nature of the effluent determines which of four common fluids should be used: dilute sulfuric acid, dilute alkali or sodium carbonate solution, laundry bleach when an oxidizing scrubber is needed, and sodium thiosulfate solution or diluted alkaline sodium borohydride when a reducing scrubber is needed. Ice should be added if an exotherm is anticipated.

Larger scale operations may require the use of a pH meter or starch/iodide test paper to ensure that the scrubbing capacity is not being exceeded.

When the operation is complete, the contents of the scrubber can be poured down the laboratory sink with a large excess (10-100 volumes) of water. If the solution is a large volume of dilute acid or base, it should be neutralized before being poured down the sink.

Liquids

Every laboratory should be equipped with a waste solvent container in which *all* waste organic solvents and solutions are collected. The contents of these containers should be periodically transferred to properly labeled waste solvent drums and arrangements made for contracted disposal in a regulated and licensed incineration facility.**

**If arrangements for incineration of waste solvent and disposal of solid chemical waste by licensed contract disposal services are not in place, a list of providers of such services should be available from a state or local office of environmental protection.

Aqueous waste containing dissolved toxic organic material should be decomposed *in situ*, when feasible, by adding acid, base, oxidant, or reductant. Otherwise, the material should be concentrated to a minimum volume and added to the contents of a waste solvent drum.

Aqueous waste containing dissolved toxic inorganic material should be evaporated to dryness and the residue handled as a solid chemical waste.

Solids

Soluble organic solid waste can usually be transferred into a waste solvent drum, provided near-term incineration of the contents is assured.

Inorganic solid wastes, particularly those containing toxic metals and toxic metal compounds, used Raney nickel, manganese dioxide, etc. should be placed in glass bottles or lined fiber drums, sealed, properly labeled, and arrangements made for disposal in a secure landfill.** Used mercury is particularly pernicious and small amounts should first be amalgamated with zinc or combined with excess sulfur to solidify the material.

Other types of solid laboratory waste including used silica gel and charcoal should also be packed, labeled, and sent for disposal in a secure landfill.

Special Note

Since local ordinances may vary widely from one locale to another, one should always check with appropriate authorities. Also, professional disposal services differ in their requirements for segregating and packaging waste.

Ronald Breslow
1931 – 2017

Ronald Breslow, one of the most influential chemists of the past 60 years, passed away October 25, 2017 in New York City at the age of 86.

Breslow was born March 14, 1931 in Rahway, New Jersey, the son of Gladyz and Alexander Breslow. His interest in chemistry was whetted in part by interactions with Max Tischler, a family friend and Director of Chemistry at Merck & Co. at the time. At the age of 17 he was a finalist in the prestigious Westinghouse Science Talent Search, and later that year entered Harvard graduating with a A.B. in chemistry in 1952. At Harvard, Breslow worked in Gilbert Stork's laboratory, coauthoring two papers that described the structure elucidation of the tricyclic sesquiterpene cedrene. Breslow initially began graduate studies in Harvard's program in medical sciences in 1952, which he left a year later with a master's degree to focus on graduate studies in chemistry. He completed his Ph. D. degree in 1955 working with R. B. Woodward on early studies directed at the structure of the macrolide antibiotic magnamycin. Breslow then spent one year in Cambridge University working with Alexander Todd on early efforts to developed chemistry for synthesizing deoxyribonucleotides and at the same time initiating his independent studies on thiamine (vitamin B1). In 1956 at the age of 25 he joined Columbia University as an Instructor in the Department of Chemistry.

Breslow's remarkable insight and broad vision of chemistry became immediately apparent when over a 10-month period in 1957–58 he authored two landmark publications in the *Journal of the American Chemical Society*. In the first, he reported the synthesis of triphenylcyclopropenyl cation, the earliest example of the simplest aromatic ring. In the second, after showing that thiazolium salts readily form zwitterions by loss of the C-2 hydrogen of the heterocyclic ring, he proposed the mechanism by which the coenzyme thiamine pyrophosphate functions in important biochemical reactions. Thiazolium zwitterions were the first examples of *N*-heterocyclic carbenes, whose chemistry – often involving Breslow intermediates – remains an active area of research today. Following up on his cyclopropenyl cation work, Breslow went on to show that cyclic conjugated ring systems having 4n π-electrons were not only not stabilized by delocalization but destabilized, molecules he termed anti-aromatic.

Ron Breslow was one of the founders of the field of bioorganic chemistry. Inspired by Nature's ability to selectively functionalize molecules not at their intrinsically most reactive position but rather by steric approximation within an enzyme-substrate complex, Breslow began in the mid 1960's systematic investigations to mimic enzymatic selectivity by the positioning of reagents and substrates. In initial investigations an oxidant was tethered to a substrate to functionalize remote C–H bonds. Later studies exploited several types of non-covalent molecular recognitions, with this approach to achieving high selectivity in chemical reaction being pursued by many researchers worldwide. He also pioneered in developing simple chemical systems that mimicked both the binding and active-site turnover steps of enzyme action. In collaboration with Paul A. Marks, at the time President and CEO of Memorial Sloan-Kettering Cancer Center, Breslow developed the first cancer therapeutics that worked by inhibiting members of the zinc-binding class of histone deacetylases. This discovery was commercialized by Merck, with Zolinda® (suberoylanilide hydroxamic acid, vorinostat) being approved by the FDA in 2006 for treating cutaneous T-cell lymphoma.

Ron Breslow's impact on the chemical enterprise extended far beyond his research accomplishments. He had leadership roles in the Chemical Section of the National Academy of Sciences and served as President of the American Chemical Society in 1996. In this latter role, he was unusually active in articulating the accomplishments and promise of chemistry to a wide audience in part through the book *Chemistry Today and Tomorrow: The Central, Useful, and Creative Science*, which was authored by Breslow and published by ACS in 1996. Ron was a member of the Editorial Board of *Organic Syntheses* during the 1960s and edited Volume 50 which appeared in 1970. In addition, Ron was a long-time member of the *Organic Syntheses* Board of Directors.

Ron Breslow's many honors and awards include the National Medal of Science, the Welch Award, the NAS Award in the Chemical Sciences, the American Institute of Chemists Gold Medal, the Swiss Chemical Society's Paracelsus Prize, and American Chemical Society Awards in Pure Chemistry, Arthur C. Cope Award, and Priestley Medal. He was a member of the National Academy of Sciences, the American Academy of Arts and Sciences and the American Philosophical Society. Breslow was also a foreign member of the Royal Society and an honorary member of many other scientific bodies around the world. In recognition of his classroom teaching skills, Columbia awarded him both its Mark Van Doren Award and its Great Teacher Award.

More than 250 Ph.D. students and postdocs trained in the Breslow research group. The fundamental nature of the training one gained working with Ron is apparent in the diverse fields his former coworkers have pursued, which include polymer chemistry, organometallic chemistry, chemical biology, organic synthesis, and drug discovery. Like many former coworkers, the author considers the time he spent in the Breslow group as indispensable to his academic endeavors. Ron coupled his love of chemistry and intensity for research with enormous personal warmth. His quick mind and intellect is legend yet was projected in a way that inspired rather than intimidated. Ron was immensely loyal and supportive of his former coworkers.

Ronald Breslow is survived by his wife Esther, professor emerita of biochemistry at Weill Cornell Medical College, their daughters Stephanie and Karen and grandchildren. Ron's friendship and wise council will be deeply missed by all who knew him.

Larry E. Overman
Irvine, California

Andrew S. Kende
1932 – 2018

Andrew S. (Andy) Kende was born in Budapest Hungary in 1932. With World War II impending, his family immigrated to the United States in 1939 and settled in the Bronx where he received his earliest schooling. His father was recruited to Chicago IL to join Marshall Fields Co. and the family settled in the suburb of Evanston IL where Andy grew up and was educated through high school. A portent of things to come, he later described being fascinated with chemistry at six years old: "It's always seemed to me an interesting way to structure information," he said. "Things relate in a sensible way."

At an early age, Andy exemplified the excellence he would later encourage in his students. At 15, he was the winner of the National Westinghouse Science Talent Search where he bested in the final another noted organic chemist, National Academy member, and former President of the American Chemical Society Ronald Breslow, who as fate would have it, would later become Andy's graduate school lab mate at Harvard. At 16, he enrolled at the University of Chicago. After earning a 2-year A.B. degree, Andy realized he was not prepared to do advanced work in chemistry so he moved south to Florida State University for advanced coursework. He moved from there to Harvard University where he received his Ph.D. degree in Organic Chemistry in 1957 working with the Nobel Laureate, Professor Robert B. Woodward. During his doctoral work he elucidated new pathways for the reactions of aliphatic diazo compounds with ketenes and led to the first spectroscopic characterization of pure cyclopropanone. With the assistance of an NRC-American Cancer Society Postdoctoral Fellowship (1956-57), Andy moved from Harvard to the UK's University of Glasgow to work with another Nobel Laureate Sir Derek H. R. Barton where he demonstrated the structure of the major photoisomerization product of dehydroergosteryl acetate.

Andy returned to the United States 1957-1958 to join Lederle Laboratories in Pearl River New York where his excursion into natural products chemistry would continue as he worked as part of the team that synthesized the antibiotic Tetracycline. He had the scientific freedom during this time to work on some of his own chemistry, which was focused on his fascination with the nascent field of organic photochemistry.

In 1968, Andy accepted an appointment as Professor of Chemistry at the University of Rochester. At Rochester, his research program focused on three principal themes: early work in organic photochemistry, pericyclic reactions, and total synthesis. During his time at Rochester, Andy's work on pericyclic reactions included the photochemistry of β,γ-enones and homoconjugated dienones, singlet oxygen chemistry, carbene reactions, rearrangements of cyclic polyenes, methylenecyclopropane isomerization and fragmentation, and the chemistry of isobenzofurans, mesoionic oxyallyl species, and phenalene derivatives. His studies in total synthesis include the construction of the antineoplastic alkaloid camptothecin from furfural, new methods for nucleophilic acylation and transition metal coupling reactions, development of selective photochemical methods in synthesis and new routes to the anthracycline antibiotics, including aklavinone which could then be modified as a less expensive, less toxic anti-tumor drug than other treatments then available. He also completed the total syntheses of numerous terpenoids including, steganacin, deoxyanisatin, and alkaloids including streptonigrin, dendrobine and sesbanine, and constructed the intricate tricyclic framework of the taxane diterpenes.

He had wide ranging scientific interests as exemplified by his collaboration with Alan Poland, an assistant professor of pharmacology and toxicology, in synthesizing TCDD, a persistent environmental contaminant and toxin resulting from the manufacture of herbicides, wood preservatives, and lubricating oils, and identifying the biological receptor for TCDD which mediated its potent toxic effects. This was an important first step in setting standards to help regulate TCDD.

Andy's research led to a 1978 Guggenheim Fellowship and to numerous invited lectures, including several Gordon Conference lectures, NSF Workshops in Natural Products Chemistry (1972 and 1974), and the International Symposium on Anthracycline Chemistry (Winnipeg, 1978), as well as plenary lectures at the Royal Society of Chemistry (Cambridge, England, July, 1983), the International Conference on Heterocyclic Chemistry (Tokyo, August, 1983), and the Medicinal Chemistry Symposium (Cambridge, England, September, 1983). In 1986, he was awarded a Japan Society for Promotion of Science Fellowship. Andy also received an ACS Cope Senior Scholar Award in 2003. Several of Andy's compounds were patented; in 1979 he was chosen inventor of the year by the Rochester Patent Law Association. That same year, he was one of 50 U.S. scientists chosen to participate in the nation's first bilateral scientific symposium with China. Other honors included the Rochester Section Award of the American Chemical Society, and a

Fellowship from the Japanese Society for the Promotion of Science. The Department of Chemistry has also created a named professorship in his honor.

Andy enjoyed teaching and was demanding instructor, but his real thrill was in mentoring and training his graduate students and participating in the research they did. He mentored over 50 postdocs and 50 students during his career. At Rochester, Andy had the reputation for demanding scientific excellence and would not settle for less than the pursuit of science at the highest level. The lessons they learned from him – about chemistry but also about working hard and achieving excellence – have remained with them over the years. Yuh-geng Tsay, an early Ph.D. student, remembers that whenever Andy returned from a business trip, "he would stop by the lab first to see how everyone was doing. This type of work ethic has inspired us not only to work hard, but to have a sense of urgency in everything you do. His teaching style empowered us to solve any technical challenge and to be independent problem solvers." Andy demanded excellence from all of those around him; students, staff, and faculty. His vision, imagination, and vast knowledge of organic chemistry were a valuable resource and set a standard that many tried to emulate. His selfless and energetic service to the field of organic chemistry was exemplary, a standard for all to strive.

Lanny Liebeskind, a Ph.D. student in the mid '70's, remembers walking into Andy's lab for the first time as a new graduate student. "I remember asking Andy when I should start my research. His succinct answer, in effect, was 'Now!'" says Liebeskind, the vice provost for strategic research initiatives and Samuel Dobbs Professor of Chemistry at Emory University. "I got the message loud and clear. It was a bit like being dropped into a professional sports team where the coach is constantly challenging you to push yourself beyond the comfort level. In doing so, you grew in ways as a scholar and person that you never would have on your own." Tsay had a similar experience. "When I toured Professor Kende's labs, I noticed there was a memo from him posted in each cubicle of his graduate students and postdocs. Two key phrases stood out that got my attention. 'When you are here, you should roll up your sleeves and work. If you cannot manage at least two experiments at the same time, you don't belong in this group.'"

During his 50-plus years as a member of the American Chemical Society, Andy served Organic Chemistry in numerous ways outside of his research, teaching, and mentoring. As department chair at Rochester from 1979 to 1983, he worked with the University's chief science librarian to introduce chemistry undergraduates to the wonders of the computer as a new way to search for articles and information "buried in the huge and growing body of scientific literature." This consisted of using an "ordinary phone" to dial a database, attaching the receiver to a portable computer terminal, typing in a request, and then "within seconds" getting a printout. Although truth be told, Andy was the very last to give up his venerable IBM Selectric typewriter and in the beginning was mystified by the new-fangled gadget called a mouse. He was noted running the mouse around on his assistant's computer monitor in her absence clicking away to no avail since his only exposure to computers, to that point, had been a touch-screen monitor at the local Wegman's supermarket.

Andy joined the editorial board of *Organic Reactions* in 1970 while William Dauben was Editor-in-Chief. During his twelve-year term on the editorial board, Organic Reactions published ten volumes. Because of Andy's avid interest in the Organic Reactions organization, he was elected to membership on the Board of Directors in 1980. He assumed the role of Editor-in-Chief in 1984 and held that position for five years before handing the mantle to Leo Paquette. Andy remained on the Board of Directors until his resignation in 2013.

During his four decades serving these various roles, Andy was responsible for the appointment of an additional Editorial Board secretary. This position was required as Organic Reactions grew from publishing roughly a single volume annually to its current multiple volumes a year.

Andy also elected as a member of the Editorial Board and was Editor of Volume *64* of *Organic Syntheses*. He subsequently joined the Board of Directors and served as President of *Organic Syntheses* Inc. from 2002-2012. In this role, he oversaw the beginning of *Organic Syntheses* transition from a purely print format to the current electronic and print publications.

He was appointed an associate editor of the ACS *Journal of Organic Chemistry*, serving for 12 years. He also was selected as a member of the Editorial Advisory Boards of *Chemical Reviews,* and *Synthetic Communications*. At the University of Rochester, Andy was a member of numerous University committees including graduate admissions, budget, tenure and privilege, and River Campus parking, served in the University Senate, and chaired the Scientific Advisory Board of the UR Cancer Center.

After his retirement from teaching 1998, he continued productive research until 2002, when Andy and his wife Fran moved from Rochester to Scottsdale, AZ. During the next 16 years, Andy and Fran continued to indulge their passion for travel and good food. Andy was a lifelong lover of good food of all types (particularly meat and potatoes but famously not fish), prepared and served well. As in everything he did, he brooked no incompetence in either area. The only "compensation" for his service to *Organic Reactions* and *Organic Syntheses* was an invitation to the annual or biannual dinners on the occasion of the meetings of the Editorial Boards and the Boards of Directors, typically at ACS national meetings. Because of Andy's well-known like of good food and wine as well as his involvement in both the *Organic Reactions* and *Organic Syntheses* organizations, there was a good-spirited competition at each annual or bi-annual meeting to determine which dinner was "best." Andy, of course, was the sole judge.

Andy and Fran were also justifiably immensely proud of their only child, their son Mark, who in his own right has gone on to considerable distinction in academia as a constitutional law scholar holding the James Madison Chair in Constitutional Law at the Drake University School of Law.

Andy rest in peace, we all will miss you on so many levels, as a person, as a scholar, as a teacher who trained your students to carry on your high standards of scientific rigor and excellence, and for your unselfish service to your science and your profession. Godspeed.

ROBERT K. BOECKMAN Jr.
Rochester, New York

John D. (Jack) Roberts
1918 – 2016

John D. Roberts, the Institute Professor of Chemistry, Emeritus, and one of the most influential chemists of the 20th century, passed away on October 29, 2016 at the age of 98 following a stroke.

John Dombrowski "Jack" Roberts was born on June 8, 1918 in Los Angeles, California. He spent most of his 98 years in Southern California and witnessed first hand its transformation from a reasonably under-populated region into one of the world's busiest metropolitan areas. In fact, Jack (or "JDR" as he was oft referred in the labs at Caltech) was born essentially right underneath what is now the famous four level interchange connecting the 101 and 110 freeways in modern day downtown LA. JDR also witnessed the growth and explosion of science and in particular chemistry over that century span. As summarized in his *J. Org. Chem.* **2009**, *74*, 4897-4917 article and numerous talks over the later part of his life, the explosion of instrumentation capabilities available to the organic chemist progressed in the course of his scientific career from no less than the melting point apparatus to some of the most advanced instruments on the planet. Without doubt, the advances most influential to JDR's monumental career in chemistry were the advent of nuclear magnetic resonance (NMR) spectroscopy and the accompanying explosion in computing. Combined, these tools greatly facilitated the insightfully designed experimentation and careful analyses that became the hallmark of JDR's career. It is clear that

Jack's thoroughgoing nature combined with his deep understanding of instrumentation and fundamental chemistry served as an inspiration to nearly four generations of scientists. Roberts was oft credited by those far senior to the author as having "taught organic chemists" the power of NMR spectroscopy. Indeed, even Jack's license plate bestowed a love for NMR that only a limited number of SoCal residents could appreciate ("N15NMR"). JDR was a stalwart organic chemist and a scientist who would never shy away from discussion (or argument) in pursuit of the truth. He was a giant of a man.

JDR attended local elementary and secondary schools in SoCal before heading to UCLA as an undergraduate in 1936, where he worked with Bill McMillan and later with William R. Crowell, Don M. Yost, and Charles Coryell (with whom he published his first two research papers). JDR earned his B. A. degree in chemistry from UCLA in 1941. Jack held a lifelong fondness for his Alma matter, once recalling to the author the scene at the UCLA football stadium where his undergraduate contemporaries Jackie Robinson and Kenny Washington connected on touchdown passes. Both of these African American men would go on to play critical roles in breaking professional sports color barriers, for which JDR was proud to have had an association through UCLA. JDR moved from UCLA to attend Penn State University for graduate school in chemistry to work with Frank Whitmore, but returned to UCLA after only one year in the midst of WWII after the bombing of Pearl Harbor, in order to be closer to family. Back in SoCal JDR performed wartime research at UCLA with William Young.

In 1942 JDR married Edith Mary Johnson, who would play a central role in his life, his family's life, and the lives of many "adopted" families in the Pasadena area. Edith and Jack would have four children, Anne, Donald, John Paul, and Allen, who have remained a strong part of the Caltech family through the decades. Jack and especially Edith played a major role in local activities wherever they were, and in their later years were particularly strong supporters of the Pasadena Symphony. In addition to chemistry, Jack was an avid photographer. In his early years, he amassed one of the largest collections of photographs of 20th Century Chemists, many of which can be found in the supplementary materials to his 2009 *J. Org. Chem.* article mentioned above.

In 1944 JDR completed his Ph. D. from UCLA and began an appointment there as an instructor. After a brief time in this position, JDR received a National Research Council Fellowship and began a position at Harvard working with Paul D. Bartlett. In 1947 JDR was offered an instructorship at MIT by Arthur C. Cope and was rapidly promoted first to Assistant Professor in 1947 and then to Associate Professor in 1950. In 1950 Roberts also initiated a consulting and advisory relationship with DuPont that would last more than five decades. In 1952, while in Los Angeles on a Guggenheim Fellowship, JDR was offered a position at the California Institute of Technology and moved in 1953 as Professor of Organic Chemistry. Jack served on the Faculty at Caltech for the remainder of his life holding many positions including Division Chair of Chemistry and Chemical Engineering (1963-1968), Institute Provost (1980-1983), Institute Professor of Chemistry (1972-1988), and Institute Professor of Chemistry Emeritus (1988-2016).

In his illustrious career, JDR received countless awards, professorships, honorary degrees and numerous other notable accolades. He was the recipient of American Chemical Society Award in Pure Chemistry (1954), the Harrison Howe Award (1957), the Roger Adams Award in Organic Chemistry from the American Chemical Society (1967), the UCLA Alumni Achievement Award (1967), the William H. Nichols Medal (1972), the Richard C. Tolman Medal (1975), the Michaelson-Morley Award (1976), the James Flack Norris Award in Physical Organic Chemistry (1979), the Linus Pauling Award (1980), the Theodore William Richards Medal (1982), the Willard Gibbs Gold Medalist (1983), the American Academy of Achievement Golden Plate Award (1984), the Priestley Medal from the American Chemical Society (1987), the Madison Marshall Award (1989), the Robert A. Welch Award in Chemistry shared with W. von E . Doering (1990), the National Medal of Science (1990), the Glenn T. Seaborg Medal (1991), Award for Achievements in Magnetic Resonance (1991), the Service to Chemistry Award (1991), the SURF 92 Dedicatee (1992), the Arthur C. Cope Award from the American Chemical Society (1994), a Chemical Pioneer Award (1994), the History Maker Award of Pasadena Historical Society (1994), *a Chemical and Engineering News* citation as one of the 75 Most Influential Chemists of the Last 75 Years (1998), the National Academy of Sciences Award in Chemical Sciences (1999), the Nakanishi Prize from the American Chemical Society (2001), the Auburn-G. M. Kosolapdoff Award (2003), the Linus Pauling Legacy Award (2006), the Reaction Mechanisms Conference - Special Session Honoree (2006), the NAS Award for Chemistry in Service to Society, and the American Institute of Chemists Gold Medal Award (2013).

Roberts was a proud member of the American Chemical Society and became an ACS Fellow in 2009. JDR was also elected a Fellow of the Royal Society of Chemistry in 2008. He had a long association with *Organic Syntheses* (1955-2016) including as a long-time member of the Board of Directors. Notably, he served as Editor-in-chief of Volume 41 of *Organic Syntheses* (1961) where his laboratory checked no less than 25% of the published procedures! Roberts was an elected member of the American Academy of Arts and Sciences (1952), American Philosophical Society (1974), and the National Academy of Sciences (1956).

JDR was a prolific author and produced over five hundred published articles in addition to twelve books including his autobiography "The Right Place at the Right Time". Worthy of particular mention is *Basic*

Principles of Organic Chemistry (W.A. Benjamin, Inc., New York, 1964), which was co-authored with Marjorie C. Caserio. This book, known in all circles simply as "Roberts and Caserio", was a highly influential work that introduced a modern approach to the teaching of organic chemistry. The work, now freely available for download as an open access e-book, has been downloaded nearly a million times as of this writing. Additionally, Roberts' small book of notes on *Molecular Orbital Calculations* (W.A. Benjamin, Inc., New York, 1961) was influential in that it introduced many organic chemists to the relatively straightforward Linear Combination of Atomic Orbitals (LCAO) methods for calculating molecular orbitals and fostered the future intimate pairing of experiment and computation in physical organic chemistry.

Roberts' contributions to the primary research literature span an amazing eight decades (1940-2014). Although there are far too many published works to be perused by the casual reader, Roberts himself provided personal perspective on his career, when late in his life he listed what he believed to be his five most significant publications, which is reproduced below. This small sampling of published works provides a glimpse of both the breadth and depth that defined Roberts' scientific style.

1. Small-Ring Compounds. IV. Interconversion Reactions of Cyclobutyl, Cyclopropylcarbinyl and Allylcarbinyl Derivatives, *J. Am. Chem. Soc.*, **73**, 2509 (1951). With R. H. Mazur.

2. Rearrangement in the Reaction of Chlorobenzene-1-C14 with Potassium Amide, *J. Am. Chem. Soc.*, **75**, 3290 (1953). With H. E. Simmons, Jr., L. A. Carlsmith, and C. W. Vaughan.

3. Nuclear Magnetic Resonance Spectroscopy. Carbon-13 Spectra of Steroids, *J. Am. Chem. Soc.*, **91**, 7445 (1969). With H. J. Reich, M. Jautelat, M. T. Messe and F. J. Weigert.

4. Applications of Natural-Abundance Nitrogen-15 Nuclear Magnetic Resonance to Large Biochemically Important Molecules, *Proc. Nat. Acad. Sci. USA*, **72**, 4696 (1975). With D. Gust and R. B. Moon.

5. Nitrogen-15 Nuclear Magnetic Resonance Spectroscopy. The State of Histidine in the Catalytic Triad of α-Lytic Protease. Implications for the Charge-Relay Mechanism of Peptide-Bond Cleavage by Serine Proteases, *J. Am. Chem. Soc.*, **100**, 8041 (1978). With W. W. Bachovchin.

Jack and his family were quintessential Caltech. His booming voice, strong opinions, and oft thunderous laugh were all characteristic of his style. JDR was no shrinking violet and indeed was a legend. His staunch support of students across all disciplines at Caltech and through the decades are equally legendary and cemented his stature at the institute.

One of his earliest actions at Caltech was to assert that a Ph.D. student working with him at MIT, Dorothy Semenow, be admitted to Caltech in order to complete her degree. At the time, the institute was male only, yet with this action, and the intervention and assistance of Linus Pauling, Caltech instantly became co-ed at the graduate ranks. Although the road toward full and equal opportunity was still years away, Roberts helped pave that path. Much later in his career, JDR with Edith by his side welcomed legions of undergraduate "SURFers" (Caltech Undergraduate Research Fellowship awardees) to his group and introduced countless students to chemical research. The Roberts' collegiality extended to all who walked Caltech's campus. In the summer of 2000, when a cadre of new organic chemists joined the Caltech faculty (including the author) it was Jack and Edith who extended an invitation to all of us and our spouses to their home in beautiful Altadena to enjoy a memorable welcome dinner and celebration lauding our division chair. It was a pleasure and an honor to know Jack over the final 16 years of his life. It was clear that JDR worried about his legacy and his life's works. Shortly after our esteemed colleague and dear friend Nelson J. Leonard had passed away, JDR lamented to me that everyone who had known him as a young man had now all died and he could only barely recall what he himself was like in his younger days. I suppose that is a challenge of longevity, but it is clear that at every stage of JDRs life, he touched people in ways that will outlive him.

John D. Roberts is survived by his four children, nine grandchildren and one great grandchild.

Brian M. Stoltz
Pasadena, California

Edwin Vedejs
1941 – 2017

Edwin Vedejs, Editor-in-Chief of Volume 65 of Organic Syntheses, passed away on December 2, 2017. He was 76 years old.

Edwin Vedejs was born in Riga, Latvia, on January 31, 1941. During World War II, he lived for six years in displaced persons camps in Germany before emigrating to the United States in 1950, where he ultimately settled with his family in Grand Rapids, MI.

Ed received a Bachelor of Science degree in chemistry from the University of Michigan (Ann Arbor) in 1962. He completed his Ph.D. in chemistry in 1966 at the University of Wisconsin, Madison, under the direction of Professor Hans Muxfeldt, and performed post-doctoral research from 1966–67 at Harvard University in the laboratory of Professor E. J. Corey.

Ed Vedejs had a long and distinguished career at his two alma maters. He began his independent career at the University of Wisconsin, Madison, in 1967 where he rose to the rank of professor of chemistry, serving as the Helfaer Professor (1991-1996) and Robert M. Bock Professor (1997-98). In 1999, he moved to the University of Michigan, Ann Arbor, as the Moses Gomberg Collegiate Professor of Chemistry, a position that he held until his retirement in 2011. In recognition of his accomplishments, the University of Michigan established the Edwin Vedejs Collegiate Professor of Chemistry Chair after Ed's retirement.

Ed Vedejs was a prolific author and an internationally recognized scholar. The American Chemical Society awarded him the Herbert C. Brown Award for Creative Research in Synthetic Methods in 2004, and in 2008 Ed was named a Fellow of the American Chemical Society.

Ed Vedejs was an expert in synthetic and mechanistic organic chemistry and was also a leader in the development of synthetic approaches to numerous natural products. Throughout his career, Ed studied a number of fundamentally important problems in the field of organic chemistry. Among his most noteworthy contributions are his seminal studies on the mechanistic and stereochemistry of the Wittig reaction. Ed was the first to demonstrate that oxaphosphetanes are stable reaction intermediates, and his careful kinetic analyses were critical to the development of a unifying mechanistic scheme that provides the now widely accepted rationale for understanding the stereoselectivity of this important reaction. His kinetic and mechanistic insights also led to practical improvements of the reaction, specifically with respect to the synthesis of (E)-alkenes from non-stabilized ylide precursors.

Ed's early contributions from the University of Wisconsin also included seminal studies in organopalladium chemistry, and the development of the MoOPH reagent for enolate hydrosylation. Additional work in sulfur ylide ring expansions, thioaldehyde cycloadditions, and silicon-mediated 1,3-dipole generation allowed for the exploration of basic questions of chemical reactivity in the context of total synthesis.

After moving to the University of Michigan in 1999, Vedejs' research focused on several unique and powerful methods for asymmetric synthesis, including the use of "fragile asymmetry" based on temporary boron-based chiral centers, asymmetric protonation of prochiral enolates, parallel kinetic resolution, chiral phosphines as acyl transfer catalysts, and the clever use of deuterium isotope effects to "protect" an acidic center. Ed's seminal contributions in these areas have had a substantial (and still growing) impact on the field of asymmetric synthesis.

During his 45-year career, Ed supervised more than 80 Ph.D. theses and numerous post-doctoral and undergraduate research projects. Many of his former students now hold prestigious positions in academia and in the biopharmaceutical industry. All testify to Ed's brilliant insights, his exceptional chemical intuition, his unwavering requirement of scientific rigor, and his passion for the chemical sciences.

In addition to his stellar contributions as a research scientist, Ed Vedejs served the organic synthesis research community with distinction. He served as associate editor of the *Journal of the American Chemical Society* from 1994–99, as chairman of the NIH Medicinal Chemistry Study Section (1990–91), and as chair of the Organic Division of the American Chemical Society (2003), among many other contributions. Ed was a member of the Organic Syntheses Board of Editors from 1980 to 1988 and served as the Editor for Volume 65.

Ed also had an unwavering commitment to the Latvian chemistry research community. He worked tirelessly to promote science in Latvia by sponsoring graduate students and professors to study at the University of Wisconsin. He helped Latvian universities to gain access to research journals, collaborated on international research grants, and taught courses at Riga Technical University. For these efforts he received many honors from Latvia including the Paul Walden Medal (1997), the Grand Medal of the Latvian Academy of Sciences (2005), the Order of the Three Stars (2006), and an honorary doctorate from Riga Technical University (2010).

Ed Vedejs was a scholar of the first magnitude with the unique ability to provide understanding and insights in areas of chemistry that ranged far outside of the problems he was addressing in his laboratory. Ed had the remarkable ability to ask penetrating questions that challenged assumptions and which always addressed the fundamental issues of the problem at hand. I can attest that I grew scientifically and learned enormously from my interactions with Ed at the University of Michigan – he was the consummate colleague, scholar and educator who enriched everyone who had the privilege of interacting with him.

Ed is survived by his loving wife, Pat Anderson; his son, Michael; his daughters Christina Mersereau, Jesikah Cordova, and Julia Vander Meer; and his former wife, Melita Vedejs. He will be deeply missed by his professional colleagues and friends in the United States, Latvia and elsewhere.

WILLIAM R. ROUSH

PREFACE

The publication of this volume of *Organic Syntheses* marks the 95th in a series of uncompromised works wherein protocols for the practicing organic chemist have been submitted and checked by experts in the field. The uniqueness of this endeavor in science cannot be understated as reproducibility is at the heart of the scientific method. For observations to count and to truly matter, they must be replicated and observed by more than just the initial discoverer. For the science of organic chemistry, Organic Syntheses plays a pivotal and crucial role in establishing reliable synthetic procedures of high scientific quality. As in the past ninety-four volumes, the current edition, Volume 95, holds within its bounds protocols from the cutting-edge of chemistry. The procedures include a number of methods for the synthesis of hetero- and carbocyclic scaffolds, the preparation of new catalysts, and the implementation of frontline catalytic methods. Asymmetric synthesis continues to be a driver in the field and the current volume is replete with methods for enantio- and diastereoselective synthesis of many important compound types. Finally, there are new examples of C−H functionalization reactions within this volume, again signifying how much organic chemistry has grown and continues to broaden with each passing year. The level of sophistication demonstrated in the current Volume 95 of Organic Syntheses would undoubtedly be remarkable to our predecessors.

The past year has also seen a number of distinguished past editors depart this earth. John D. (Jack) Roberts, Ronald Breslow, Andrew S. Kende, Ichiro Shinkai, and Edwin Vedejs all recently passed away. It cannot be understated the role these great scientists played in the development of the field of chemical synthesis and organic chemistry more broadly. Bound in the current Volume are biographies of this distinguished group, and as a memorial to their service and contributions, I would like to dedicate this volume to them.

Enterprises such as *Organic Syntheses* and outstanding compendia, akin to Volume 95, require an enormous collective effort on the part of many individuals. The process of bringing together thoughtful scientific contributors, a diligent Board of Editors and their co-workers, a highly supportive Board of Directors and corporation, and a stellar Associate Editor (Chuck Zercher) is the primary undertaking of our Editor-in-Chief, Rick Danheiser. Rick and Chuck serve to keep their Board of Editors in line and on task, yet always do so in a pleasant way and with a patience that cannot be overstated. I, for one, thank them for their patience with me as both a member of the board and as the Editor of Volume 95. The experience of serving on the Board of Editors for the past eight years was an enjoyable and intellectually stimulating one, and I am grateful to all of the past and current members of the Board for providing such a pleasant atmosphere at Organic Syntheses.

BRIAN M. STOLTZ

TABLE OF CONTENTS

Wei Wen Tan, Bin Wu, Ye Wei, and Naohiko Yoshikai

Xinpeng Cheng, Yanzhao Wang, and Liming Zhang

Kip A. Teegardin and Jimmie D. Weaver

Preparation of Cyclopent-2-enone Derivatives via the Aza-Piancatelli Rearrangement   46

Meghan F. Nichol, Luis Limon, and Javier Read de Alaniz

Catalytic, Metal-Free Oxidation of Primary Amines to Nitriles   60

Kyle M. Lambert, Sherif A. Eldirany, James M. Bobbitt, and William F. Bailey

Copper-Catalyzed Enantioselective Hydroamination of Alkenes   80

Richard Y. Liu and Stephen L. Buchwald

Preparation of 6H-Benzo[c]chromen-6-one

Yang Wang, Yi Shi, and Vladimir Gevorgyan

Synthesis of N-Acyl Pyridinium-N-Aminides and Their Conversion to 4-Aminooxazoles via a Gold-Catalyzed Formal (3+2)-Dipolar Cycloaddition

Matthew P. Ball-Jones and Paul W. Davies

Preparation of Solid Organozinc Pivalates and Their Reaction in Pd-Catalyzed Cross-Couplings

Mario Ellwart, Yi-Hung Chen, Carl Phillip Tüllmann, Vladimir Malakhov, and Paul Knochel

Emerson Teixeira da Silva, Adriana Marques Moraes, Adriele da Silva Araújo, and Marcus Vinícius Nora de Souza

Marvin Kischkewitz and Armido Studer

Hiroto Yoshida, Yuya Murashige, and Itaru Osaka

(pin)B–B(dan)

Discussion Addendum for:
Preparation of a Carbazole-Based Macrocycle Via Precipitation-Driven Alkyne Metathesis

Christopher C. Pattillo, Morgan M. Cencer, and Jeffrey S. Moore

Hydrodecyanation by a Sodium Hydride-Iodide Composite

Guo Hao Chan, Derek Yiren Ong, and Shunsuke Chiba

Indole-Catalyzed Bromolactonization: Preparation of Bromolactone 256 in Lipophilic Media

Zhihai Ke, Tao Chen, and Ying-Yeung Yeung

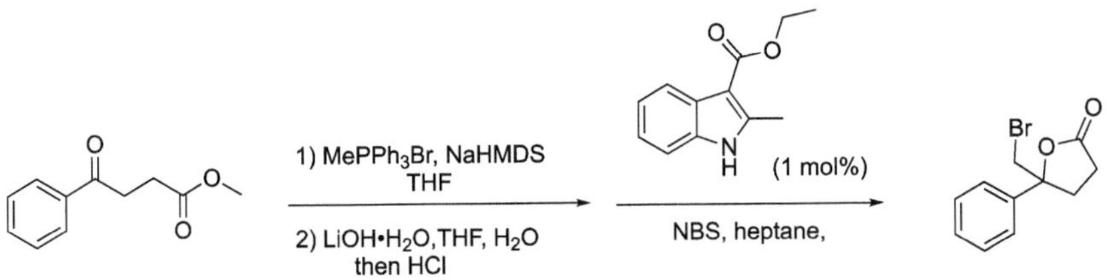

1) MePPh₃Br, NaHMDS
THF

2) LiOH•H₂O, THF, H₂O
then HCl

(1 mol%)

NBS, heptane,

Discussion Addendum for:
Stereoselective Synthesis of 3-Arylacrylates by Copper-Catalyzed Syn 267 Hydroarylation [(E)-Methyl 3-phenyloct-2-enoate]

Yoshihiko Yamamoto

$n\text{-}C_5H_{11}$——CO_2Me + ⬡—$B(OH)_2$

cat. CuOAc

MeOH

$n\text{-}C_5H_{11}$ CO_2Me

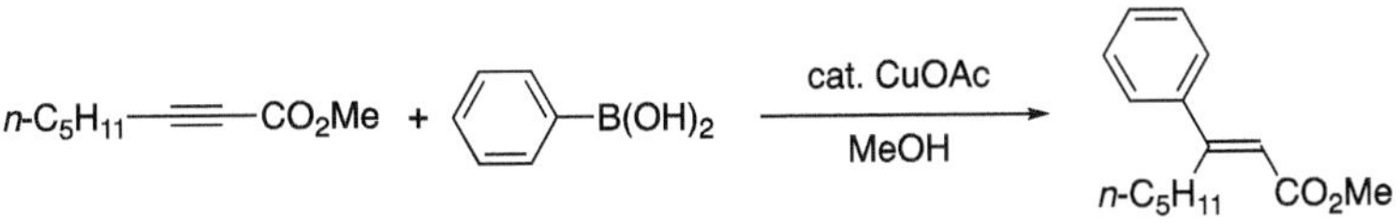

Modified McFadyen-Stevens Reaction for a Versatile Synthesis of 276 Aromatic Aldehydes

Yuri Iwai and Jun Shimokawa

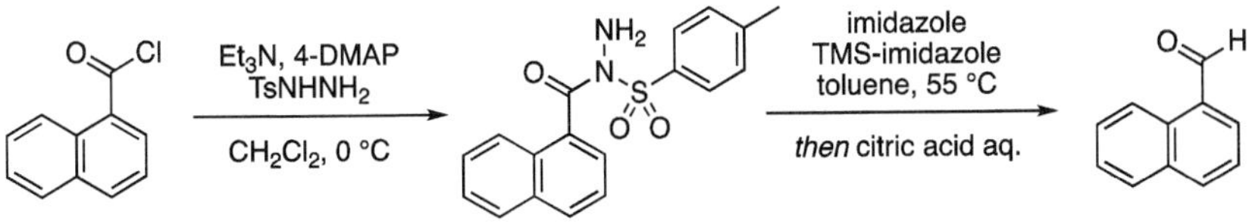

Et₃N, 4-DMAP
TsNHNH₂

CH₂Cl₂, 0 °C

imidazole
TMS-imidazole
toluene, 55 °C

then citric acid aq.

Facile Syntheses of Aminocyclopropanes: **289**
N,N-**Dibenzyl-***N*-**(2-ethenylcyclopropyl)amine [(Benzenemethanamine,**
N-**(2-ethenylcyclopropyl)-***N*-**(phenylmethyl))**

Armin de Meijere and Sergei I. Kozhushkov

$$Ti(OiPr)_4 \ + \ TiCl_4 \ \xrightarrow[\text{ether}]{\text{MeLi}} \ MeTi(OiPr)_3$$

(*R*)-2,2,2-Trichloro-1-phenylethyl (methylsulfonyl)oxycarbamate   **310**

Hélène Lebel, Henri Piras, and Johan Bartholoméüs

Large Scale Synthesis of Enantiomerically Pure   **328**
(*S*)-3-(4-Bromophenyl)- butanoic Acid

J. Craig Ruble, H. George Vandeveer, and Antonio Navarro

Preparation of Tributyl(iodomethyl)stannane

Michael U. Luescher, Chalupat Jindakun, and Jeffrey W. Bode

Bu₃SnH → [DIPA, *n*-BuLi, (CH₂)ₙO, THF; then CH₃SO₂Cl] → Bu₃Sn—CH₂—Cl → [NaI / acetone] → Bu₃Sn—CH₂—I

Stannylamine Protocol (SnAP) Reagents for the Synthesis of C–Substituted Morpholines from Aldehydes

Michael U. Luescher, Chalupat Jindakun, and Jeffrey W. Bode

Trimethylsilyldiazo[^{13}C]methane: A Versatile ^{13}C-Labelling Reagent 374

Chris Nottingham and Guy C. Lloyd-Jones

^{13}CH$_3$OH → (Tosyl Chloride / NaOH / THF:Water / 0 °C to rt) → ^{13}CH$_3$OTs **1**

NH, Ph, Ph → (i) nBuLi, THF, -78 °C (ii) **1**, THF -78 °C to rt → ^{13}CH$_3$, N, Ph, Ph → LDA, TMSCl, THF, -78 °C → SiMe$_3$, ^{13}CH$_2$, N, Ph, Ph

SiMe$_3$, ^{13}CH$_2$, N, Ph, Ph → (i) H$_2$, Pd/C (5 mol% Pd), MeOtBu (ii) HCl in Et$_2$O → Cl$^-$ H$_3$N$^+$, ^{13}CH$_2$, SiMe$_3$ → NaOH, Et$_2$O → H$_2$N, ^{13}CH$_2$, SiMe$_3$

H$_2$N, ^{13}CH$_2$, SiMe$_3$ → (ONO...ONO) / 3-NO$_2$-Phenol (10 mol%) / 2-MeTHF/Et$_2$O, rt → N$_2$, ^{13}CH, SiMe$_3$

Stereoretentive Iron-catalyzed Cross-coupling of an Enol Tosylate with MeMgBr 403

Takeshi Tsutsumi, Yuichiro Ashida, Hiroshi Nishikado, and Yoo Tanabe

Ph, CO$_2$Me → HCO$_2$Me, NaOt-Bu, THF, 0 – 10 °C → [Ph, O$^-$Na, O, OMe] → TsCl, THF, 0 – 10 °C → Ph, OTs, CO$_2$Me

Ph, OTs, CO$_2$Me → MeMgBr, Fe(acac)$_3$ (cat), TMEDA, AcOEt – THF, 20 – 25 °C → Ph, Me, CO$_2$Me

Synthesis of Methyl 2-Bromo-3-oxocyclopent-1-ene-1-carboxylate   **425**

Rama Rao Tata and Michael Harmata

Discussion Addendum for:
Preparation of (*S*)-*tert*-ButylPHOX and (*S*)-2-Allyl-2-   **439**
Methylcyclohexanone

Alexander W. Sun and Brian M. Stoltz

Synthesis of Acyl Derivatives of Cotarnine   **455**

Laxmidhar Rout, Bibhuti Bhusan Parida, Ganngum Phaomei, Bertounesque Emmanuel, and
Akhila Kumar Sahoo

Carbonyl-Olefin Metathesis for the Synthesis of Cyclic Olefins

Marc R. Becker, Katie A. Rykaczewski, Jacob R. Ludwig, and Corinna S. Schindler

Hexafluoro-2-propanol-promoted Intramolecular Friedel-Crafts Acylation

Rakesh H. Vekariya, Matthew C. Hortonand Jeffrey Aubé

Discussion Addendum for:
Preparation of the COP Catalysts: [(S)-COP-OAc]₂, [(S)-COP-Cl]₂,   **500**
and (S)-COP-hfacac

Jeffrey S. Cannon and Larry E. Overman

Copper and Secondary Amine-Catalyzed Pyridine Synthesis from *O*-Acetyl Oximes and α,β-Unsaturated Aldehydes

Wei Wen Tan, Bin Wu, Ye Wei, and Naohiko Yoshikai[1]*

Division of Chemistry and Biological Chemistry, School of Physical and Mathematical Sciences, Nanyang Technological University, Singapore 637371, Singapore

Checked by Suttipol Radomkit and Chris H. Senanayake

A.

1) NH$_2$OH·HCl, pyridine
ethanol, 60 °C
2) Ac$_2$O, DMAP, pyridine, rt

Ph—C(=O)—Me → Ph—C(=NOAc)—Me **1**

B.

Ph—C(=NOAc)—Me **1** + H—C(=O)—CH=CH—Ph

CuI (20 mol%)
i-Pr$_2$NH (2 equiv)

DMSO, 60 °C, 16 h

→ Ph—(pyridine)—Ph **2**

Procedure (Note 1)

A. *Acetophenone O-acetyl oxime (1).* A 250-mL, two-necked round-bottomed flask equipped with a Teflon-coated magnetic stir bar (oval, 20 x 10 mm) is charged with acetophenone (9.33 mL, 80.0 mmol, 1 equiv) (Note 2), pyridine (18 mL, 223 mmol, 2.8 equiv) (Note 3), ethanol (40 mL) (Note 4), and hydroxylamine hydrochloride (8.33 g, 120 mmol, 1.5 equiv) (Note 5). The resulting mixture is stirred at 60 °C for 75 min (Notes 6 and 7) (Figure 1) and then cooled to 22 °C. Water (80 mL) is added, and the resulting biphasic mixture is transferred to a 500-mL separatory funnel. The mixture is partitioned, and the aqueous layer is extracted with ethyl acetate

(2 × 80 mL). The organic extracts are combined, washed successively with 1 M HCl (50 mL) and brine (50 mL), and dried over anhydrous MgSO₄ (10 g) for 15 min. The extracts are gravity filtered through a filter paper and

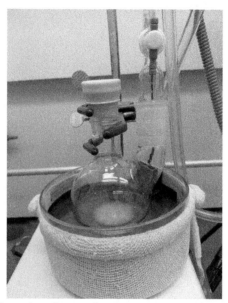

Figure 1. Reaction Setup for Step A (acetophenone oxime)

concentrated on a rotary evaporator (35 °C, 30 mmHg). The residue is transferred to a 250-mL, one-necked round-bottomed flask and dried further for 1 h under reduced pressure (1.5 mmHg, 22 °C) (Notes 8 and 9). A Teflon-coated stir bar (15 x 6 mm) is placed in the flask. To the flask are added *N,N*-dimethylaminopyridine (20 mg, 0.16 mmol, 0.002 equiv) (Note 10), pyridine (35 mL) (Note 2), and acetic anhydride (15 mL, 160 mmol, 2 equiv) (Note 11). The resulting mixture is stirred at room temperature (24 °C) for 1 h (Note 12). The volatile materials are removed on a rotary evaporator (45 °C, 35 mmHg). To the residue is added water (50 mL), and the mixture is transferred to a 250-mL separatory funnel. The mixture is partitioned, and the aqueous layer is extracted with ethyl acetate (2 x 50 mL). The organic extracts are combined, washed successively with 1 M HCl (50 mL) and brine (50 mL), and dried over anhydrous MgSO₄ (10 g) for 15 min. The extracts are filtered through a filter paper and concentrated on a rotary evaporator (35 °C, 30 mmHg). Recrystallization of the residual white solid from ethyl acetate (Note 13) affords (*E*)-acetophenone *O*-acetyl oxime

(**1**) as white crystals (9.41 g, 53.1 mmol, 66%) (Figure 2) (Notes 14, 15, and 16).

Figure 2. (*E*)-Acetophenone *O*-acetyl oxime (1)

B. *2,4-Diphenylpyridine (2)*. An oven-dried, 500-mL, three-necked round-bottomed flask is equipped with a Teflon-coated magnetic stir bar (oval, 30 x 16 mm) and the side neck is fitted with a glass gas inlet adapter connected to a vacuum/nitrogen manifold. The other side neck is capped with a glass stopper, and the center with a rubber septum (Figure 3). The flask is charged with **1** (6.19 g, 34.9 mmol) and copper(I) iodide (1.33 g, 6.98 mmol, 0.2 equiv) (Note 17). The flask is evacuated and backfilled with N_2 three times. To the flask are added cinnamaldehyde (6.6 mL, 52 mmol, 1.5 equiv) (Note 18), diisopropylamine (9.8 mL, 70 mmol, 2 equiv) (Note 19), and DMSO (175 mL) (Note 20) via syringe. The center neck is capped with a glass stopper, and the resulting mixture is stirred at 60 °C for 16 h under a gentle stream of nitrogen (Notes 21 and 22) (Figure 1). Upon cooling to room temperature, the reaction mixture is diluted with ethyl acetate (175 mL) and water (100 mL). The mixture is filtered through Celite (10 g) on a glass filter (Note 22) while washing with ethyl acetate (175 mL). The filtrate is transferred to a 1-L separatory funnel. The mixture is partitioned, and the organic layer is washed successively with water (150 mL) and brine

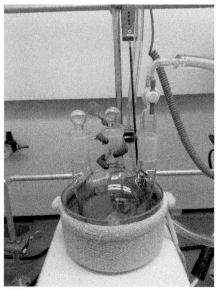

Figure 3. Reaction Setup for Step B

(50 mL), and dried over anhydrous MgSO₄ (15 g) for 30 min. The organic layer is filtered through Celite (10 g) on a glass filter (Note 23) and concentrated on a rotary evaporator (35 °C, 30 mmHg). The residue is transferred to a 25-mL flask and subjected to Kugelrohr distillation (Note 24) to remove excess cinnamaldehyde. The residue is subjected to column chromatography on silica gel (Note 25). The fractions containing the product are combined and concentrated on a rotary evaporator (35 °C, 30 mmHg), and the product is dried under reduced pressure at room temperature (1.5 mmHg, 2 h) to yield 2,4-diphenylpyridine (**2**) as a brown solid (5.17 g, 22.4 mmol, 64%) (Figure 4) (Notes 26, 27 and 28).

Figure 4. 2,4-Diphenylpyridine (2)

Notes

1. Prior to performing each reaction, a thorough hazard analysis and risk assessment should be carried out with regard to each chemical substance and experimental operation on the scale planned and in the context of the laboratory where the procedures will be carried out. Guidelines for carrying out risk assessments and for analyzing the hazards associated with chemicals can be found in references such as Chapter 4 of "Prudent Practices in the Laboratory" (The National Academies Press, Washington, D.C., 2011; the full text can be accessed free of charge at https://www.nap.edu/catalog/12654/prudent-practices-in-the-laboratory-handling-and-management-of-chemical. See also "Identifying and Evaluating Hazards in Research Laboratories" (American Chemical Society, 2015) which is available via the associated website "Hazard Assessment in Research Laboratories" at https://www.acs.org/content/acs/en/about/governance/committees/chemicalsafety/hazard-assessment.html. In the case of this procedure, the risk assessment should include (but not necessarily be limited to) an evaluation of the potential hazards associated with acetophenone, pyridine, ethanol, hydroxylamine hydrochloride, ethyl acetate, hydrochloric acid, magnesium sulfate, *N,N*-dimethylaminopyridine, acetic anhydride, copper(I) iodide, cinnamaldehyde, diisopropylamine, dimethylsulfoxide, and Celite, as well as the proper procedures for operation of a Kugelrohr apparatus.

2. Acetophenone (99%) was purchased from Alfa Aesar and used as received.

3. Anhydrous pyridine (99.8%) was purchased from Sigma-Aldrich and used as received.

4. Anhydrous, denatured ethanol was purchased from Sigma-Aldrich and used as received.

5. Hydroxylamine hydrochloride (99%) was purchased from Alfa Aesar and used as received.

6. A 1.2-L oil bath was used, and stirring was performed at 450 rpm. After 5 min of heating, the white solids of hydroxylamine hydrochloride were completely dissolved and a colorless solution was obtained. No color change was observed over the course of the reaction.

7. The reaction progress was monitored by TLC analysis on Merck silica gel 60 F_{254} plates with hexane/ethyl acetate (9:1) as the eluent. The TLC

plates were visualized under 254 nm UV lamp. The R_f values of acetophenone and acetophenone oxime are 0.31 and 0.20, respectively.

8. A Schlenk line equipped with an oil pump was used. A colorless solid (ca. 12 g) containing some residual solvents, which did not interfere with the next step, was obtained.

9. The characterization data of the non-purified acetophenone oxime are as follows: ATR-FTIR: 3236, 1497, 1370, 1302, 1005, 925 cm⁻¹; ¹H NMR (500 MHz, CDCl₃) δ: 2.34 (s, 3H), 7.41-7.43 (m, 3H), 7.65-7.66 (m, 2H), 9.79 (brs, 1H); ¹³C NMR (125 MHz, CDCl₃) δ: 12.5, 126.2, 128.7, 129.4, 136.6, 156.1; HRMS (ESI) *m/z* calcd for C₈H₁₀NO [M+H]⁺ 136.0762, found 136.0765.

10. 4-(Dimethylamino)pyridine (99%) was purchased from Alfa Aesar and used as received.

11. Acetic anhydride (99+%) was purchased from Acros and used as received (checkers). Acetic anhydride (98%) was purchased from Merck KGaA and used as received (submitters).

12. The reaction progress was monitored by TLC analysis on Merck silica gel 60 F₂₅₄ plates with hexane/*tert*-butyl methyl ether (4:1) as the eluent. The TLC plates were visualized under 254 nm UV lamp. Acetophenone oxime and **1** show the R_f value of 0.35 and 0.2, respectively.

13. Recrystallization was performed as follows: Ethyl acetate (5 mL) was added to the crude residue and the flask equipped with a condensor. The mixture was gently heated until all the solids were dissolved. The solution was left to cool to room temperature and then placed in an ice bath to grow white crystalline solids. The solids were collected by filtration through a Büchner funnel and washed with ice-cold hexane (100 mL).

14. Acetophenone *O*-acetyl oxime (**1**) has the following physical properties: mp = 56–57 °C; ATR-FTIR: 1758, 1615, 1571, 1445, 1361, 1308, 1203, 1002, 978, 934, 892 cm⁻¹; ¹H NMR (500 MHz, CDCl₃) δ: 2.26 (s, 3H), 2.38 (s, 3H), 7.38-7.45 (m, 3H), 7.73-7.74 (m, 2H); ¹³C NMR (125 MHz, CDCl₃) δ: 14.5, 19.9, 127.1, 128.7, 130.7, 135.0, 162.5, 169.0; HRMS (ESI) *m/z* calcd for C₁₀H₁₂NO₂ [M+H]⁺ 178.0868, found 178.0866. The compound was evaluated by DSC (Differential Scanning Calorimetry), which indicates the compound should never be heated above ~150 °C.

15. The purity was determined to be >99% wt. by quantitative ¹H NMR spectroscopy in CDCl₃ using 31.5 mg of the compound **1** and 16.1 mg of dimethyl fumarate as an internal standard.

16. A second reaction on the same scale provided 9.88 g (70%) of the product.
17. CuI (98%) was purchased from Strem Chemicals and used as received.
18. *trans*-Cinnamaldehyde (98%) was purchased from Alfa Aesar and purified by Kugelrohr distillation (120 °C, 2.3 mmHg).
19. Diisopropylamine (99.5%) was purchased from Aldrich and purified by distillation over CaH_2 (checkers). Diisopropylamine (99+%) was purchased from Alfa Aesar and purified by distillation over CaH_2 (submitters).
20. Anhydrous DMSO was purchased from Alfa Aesar and used as received.
21. A 1.2-L oil bath was used, and stirring was performed at 450 rpm. During the initial 15 min of heating, the reaction turned from light green to dark green to black.
22. The reaction progress was monitored by TLC analysis on Merck silica gel 60 F_{254} plates with hexane/ethyl acetate (9:1) as the eluent. The TLC plates were visualized under 254 nm UV lamp and then with 2,4-dinitrophenylhydrazine stain. R_f values of **1**, **2**, and cinnamaldehyde are 0.20, 0.32 and 0.32 respectively. Only cinnamaldehyde turns red/orange immediately after dipping into 2,4-dinitrophenylhydrazine stain.
23. The specifications of the glass filter are as follows: Volume 150 mL, OD 60 mm, porosity 40-60 μm.
24. Kugelrohr distillation was performed at 120 °C under 0.75 mmHg for 90 min.
25. The crude material is dissolved in dichloromethane (4 mL) and then charged onto a column (diameter = 7 cm, height (silica height) = 11 cm) of 190-gram silica gel. The column is first eluted with hexane (100 mL), then (sequentially) eluted with hexane/ethyl acetate = 95:5 (500 mL), hexane/ethyl acetate = 92:8 (1000 mL), hexane/ethyl acetate = 90:10 (500 mL), hexane/ethyl acetate = 88:12 (500 mL). The first 500 mL eluent was not collected, and then 13-14-mL fractions were collected. Fractions 71-108 were combined and concentrated on a rotary evaporator (35 °C, 30 mmHg).
26. 2,4-Diphenylpyridine (**2**) has the following physical properties: mp = 66–67 °C; ATR-FTIR: 1604, 1592, 1577, 1540, 1493, 1468, 1447, 1388, 1074, 988, 885, 842 cm^{-1}; ^1H NMR (500 MHz, $CDCl_3$) δ: 7.43-7.53 (m, 7H), 7.68 (d, *J* = 7.5 Hz, 2H), 7.93 (s, 1H), 8.10 (d, *J* = 8.0 Hz, 2H), 8.75 (d, *J* = 5.5 Hz, 1H); ^{13}C NMR (125 MHz, $CDCl_3$) δ: 118.9, 120.4, 127.16, 127.2,

128.9, 129.14, 129.15, 129.2, 138.7, 139.6, 149.4, 150.2, 158.2; HRMS (ESI) m/z calcd for $C_{17}H_{14}N$ [M+H]$^+$ 232.1126, found 232.1120.

27. The purity was determined to be >99% wt. by quantitative ^1H NMR spectroscopy in CDCl$_3$ using 23.5 mg of the compound **2** and 12.4 mg of dimethyl fumarate as an internal standard.

28. A second reaction on the same scale provided 5.41 g (67%) of the product.

Working with Hazardous Chemicals

The procedures in *Organic Syntheses* are intended for use only by persons with proper training in experimental organic chemistry. All hazardous materials should be handled using the standard procedures for work with chemicals described in references such as "Prudent Practices in the Laboratory" (The National Academies Press, Washington, D.C., 2011; the full text can be accessed free of charge at http://www.nap.edu/catalog.php?record_id=12654). All chemical waste should be disposed of in accordance with local regulations. For general guidelines for the management of chemical waste, see Chapter 8 of Prudent Practices.

In some articles in *Organic Syntheses*, chemical-specific hazards are highlighted in red "Caution Notes" within a procedure. It is important to recognize that the absence of a caution note does not imply that no significant hazards are associated with the chemicals involved in that procedure. Prior to performing a reaction, a thorough risk assessment should be carried out that includes a review of the potential hazards associated with each chemical and experimental operation on the scale that is planned for the procedure. Guidelines for carrying out a risk assessment and for analyzing the hazards associated with chemicals can be found in Chapter 4 of Prudent Practices.

The procedures described in *Organic Syntheses* are provided as published and are conducted at one's own risk. *Organic Syntheses, Inc.*, its Editors, and its Board of Directors do not warrant or guarantee the safety of individuals using these procedures and hereby disclaim any liability for any injuries or damages claimed to have resulted from or related in any way to the procedures herein.

Discussion

The efficient and regioselective synthesis of substituted pyridines represents one of the most important subjects in heterocyclic chemistry, and is relevant to medicinal and materials chemistry because of the prevalence of the pyridine core in pharmaceuticals and other organic functional materials. Besides conventional methods based on carbonyl condensation chemistry, a variety of alternative approaches, those catalyzed by transition metals in particular, have been extensively explored to date.[2] In this context, oxime derivatives have emerged as readily available and versatile starting materials for the synthesis of pyridines and other nitrogen-containing heterocycles under transition metal catalysis.[3] For example, Liebeskind developed [4+2]-type pyridine synthesis via copper-catalyzed C–N coupling between alkenylboronic acid and α,β-unsaturated oxime O-pentafluorobenzoate and subsequent electrocyclization.[4] α,β-Unsaturated oxime derivatives have also been exploited for [4+2]-type pyridine synthesis via rhodium-catalyzed, oxime-directed C–H activation and annulation with alkynes[5] or alkenes.[6]

The pyridine synthesis described here is based on our previous work on the [3+3]-type condensation of oxime acetates having α-protons with α,β-unsaturated aldehydes in the presence of a copper(I) salt and a secondary amine (or ammonium salt).[7] We developed two viable catalytic systems for this transformation. One comprises catalytic amounts (20 mol% each) of CuI and pyrrolidinium perchlorate (denoted as method A), while the other employs catalytic CuI (20 mol%) in combination with a superstoichiometric amount (2 equiv) of diisopropylamine (denoted as method B). As briefly summarized in Table 1, both methods A and B are effective for the condensation of oximes derived from acetophenone and related aryl ketones with cinnamaldehyde derivatives (see the products **2–8**). On the other hand, only method B is practicable for oximes derived from other types of ketones (see the products **9–12**). In addition, there remains room for improvement for the reaction of unsaturated aldehydes other than cinnamaldehyde derivatives (see the products **15–17**).

In this report, the synthesis of 2,4-diphenylpyridine on a multigram scale is described in detail. Here, method B was chosen because diisopropylamine would be in the list of commodity chemicals of most synthetic laboratories. While the synthesis of 2,4-diphenylpyridine itself was achieved in many different ways (e.g., cross-coupling between 2,4-

Table 1. Selected pyridines obtained by the copper/amine-catalyzed condensation (previous work)[a]

method A: CuI (20 mol%), pyrrolidinium perchlorate (20 mol%)
method B: CuI (20 mol%), i-Pr₂NH (2 equiv)

2 75% (A)[b] 84% (B)[b]	**3** 78% (A) 66% (B)	**4** 49% (A) 72% (B)	**5** 42% (A) 79% (B)
6 85% (A) 78% (B)	**7** 80% (A) 92% (B)	**8** 80% (A) 92% (B)	**9** 20% (A)[c] 55% (B)
10 <10% (A)[c] 65% (B)	**11** <10% (A)[c] 79% (B)	**12** <10% (A)[c] 52% (B)	**13** 75% (A) 76% (B)
14 79% (A)[d] 80% (B)	**15** 36% (B)[e]	**16** 23% (A)[f] 72% (B)[f]	**17** 36% (A)[g]

[a] Unless otherwise noted, the reaction was performed using 0.2 mmol of oxime acetate and 0.3 mmol (1.5 equiv) of α,β-unsaturated aldehyde. [b] 10 mmol-scale reaction. [c] Determined by GC. [d] 40 mol% of CuI was used. [e] 5 equiv of acrolein was used. [f] 2 equiv of methacrolein was used. [g] 3 equiv of aldehyde was used.

dihalopyridine and phenylmetal reagents), there had been practically no general method for the synthesis of diversely substituted 2,4-diarylpyridines from simple starting materials. Indeed, many of the pyridine products described in our previous work[7] had not been reported before.

From a mechanistic point of view, the present condensation reaction capitalizes on the ability of copper(I) to reduce the oxime N–O bond[8] as well as that of secondary amine (or ammonium) to activate an aldehyde via iminium formation. Thus, the copper catalyst would reduce the oxime acetate via sequential single-electron transfer to generate an iminylcopper(II), which would then tautomerize to a nucleophilic copper(II) enamide species.[9] A Michael addition of the copper enamide to the α,β-unsaturated iminium and subsequent cyclocondensation would give a dihydropyridine intermediate while liberating the secondary amine. The dihydropyridine would be oxidized by copper(II) to furnish the pyridine and regenerate copper(I). The overall reaction mechanism may be understood in terms of synergistic catalysis of copper(I) and amine.[10]

Since our report, a few variants of the present reaction have been reported. Cui et al. developed a copper-catalyzed three-component condensation of an oxime pivalate, an aldehyde, and malononitrile to afford a 2-aminonicotinonitrile derivative,[11] while Jiang et al. developed an analogous three-component condensation of an oxime acetate, an aldehyde, and a β-keto ester to afford a multi-substituted pyridine.[12] These reactions most likely involve the formation of an activated Michael acceptor via Knoevenagel condensation. Huang, Deng, and coworkers reported a metal-free system employing iodine and triethylamine, as an alternative to the present copper/amine system, for the condensation of an oxime acetate and an α,β-unsaturated aldehyde.[13] Despite the simplicity, the system requires a higher reaction temperature (120 °C) and appears to have a narrower scope than that of the present method, as it is not applicable to oximes derived from dialkylketones. Besides these reports, several other transition metal-catalyzed pyridine-forming reactions employing oxime acetates bearing α-protons have been reported.[14]

In conclusion, the copper/amine-catalyzed method developed by us allows for convenient preparation of certain types of substituted pyridines, some of which may not be readily accessible by conventional methods. The method shows excellent functional group compatibility and tolerates scale-up.

References

1. Division of Chemistry and Biological Chemistry, School of Physical and Mathematical Sciences, Nanyang Technological University, Singapore 637371 (Singapore). E-mail: nyoshikai@ntu.edu.sg.
2. For selected general reviews, see: (a) Allais, C.; Grassot, J. M.; Rodriguez, J.; Constantieux, T. *Chem. Rev.* **2014**, *114*, 10829–10868. (b) Gulevich, A. V.; Dudnik, A. S.; Chernyak, N.; Gevorgyan, V. *Chem. Rev.* **2013**, *113*, 3084–3213. (c) Hill, M. D. *Chem. Eur. J.* **2010**, *16*, 12052–12062. (d) Henry, G. D. *Tetrahedron* **2004**, *60*, 6043–6061.
3. (a) Huang, H.; Ji, X.; Wu, W.; Jiang, H. *Chem. Soc. Rev.* **2015**, *44*, 1155–1171. (b) Huang, H.; Cai, J.; Deng, G.-J. *Org Biomol Chem* **2016**, *14*, 1519–1530.
4. Liu, S.; Liebeskind, L. S. *J. Am. Chem. Soc.* **2008**, *130*, 6918–6919.
5. (a) Parthasarathy, K.; Jeganmohan, M.; Cheng, C.-H. *Org. Lett.* **2008**, *10*, 325–328. (b) Hyster, T. K.; Rovis, T. *Chem. Commun.* **2011**, *47*, 11846–11848. (c) Too, P. C.; Noji, T.; Lim, Y. J.; Li, X.; Chiba, S. *Synlett* **2011**, 2789–2794. (d) Martin, R. M.; Bergman, R. G.; Ellman, J. A. *J. Org. Chem.* **2012**, *77*, 2501–2507.
6. (a) Neely, J. M.; Rovis, T. *J. Am. Chem. Soc.* **2013**, *135*, 66–69. (b) Neely, J. M.; Rovis, T. *J. Am. Chem. Soc.* **2014**, *136*, 2735–2738.
7. Wei, Y.; Yoshikai, N. *J. Am. Chem. Soc.* **2013**, *135*, 3756–3759.
8. (a) Kitamura, M.; Narasaka, K. *Chem. Rec.* **2002**, *2*, 268–277. (b) Narasaka, K.; Kitamura, M. *Eur. J. Org. Chem.* **2005**, 4505–4519.
9. Takai, K.; Katsura, N.; Kunisada, Y. *Chem. Commun.* **2001**, 1724–1725.
10. (a) Allen, A. E.; MacMillan, D. W. C. *Chem. Sci.* **2012**, *3*, 633–658. (b) Shao, Z.; Zhang, H. *Chem. Soc. Rev.* **2009**, *38*, 2745–2755.
11. Wu, Q.; Zhang, Y.; Cui, S. *Org. Lett.* **2014**, *16*, 1350–1353.
12. Jiang, H.; Yang, J.; Tang, X.; Li, J.; Wu, W. *J. Org. Chem.* **2015**, *80*, 8763–8771.
13. Huang, H.; Cai, J.; Tang, L.; Wang, Z.; Li, F.; Deng, G.-J. *J. Org. Chem.* **2016**, *81*, 1499–1505.
14. (a) Ren, Z.-H.; Zhang, Z.-Y.; Yang, B.-Q.; Wang, Y.-Y.; Guan, Z.-H. *Org. Lett.* **2011**, *13*, 5394–5397. (b) Zhao, M.-N.; Hui, R.-R.; Ren, Z.-H.; Wang, Y.-Y.; Guan, Z.-H. *Org. Lett.* **2014**, *16*, 3082–3085. (c) Zhao, M.-N.; Ren, Z.-H.; Yu, L.; Wang, Y.-Y.; Guan, Z.-H. *Org. Lett.* **2016**, *18*, 1194–1197. (d) Zheng, M.; Chen, P.; Wu, W.; Jiang, H. *Chem. Commun.* **2016**, *52*, 84–87.

Appendix
Chemical Abstracts Nomenclature (Registry Number)

Acetophenone: Ethanone, 1-phenyl-; (98-86-2)
Hydroxylamine hydrochloride: Hydroxylamine, hydrochloride (1:1); (5470-11-1)
Pyridine: Pyridine; (110-86-1)
N,N-Dimethylaminopyridine: 4-Pyridinamine, *N,N*-dimethyl-; (1122-58-3)
Acetic anhydride: Acetic acid, 1,1'-anhydride; (108-24-7)
Copper(I) iodide: Copper iodide; (7681-65-4)
Diisopropylamine: 2-Propanamine, *N*-(1-methylethyl)-; (108-18-9)
trans-Cinnamaldehyde: 2-Propenal, 3-phenyl-, (2*E*)-; (14371-10-9)

Wei Wen (Simon) Tan was born in Singapore. He received his B.Sc. (1st class honors) degree in chemistry and biological chemistry from Nanyang Technological University in 2013. He received his Ph.D. degree in chemistry from Nanyang Technological University in 2017 under the supervision of Professor Naohiko Yoshikai, working on the development of novel copper-catalyzed condensation methods for the synthesis of heteroarenes. He then joined Mondelēz International as a flavor scientist.

Bin Wu received his B.Sc. degree in 2009 and M.Sc. degree in 2012 (both in chemistry) from Soochow University. He subsequently moved to Singapore and obtained his Ph.D. degree under the guidance of Prof. Naohiko Yoshikai from Nanyang Technological University in 2017. After a brief stint as a postdoctoral associate in the same group, he joined Novartis Pharma in Suzhou (China) as a Senior Scientist. His current research focuses on the process research and development of drug substances.

Ye Wei was born in 1982 in China. He earned his B.Sc. degree in 2005 from Huaqiao University and his Ph.D. degree in 2010 from Fujian Institute of Research on the Structure of Matter, Chinese Academy of Sciences, under the guidance of Prof. Weiping Su. After a two-year postdoctoral appointment with Prof. Naohiko Yoshikai at Nanyang Technological University, he went back to China to join the College of Pharmacy at Third Military University. He was promoted to Professor in 2015. His research interests are primarily focused on transition metal-mediated novel organic transformations to construct heterocyclic scaffolds.

Naohiko Yoshikai was born in 1978 and raised in Tokyo. He received his B.Sc. (2000), M.Sc. (2002), and Ph.D. (2005) degrees from the University of Tokyo under the guidance of Professor Eiichi Nakamura. He then served as an Assistant Professor at the same institute (2005-2009). In 2009, he moved to Singapore to join the faculty of Nanyang Technological University as an Assistant Professor and a Research Fellow of the Singapore National Research Foundation. He has been an Associate Professor since 2016. His research interests revolve around the development and mechanistic study of novel transition metal-catalyzed reactions and their synthetic applications.

Suttipol Radomkit was born in 1986 in Thailand. He received his B.S. (2008) in Chemistry from Kasetsart University and M.S. (2010) in Chemical Biology from Chulabhorn Graduate Institute, Thailand. He obtained his Ph.D. under the guidance of Professor Amir H. Hoveyda from Boston College in 2016. He then joined Boehringer Ingelheim Pharmaceuticals Inc. in Ridgefield, CT where he is currently a Senior Scientist. His research focuses on the design and development of practical synthetic methods for drug candidates.

An Au/Zn-catalyzed Synthesis of N-Protected Indole via Annulation of *N*-Arylhydroxamic Acid and Alkyne

Xinpeng Cheng, Yanzhao Wang, and Liming Zhang[1*]

Department of Chemistry and Biochemistry, University of California, Santa Barbara, California 93106, United States

Checked by Lucas Morrill, Junyong Kim, and Neil K. Garg

A.

$$\left(tBu - \underset{tBu}{\underset{|}{\bigcirc}} - O \right)_3 P \quad \xrightarrow[\text{CH}_2\text{Cl}_2]{\underset{\textbf{2}}{S-Au-Cl}} \quad \left(tBu - \underset{tBu}{\underset{|}{\bigcirc}} - O \right)_3 P-Au-Cl$$

1   **3**

B.

5

3 (0.5 mol%)
AgNTf₂ (0.5 mol%)
Zn(OTf)₂ (5 mol%)
toluene, 60 °C

4   **6**

Procedure (Note 1)

*A. Chloro[tris(2,4-di-tert-butylphenyl)phosphite]gold (I) (**3**).* A one-necked (B14, diameter: 4.5 cm) 25 mL round-bottomed flask is open to air, equipped with a 3 × 10 mm egg shaped magnetic stirring bar, and charged with tris(2,4-di-*tert*-butylphenyl)phosphite (**1**) (0.647 g, 1.00 mmol) and chloro(dimethyl sulfide)gold (**2**) (0.295 g, 1.00 mmol, 1.0 equiv) (Note 2). Dichloromethane (5 mL) is added via syringe (Note 3) and the flask is fitted with a glass stopper.

The resulting colorless solution (Figure 1) is stirred (800 rpm) at 23 °C for 1 h. The volatiles are removed by rotatory evaporation (300 mmHg, 30 °C bath temperature) and then, under a higher vacuum (1 mmHg) for 24 h to afford gold(I) chloride complex **3** (0.879 g, quantitative yield) as a white solid (Note 4) (Figure 2).

Figure 1. Colorless solution   **Figure 2. White solid**

B. 1-(2-Butyl-1H-indol-1-yl)ethanone (**6**). A two-necked (B24, diameter: 8 cm) 250 mL round-bottomed flask is open to air, equipped with an egg shaped magnetic stirring bar (2.5 x 1.0 cm) and a thermometer (–10 °C – 250 °C), and charged with toluene (80 mL) (Note 5). *N*-Hydroxy-*N*-phenylacetamide (**4**) (6.046 g, 40.0 mmol), 1-hexyne (**5**) (6.50 mL, 4.60 g, 56.0 mmol, 1.4 equiv), zinc trifluoromethanesulfonate (0.731 g, 2.0 mmol, 0.05 equiv), chloro[tris(2,4-di-*tert*-butylphenyl)phosphite]gold (I) (**3**) (176 mg, 0.2 mmol, 0.005 equiv), and silver bis(trifluoromethanesulfonyl)imide (78 mg, 0.2 mmol, 0.005 equiv) are successively added (Note 6). The color of the solution changes from colorless to yellow upon the addition of silver bis(trifluoromethanesulfonyl)imide into the solution, while white silver chloride precipitation is observed. (Figure 3)

Both necks of the flask are fitted with septa. The reaction mixture is heated in a 60 °C oil bath and stirred for 24 h. During its course, the reaction turns progressively from yellow to orange and finally to black (Figure 4).

Figure 3. Apparatus Assembly in Step B

Figure 4. Progression of color from yellow to black in Step B

A 150 mL Büchner funnel with fritted disc (diameter: 7 cm) is mounted on the top of a 250 mL one-necked round-bottomed flask and charged with 10.5 g celite (Note 7). While the funnel is connected to a vacuum source (375 mmHg), THF (10 mL) (Note 8) is poured into the funnel, followed by the reaction mixture, and THF (2 x 10 mL) (Note 8). The filtrate is washed with 50 mL saturated sodium bicarbonate solution (Note 9), and the aqueous solution is extracted with dichloromethane (3 x 10 mL) (Note 10). The combined organic layers are dried for 20 min over Na_2SO_4 (15 g) (Note 11). The volatiles are removed by rotatory evaporation (30 mmHg, 40 °C bath temperature), and then under a higher vacuum for 4 h (1 mmHg) (Note 12). The resulting solid is dissolved in dichloromethane (20 mL) (Note 10), and 10 g silica gel (Note 13) is added into the solution. After removing the solvent by rotatory evaporation (300 mmHg, 30 °C bath temperature), the silica gel with adsorbed crude material is charged on a silica gel (Note 13) column (5 cm diameter x 15 cm height). The column is packed with hexanes (800 mL) and eluted with hexanes/ethyl acetate (20/1, 1.3 L) using compressed air (2 atm) (Notes 14 and 15). Fractions 42-58 (Note 16) containing the pure product are concentrated by rotary evaporation (45 mmHg, 30 °C bath temperature). The resultant solid is dried under high vacuum (1.0 mmHg) for 10 h to afford indole **6** (7.77 g, 90% yield) as a white solid (Note 17) (Figure 5).

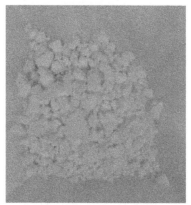

Figure 5. White solid produced in Step B

Notes

1. Prior to performing each reaction, a thorough hazard analysis and risk assessment should be carried out with regard to each chemical substance and experimental operation on the scale planned and in the context of the laboratory where the procedures will be carried out. Guidelines for carrying out risk assessments and for analyzing the hazards associated with chemicals can be found in references such as Chapter 4 of "Prudent Practices in the Laboratory" (The National Academies Press, Washington, D.C., 2011; the full text can be accessed free of charge at https://www.nap.edu/catalog/12654/prudent-practices-in-the-laboratory-handling-and-management-of-chemical. See also "Identifying and Evaluating Hazards in Research Laboratories" (American Chemical Society, 2015) which is available via the associated website "Hazard Assessment in Research Laboratories" at

https://www.acs.org/content/acs/en/about/governance/committees/chemicalsafety/hazard-assessment.html. In the case of this procedure, the risk assessment should include (but not necessarily be limited to) an evaluation of the potential hazards associated with tris(2,4-di-*tert*-butylphenyl)phosphite, chloro(dimethyl sulfide)gold, dichloromethane, toluene, *N*-hydroxy-*N*-phenylacetamide, 1-hexyne, zinc trifluoromethanesulfonate, chloro[tris(2,4-di-*tert*-butylphenyl)phosphite]gold (I), silver bis(trifluoromethanesulfonyl)-imide, celite, tetrahydrofuran, sodium bicarbonate, silica gel, hexanes, ethyl acetate, and sodium sulfate.

2. Tris(2,4-di-*tert*-butylphenyl)phosphite (98%) was purchased from Sigma-Aldrich and used as received. Chloro(dimethyl sulfide)gold (I) was purchased from Strem Chemicals and used as received.

3. Dichloromethane (unstabilized HPLC grade, ≥99.9%) was purchased from Fisher Scientific and passed over columns of activated alumina prior to use.

4. A second run on the same scale provided 0.878 g (quantitative yield) of the same products. The physical and spectroscopic properties of gold(I) chloride complex (**3**) are as follows: mp = 237–238 °C. ^{1}H NMR (500 MHz, CDCl$_3$) δ: 1.29 (s, 9H), 1.44 (s, 9H), 7.13 (dd, J = 8.5, 2.4 Hz, 1H), 7.38–7.44 (m, 2H). ^{13}C NMR (125 MHz, CDCl$_3$) δ: 30.7, 31.52, 34.8, 35.2, 119.25, 119.32, 124.3, 125.5, 139.2, 139.3, 147.35, 147.40, 148.2. ^{31}P NMR (121 MHz, CDCl$_3$) δ: 100.65. IR (film): 2960, 2907, 2870, 1490, 1176, 1075, 928 cm^{-1}. HRMS–APCI (m/z) calculated for [C$_{42}$H$_{63}$AuO$_3$P]$^+$: 843.41749, found: 843.41350. The purity was determined to be >99% wt. by quantitative ^{1}H NMR spectroscopy in CDCl$_3$ using 20.2 mg of the compound and 5.1 mg of HMB(hexamethylbenzene) as an internal standard.

5. Toluene (Certified ACS grade, ≥99.5%) was purchased from Fisher Scientific and passed over columns of activated alumina prior to use.

6. *N*-Hydroxy-*N*-phenylacetamide was prepared according to literature procedure.[3] 1-Hexyne (98%) was purchased from GFS Chemicals (Ref. 3193). Zinc trifluoromethanesulfonate (98%) was purchased from Acros Organics. Both were used as received. Silver bis(trifluoromethanesulfonyl)imide was purchased from sSigma-Aldrich and used as received.

7. Celite (545, Filter agent) was purchased from Sigma-Aldrich and used as received.

8. Tetrahydrofuran (Certified) was purchased from Fisher Scientific and used as received.

9. A gray solid is formed upon washing.

10. Dichloromethane (Certified ACS grade, ≥99.5 %) was purchased from Fisher Scientific and used as received.

11. Sodium sulfate anhydrous (Low nitrogen grade) was purchased from EMD Millipore and used as received.

12. The crude product solidifies under vacuum after 30 min.

13. Silica gel (SiliaFlash P60, pore size 60Å, 230-400 mesh particle size, 40-63 μm particle size) was purchased from SiliCycle Inc.

14. Hexanes (Certified ACS grade), ethyl acetate (Certified ACS grade) were purchased from Fisher Scientific and used as received.

15. Column fractions were checked by TLC analysis on silica gel 60 F254 TLC plate (SiliaPlate™ TLC Plates), using hexanes/ethyl acetate (20/1) as eluent. Visualization is accomplished with 254 nm UV light. The starting material N-hydroxy-N-phenylacetamide (**4**) stays at base line whereas the indole product (**6**) has R$_f$ = 0.30.

s

Figure 6. TLC analysis of the reaction mixture

16. Fractions were collected by test tubes (diameter: 1.8 cm, height: 15 cm). When indole product (**6**) starts to elute from the column, a white solid will form on the edge of test tubes due to evaporation of solvent.

17. A second run on the same scale provided 7.50 g (87%) of the same products. The physical and spectroscopic properties of indole (**6**) are as follows: mp = 56-57 °C. ^{1}H NMR (500 MHz, CDCl$_3$) δ: 0.98 (t, *J* = 7.4 Hz, 3H), 1.48 (sextet, *J* = 7.4 Hz, 2H), 1.71 (quintet, *J* = 7.7 Hz, 2H), 2.76 (s, 3H), 3.01 (t, *J* = 7.5 Hz, 2H), 6.42 (m, 1H), 7.19–7.28 (m, 2H), 7.48 (m, 1H), 7.84 (m, 1H). ^{13}C NMR (125 MHz, CDCl$_3$) δ: 14.1, 22.7, 27.8, 30.4, 31.2,

108.3, 114.9, 120.3, 123.10, 123.52, 130.1, 136.5, 143.1, 170.5. IR (film): 2956, 2869, 1703, 1454, 11369, 1302, 1197, 748 cm^{-1}. HRMS–APCI (*m/z*) calculated for $C_{14}H_{18}NO$ [M + H]$^+$ 216.13829, found 216.13802. The purity was determined to be >99% wt. by quantitative ^1H NMR spectroscopy in CDCl$_3$ using 99.1 mg of the compound and 16.1 mg of HMB (hexamethylbenzene) as an internal standard.

Working with Hazardous Chemicals

The procedures in *Organic Syntheses* are intended for use only by persons with proper training in experimental organic chemistry. All hazardous materials should be handled using the standard procedures for work with chemicals described in references such as "Prudent Practices in the Laboratory" (The National Academies Press, Washington, D.C., 2011; the full text can be accessed free of charge at http://www.nap.edu/catalog.php?record_id=12654). All chemical waste should be disposed of in accordance with local regulations. For general guidelines for the management of chemical waste, see Chapter 8 of Prudent Practices.

In some articles in *Organic Syntheses*, chemical-specific hazards are highlighted in red "Caution Notes" within a procedure. It is important to recognize that the absence of a caution note does not imply that no significant hazards are associated with the chemicals involved in that procedure. Prior to performing a reaction, a thorough risk assessment should be carried out that includes a review of the potential hazards associated with each chemical and experimental operation on the scale that is planned for the procedure. Guidelines for carrying out a risk assessment and for analyzing the hazards associated with chemicals can be found in Chapter 4 of Prudent Practices.

The procedures described in *Organic Syntheses* are provided as published and are conducted at one's own risk. *Organic Syntheses, Inc.*, its Editors, and its Board of Directors do not warrant or guarantee the safety of individuals using these procedures and hereby disclaim any liability for any injuries or damages claimed to have resulted from or related in any way to the procedures herein.

Discussion

The annulation between an arylhydrazine and a ketone to construct an indole,[3] also called the Fischer indole synthesis,[4] is one of the most important reactions in organic synthesis. This method has been used extensively in the construction of various indole alkaloids[5] since the first report by Fischer[6] in 1883. While it has been subjected to various modifications/improvements[7] over the years, there are still notable drawbacks including the poor regioselectivities with non-symmetric ketones and strong acidic reaction conditions. Furthermore, 2-alkenylindoles cannot be prepared via this method except for a few special cases.[8]

In this context, we developed a gold-catalyzed addition of N-arylhydroxylamine to aliphatic terminal alkynes to access 2-alkylindoles with regiosepecificity.[9] As shown in Scheme 1, the reaction mechanism likely entails the addition of the OH group of an N-arylhydroxylamine onto a terminal aliphatic alkyne in the presence of a gold catalyst, followed by one-pot sequential 3,3-rearrangement and dehydrative cyclization reaction. Despite the exceptionally mild reaction conditions (ambient temperature), the utility of this chemistry is largely limited by the moderate thermostability of the N-arylhydroxylamines, which prohibits the extension of this chemistry beyond kinetically facile terminal aliphatic alkyne reaction partners due to the inability to perform the reaction at elevated temperatures.

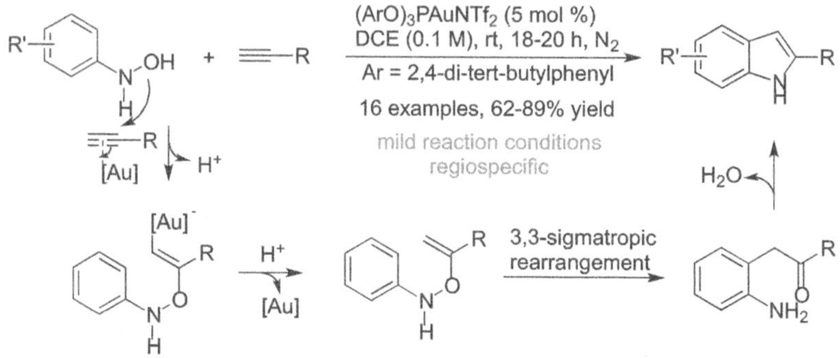

Scheme 1: Regiospecific formation of 2-alkylindole from aliphatic terminal alkyne via gold-catalyzed O-addition of hydroxylamine to C-C triple bond

To substantially improve this indole synthesis, we opted to use thermally much more stable hydroxamic acids or *N*-hydroxycarbamates in place of *N*-hydroxylamines. However, this modification is at the expense of the nucleophilicity of the *N*-hydroxyl group and hence the reaction rates are substantially reduced. Although increasing the reaction temperature could compensate the rate loss, reaction optimizations along this line led to little success.

A: In some metalloenzymes
(M: Zn, Mn)

B: metalloenzyme Mimicking

$$H_2O + [M]^+ \rightleftharpoons [M]-OH + H^+$$
more nucleophilic
than H_2O

7 + [M]$^+$ ⇌ 7-M + H$^+$

Y = OR or R

7-M
more nucleophilic
than 7

Scheme 2. Enhancing the nucleophilicity of 7 by a metal salt

Inspired by the role of metal ions in enhancing nucleophilicity of H_2O by forming metal hydroxides in metalloenzyme catalysis (Scheme 2A),[10] we reason that the nucleophilicity of hydroxamic acid or *N*-hydroxycarbamate would be enhanced in a similar manner. As shown in Scheme 2B, **7** could react with a metal ion to form metal chelate **7-M** and proton reversibly. As in metalloenzyme catalysis, **7-M** should be more nucleophilic than **7** due to the increased negative charge on the deprotonated oxygen. After screening various metal salts, we identified Zn(OTf)$_2$ as the most effective. Even with only 5 mol% of this salt, the desired annulation of **7** with alkynes can be achieved at 60 °C in toluene in the presence of a gold catalyst (5 mol%), affording *N*-protected indoles in generally good yields. Scheme 3 shows a general reaction with the proposed mechanism, which entails cooperative catalysis by Zn(OTf)$_2$ and LAu$^+$.

Some representative cases from the scope study[11] are shown in Table 1. With terminal alkynes as substrates, LAuNTf$_2$ with L = **1** is the preferred gold catalyst due to its higher acidity/electrophilicity, and the reactions finish in 4-8 h. On the other hand, IPrAuOTf is a better catalyst when less reactive internal alkynes are employed, owing to its higher themostability, and the reactions require 18-30 h to go to completion. Compared with the *N*-hydroxylamine-based indole synthesis, this modified strategy dramatically expands the scope of alkynes to include internal alkynes and enynes as well as the scope of the *N*-aryl group.

Scheme 3. The general reaction and a plausible reaction mechanism

For the scaled-up procedure reported here, in order to minimize the reaction cost, we were able to lower down the loading of the gold catalyst from 5 mol% used in our original report to 0.5 mol% without affecting the reaction yield. Moreover, the active catalyst LAuNTf$_2$ (L = **1**) can be generated in situ from its chloride salt **3** upon the treatment of it with an equivalent of AgNTf$_2$.

Table 1. Representative N-protected indoles synthesized in the reported scope study

84%
(1)AuNTf₂ as catalyst

87%
(1)AuNTf₂ as catalyst

83%
(1)AuNTf₂ as catalyst

94%
(1)AuNTf₂ as catalyst

80%
(1)AuNTf₂ as catalyst

80%
(1)AuNTf₂ as catalyst

83%
(1)AuNTf₂ as catalyst

77%
(1)AuNTf₂ as catalyst

75%
(1)AuNTf₂ as catalyst

67%
regioselectivity >19 :1
IPrAuOTf as catalyst

80%
regioselectivity, 16.6 : 1
IPrAuOTf as catalyst

66%
regioselectivity >20 : 1
IPrAuOTf as catalyst

References

1. Department of Chemistry and Biochemistry, University of California, Santa Barbara, California 93106, United States. E-mail: zhang@chem.ucsb.edu. We thank NSF CHE-1301343 for financial support and for NIH shared instrument grant S10OD012077 for the purchase of a 400 MHz NMR spectrometer.

2. Oxley, P.W.; Adger, B. M.; Sasse, M. J.; Forth, M. A. *Org. Synth.* **1989**, *67*, 187–189.

3. Humphrey, G. R.; Kuethe, J. T. *Chem. Rev.* **2006**, *106*, 2875–2911.

4. (a) Robinson, B., *The Fischer Indole Synthesis*, Wiley, Chichester, **1982**. (b) Downing, R. S. and Kunkeler, P. J. in *Fine chemicals through heterogeneous catalysis*, eds. Sheldon, R. A. and Bekkum, H. Wiley-VCH, Weinheim, New York, **2001**, pp. 178–183; (c) Hughes, D. L. *Org. Prep. Proced. Int.* **1993**, *25*, 607–632.

5. (a) Bonjoch, J.; Catena, J.; Valls, N. *J. Org. Chem.* **1996**, *61*, 7106–7115. (b) Iyengar, R.; Schildknegt, K; Aubé, J. *Org. Lett.* **2000**, *2*, 1625–1627. (c) Roberson, C. W.; Woerpel K. A. *J. Am. Chem. Soc.* **2002**, *124*, 11342–11348. (d) Gan, T.; Liu, R.; Yu, P.; Zhao, S.; Cook, J. M. *J. Org. Chem.* **1997**, *62*, 9298–9304.

6. Fischer, E.; Jourdan, F. *Ber. Dtsch. Chem. Ges.* **1883**, *16*, 2241–2245.

7. (a) Lipin'ska, T. *Chem. Heterocycl. Compd.* **2001**, *37*, 231–236. (b) Chen, C.-Y.; Senanayake, C. H.; Bill, T. J.; Larsen, R. D.; Verhoeven, T. R.; Reider, P. J. *J. Org. Chem.*, **1994**, *59*, 3738–3741. (c) Katritzky, A. R.; Rachwal, S.; Bayyuk S. *Org. Prep. Proced. Int.* **1991**, *23*, 357–363. (d) Inman, M.; Moody, C. J. *Chem. Commun.* **2011**, *47*, 788–790. (e) Narayana B.; Ashalatha B. V.; Vijaya Raj K. K.; Fernandes J.; Sarojini B. K. *Bioorg. Med. Chem.* **2005**, *13*, 4638–4644.

8. Bergman, J.; Pelcman, B. *Tetrahedron* **1988**, *44*, 5215–5228.

9. Wang, Y.; Ye, L.; Zhang, L. *Chem. Commun.* **2011**, *47*, 7815–7817.

10. Christianson, D. W.; Cox, J. D. *Annu. Rev. Biochem.* **1999**, *68*, 33–57.

11. Wang, Y.; Liu, L.; Zhang, L. *Chem. Sci.* **2013**, *4*, 739–746.

Appendix
Chemical Abstracts Nomenclature (Registry Number)

Tris(2,4-di-*tert*-butylphenyl)phosphite: Phenol, 2,4-bis(1,1-dimethylethyl)-, 1,1',1''-phosphite; (31570-04-4)

Chloro(dimethylsulfide)gold(I): Gold, chloro[thiobis[methane]]-; (29892-37-3)

Chloro[tris(2,4-di-*tert*-butylphenyl)phosphite]gold(I): Gold, chloro[tris[2,4-bis(1,1-dimethylethyl)phenyl] phosphite-κP]-; (915299-24-0)

N-Hydroxy-*N*-phenylacetamide: Acetamide, *N*-Hydroxy-*N*-phenyl-; (1795-83-1)

1-Hexyne: 1-Hexyne; (693-02-7)

1-(2-Butyl-1*H*-indol-1-yl)ethanone: Ethanone, 1-(2-butyl-1*H*-indol-1-yl)-; (116491-55-5)

Organic
Syntheses

Xinpeng Cheng was born in Nanchang (China) in 1994. He graduated with a B.S. degree in Chemistry from Zhejiang University in 2016. In the same year, he joined the group of Prof. Liming Zhang at the University of California, Santa Barbara (UCSB) to perform his Ph.D. studies.

Liming Zhang was born in Pingxiang, China in 1972. He received his B.S. degree in chemistry from Nanchang University in 1993, his first M.S. degree in organometallic chemistry with Professor Zhengzhi Zhang from Nankai Univeristy in 1996, and his second M.S. degree in organic chemistry with Professor Michael P. Cava from the University of Alabama in 1998. He obtained his Ph.D. degree with Professor Masato Koreeda from the medicinal chemistry program at the University of Michigan in 2003 and then carried out a post-doctoral study with Professor Sergey A. Kozmin at the University of Chicago. He started his independent academic career at the University of Nevada, Reno in 2005 and moved to the University of California, Santa Barbara in 2009. He is currently a Professor in Organic Chemistry. His research interests include late transition metal-catalyzed reactions, natural product synthesis and medicinal chemistry.

Lucas Morrill received his B.A. in Chemistry from Carleton College in Northfield, MN, where he performed undergraduate research with Professors David Alberg and Gretchen Hofmeister. He is currently a fourth-year graduate student in Professor Neil K. Garg's laboratory at the University of California, Los Angeles. His graduate studies are focused on the total synthesis of complex natural products.

Junyong Kim was born in Seoul, South Korea in 1988. He received his B.S. in Chemistry from Seoul National University, where he performed undergraduate research with Professor David Y.-K. Chen on the total synthesis of dendrobine. In the summer of 2013, he began his graduate studies at the University of California, Los Angeles. He is currently a fourth-year graduate student in the laboratory of Professor Neil K. Garg, pursuing the total synthesis of tubingensin natural products.

Preparation of *Fac*-Tris(2-Phenylpyridinato) Iridium(III)

Kip A. Teegardin and Jimmie D. Weaver[1*]

Department of Chemistry, Oklahoma State University, Stillwater, OK 74078

Checked by Hao Wu, Alexander Sienkiewicz, and Chris Senanayake

Procedure (Note 1)

Fac-tris(2-phenylpyridinato) iridium(III) *(A).* Iridium (III) chloride anhydrous (0.65 g , 2.18 mmol, 1 equiv) (Note 2), 2-phenylpyridine (3.74 mL, 26.1 mmol, 12.0 equiv) (Note 3), and 0.65 L of DI water (0.003 M with sssrespect to IrCl₃) is added to a 1 L Parr reactor (Figure 1) (Notes 4, 5, and 6). The reaction mixture is pressurized with argon (10.0 psi), stirred and sthen depressurized three times, and finally charged again with argon before sealing (Note 7). The reaction mixture is heated to 205 °C for 48 h. Then the reactor is cooled to 20 °C with internal cooling coils. At the end of the experiment the reactor was left in the stand and the contents cooled to 20 °C using cold water. After cooling, the reactor is opened revealing an insoluble yellow solid on the surfaces (Figure 2) and dispersed in the aqueous phase (Note 8). All contents are transferred slowly to a 6 L separatory funnel aided by a large 5 cm glass funnel. Then the interior of the reactor is mechanically scraped (to extract the yellow material), with metal tongs, cotton balls (25 in total), and 500 mL of dichloromethane (DCM) from a spray bottle, and again all contents are added to the separatory funnel (Notes 9 and 10).

Org. Synth. **2018**, *95*, 29-45
DOI: 10.15227/orgsyn.095.0029

Published on the Web 1/12/2018
© 2018 Organic Syntheses, Inc.

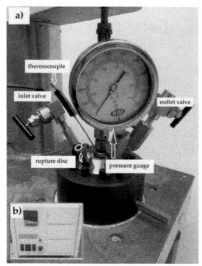

Figure 1. a) 1 L Parr stirred reactor used for the reaction b) 4843 controller

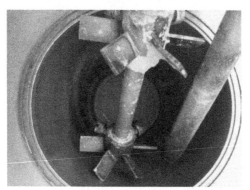

Figure 2. An example of the insoluble yellow solid found on the internal components after a successful reaction

While still in the funnel, the cotton is rinsed with 25 mL of DCM from a spray bottle and evenly pressed with tongs to release the yellow material from the cotton (Note 9). After removing the cotton, the solution is then diluted with 2.5 L of DCM. The separatory funnel is shaken vigorously, allowed to settle and again shaken, and the organic layer is then slowly separated from the aqueous layer (Notes 11, 12 and 13) and the aqueous layer is further extracted with more DCM (3 x 10 mL), and the organic layers are combined (Note 14). The aqueous layers are kept for future ligand recovery. The combined organic layer is washed with a 1 M HCl solution, with vigorous mixing prior to separation (3 x 900 mL). Each HCl wash is then back extracted with DCM (3 x 10 mL) to insure complete recovery of the product. After the final wash, the organic layer is filtered slowly (20 min) through a Celite (35 g) pad on top of a 150 mL medium porosity sintered glass funnel, into a 3 L round-bottomed flask, and then dried with 30 g of MgSO$_4$. After filtering the drying reagent using a 4 L Erlenmyer flask fitted with a 5 cm funnel/cotton plug, a homogenous aliquot is removed for NMR analysis (Note 15). Finally, the solvent is removed in batches by transferring to a 2.5 L round-bottomed flask by rotary evaporation (35 °C, 30 mm Hg, 150 rpm) to afford 1.35 g (94%) of Ir(ppy)$_3$ as a bright yellow solid (Note 16).

Further purification of Ir(ppy)$_3$ can be performed by adding the yellow solid to a 1 L round-bottomed flask and adding 600 mL of distilled hexanes. The solid material is then sonicated until a uniform slurry is achieved, and 5 mL of dichloromethane is added (Note 17). The liquid is swirled giving a slight yellow tint to the solution indicating successful dissolution of a colored compound, and selective extraction of the impurities. Then the slurry is slowly poured through a 50 mL fine porosity sintered glass funnel to collect the yellow solid, and the filtrate is collected into a 1 L Erlenmeyer flask. The yellow solid is air dried on the filter to afford 1.29 g (91%) (Notes 18 and 19) of the product in >97% purity (Notes 20 and 21).

Notes

1. Prior to performing each reaction, a thorough hazard analysis and risk assessment should be carried out with regard to each chemical substance and experimental operation on the scale planned and in the context of the laboratory where the procedures will be carried out. Guidelines for carrying out risk assessments and for analyzing the hazards associated with chemicals can be found in references such as Chapter 4 of "Prudent Practices in the Laboratory" (The National Academies Press, Washington, D.C., 2011; the full text can be accessed free of charge at https://www.nap.edu/catalog/12654/prudent-practices-in-the-laboratory-handling-and-management-of-chemical). See also "Identifying and Evaluating Hazards in Research Laboratories" (American Chemical Society, 2015) which is available via the associated

website "Hazard Assessment in Research Laboratories" at https://www.acs.org/content/acs/en/about/governance/committees /chemicalsafety/hazard-assessment.html. In the case of this procedure, the risk assessment should include (but not necessarily be limited to) an evaluation of the potential hazards associated with iridium (III) chloride anhydrous, 2-phenylpyridine, argon, dichloromethane, Celite, hydrochloric acid solution, and hexanes, as well as the proper procedures for operating a Paar apparatus. This reaction is run well above the boiling point of water and reaches high pressures. The reaction should be cooled to room temperature before venting and opening. Only reactors designed to handle this pressure should be used and shields used as appropriate.

2. IrCl₃ was obtained from Beantown Chemicals (BTC), available through VWR.

3. 2-Phenylpyridine was obtained from AK scientific and used as received.

4. A 2 L Parr stirred reactor was fitted with an inlet and outlet valve, a Duro United pressure gauge (0-5000 psi), and a 526HCPH rupture disc was used for the reaction (Figure 1a). In series with a 4843 controller the reactor was heated with a Parr heater assembly (model no. A1445HC3EE) controlled by a type J thermocouple (15.5 in. length) (Figure 1b). We have found that the stirring and glass sleeve options are unnecessary for the success of the reaction. The Checkers used a 1 L Parr series 4520 stirred reactor, material of construction 316L, Maximum Allowable Working Pressure 1900 psi at 350 °C. A Pressure relief valve was installed, Parr# A140VB2PC and set to 600 psi for safety. In addition we installed a Swagelok Pressure transducer, model# PTI-S-NG1500-11AQ with a frange of 0-1500 psi so that we could log the internal pressure. A second JKem model J thermocouple 1/16 x 12 inches was installed for logging the internal temperature during the reaction. The heating and cooling was controlled with a Parr model 4843 controller along with a magnetic solenoid valve on the cold water loop. We also discovered that the headspace in the reactor was very important to maintaining operable pressures. A headspace of 25% is needed. We found if the headspace in the same 2 L Parr reactor was decreased to 12.5% (1.75 L water), the reactor leaked and displayed significant pressure. Again, further safety precautions must be taken when changing the headspace to water ratio. With a lower fill in the reactor the system is operating as saturated steam, gas phase and liquid phase in an equilibrium, so the pressure should be pretty constant.

However as the amount of water is increased, and it expands upon heating there is no longer head space for steam to form. Once there is no room for steam, there is no equilibrium, and you are in a situation where hydraulic pressure from the water is pushing your reactor apart.

5. Inspecting the equipment, and maintaining it prior to starting, in addition to doing a pressure test before starting should eliminate most equipment failure. Having equipment failure with water at 200 °C can be hazardous. The experiment was carried out in a closed hood for safety. In the pressure test, the reactor was charged with 0.65 L water only. The internal temperature and pressure were monitored using the Parr controller. The temperature and pressure were found very constant over 48 h. The primary cause for incomplete or failed reactions stems from equipment failure. Since the reaction is highly dependent on temperature it is important to insure probes and controllers are working properly, and leak checks are performed on the entire system before every reaction. Insure all valves are free of obstruction and all areas are clean. It is imperative that the reactor temperature stays above 200 °C. The reactor used in the above reactions had shown significant temperature deviation +/–5 °C. So we set our controller to 205 °C to insure the reactor stays above 200 °C.

6. See the attached plot (generated by the Checkers) for expansion of pure water One must be careful to ensure that there is room for steam to form or else it is possible to have a catastrophic failure.

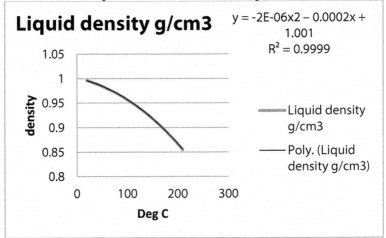

7. If the Parr reactor is not equipped with a stirring mechanism, after pressurizing the reactor with argon one can gently rock the reactor back and forth to mix the contents

8. Suspended dull yellow colored water mixtures is indicative of an unsuccessful reaction, while insoluble yellow colored (Figure 2) water mixtures are indicative of a successful reaction. For the Ir(ppy)₃ catalyst, the primary contaminant after the workup is the chloro-bridging dimer which is a dull yellow color intermediate en-route to the desired product. This was determined by NMR analysis reaction mixtures after work up.[2]

9. Yellow solid can accumulate around the walls, probe shafts, and stirring impellers and should be examined for yellow residue, and any residue should be scraped or extracted with DCM/cotton to insure high yields (Figure 3). The cotton was placed directly into the separatory funnel only to cut process time. Alternatively, the cotton can be placed into a beaker and then the cotton material can be extracted with DCM, then the DCM extracts can be added to the separatory funnel.

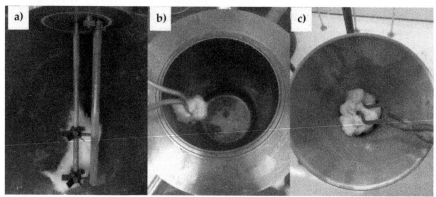

Figure 3. a) yellow solid on stirring impellers b) mechanical scraping of yellow solid off the reactor walls c) extracting compound from cotton with tongs and solvent directly into funnel

10. There may be yellow solid that did not dissolve completely into the DCM. This is normal; as the 1 M HCl washes are performed and ligand is removed the catalyst will eventually partition fully into the organic layer.

11. It is important to drain to the organic layer slowly to avoid disrupting the phase separation. In this large scale work up the aqueous layer

tends to swirl into the organic layer as the organic is dispersed through the separatory funnel, if separation is done quickly. It is important to insure that the separation of the organic layer from the aqueous layer is performed slowly in all phase separations.

12. There tends to be a small emulsion layer in between the organic and aqueous layers during the work up which contains some black material. Care should be taken to avoid the inclusion of this emulsion layer. The extra DCM rinses serve to fully extract the product that remains in the emulsion and on the walls of the separatory funnel.

13. Because it builds up pressure, it is important to vent the separatory funnel to avoid loss of product and to avoid injury. Some loss of product is apparent when venting the large 6 L separatory funnel through the stopcock, while not the standard technique, venting was performed by securing the separatory funnel back to the ring stand and carefully removing the stopper.

14. Thin-layer chromatography (TLC) was used to insure all ligand was extracted from the organic layer. TLC was performed using Sorbent Technologies Silica Gel XHL normal phase plates and utilizing an eluent system of hexanes/EtOAc (9:1). Rf values are measured using thin layer chromatography (TLC), (obtained from sorbent technology Silica XHL TLC Plates, w/UV254, glass backed, 250 μm, 20 x 20 cm) see Table 1.

Table 1. Rf values of the photocatalyst and ligand

Compound	R_f: 1/9 EtOAc/Hexanes	R_f: 1/1 DCM/Hexanes
Fac-tris(2-phenylpyridinato) iridium(III)	0.028	0.43
2-Phenylpyridine	0.31	0.19

15. The solubility of the catalysts is relatively low in dichloromethane (thus a large volume of solvent was used to dilute the compound), yet the impurities (the chloro-bridging dimer and the excess ligand) are significantly more soluble. Removing an aliquot while the entire mixture is homogenous will insure an accurate NMR analysis.

16. Since the solubility of the catalyst is low in most volatile solvents transferring the material to a smaller container can be laborious and require copious amounts of solvent. It is recommended to scrape the

sides of the round bottom meticulously and transfer all the material possible as a solid. Then any remaining solid material can be dissolved with solvent and transferred to the smaller container, thus, minimizing the solvent use.

17. To insure high yields of product, the purification step must be performed with care. The amount of DCM that is added to the hexanes can vary depending on the purity of the sample. Lower amounts of DCM are added for purer samples and larger amounts are used for compounds of lower purity. The ratio for the purification technique is typically 1/120 (DCM/hexanes).

18. Characterization of *fac*-tris(2-phenylpyridinato) iridium(III): ¹H NMR (500 MHz, CD₂Cl₂) (5.32 ppm for CH₂Cl₂ in CD₂Cl₂) δ: 6.78 (dt, *J* = 16.1, 7.4 Hz, 6H), 6.90 (dt, *J* = 18.9, 6.4 Hz, 6H), 7.57 (d, *J* = 5.3 Hz, 3H), 7.64–7.68 (m, 6H), 7.92 (d, *J* = 8.2 Hz, 3H). ¹³C NMR (400 MHz, C₆D₆ with NMP) δ: 119.1, 120.3, 122.2, 124.6, 130.1, 136.5, 137.7, 144.4, 147.3, 161.9, 167.1. IR (film): 3034, 2923, 1599, 1580, 1560, 1470, 1413, 1261, 1159, 752, 732 cm⁻¹.

19. A second reaction on the same scale provided 1.30 g (91%) of the product in >98% purity.

20. The weight percent was determined by quantitative NMR using trimethoxybenzene as the internal standard. The solubility of Irppy₃ in most deuterated solvents is low, which makes weighing out the necessarily small amounts of dissolvable compound challenging. It is best to accurately weigh out a larger sample (>100 mg of the standard or compound) and dissolve the compound into protio-dichloromethane using a volumetric flask to give an accurate molar solution. Then appropriate volumes of the standard and compound solutions can be mixed. Next, the protio-solvent removed and the mixture redissolved in the appropriate deuterated solvent. A proton NMR spectrum is thus obtained for the mixture with a minimum relaxation delay of 30 s.

21. If using more expensive ligands it may be economical to recover the valuable ligand. Below is an example of the *2-phenylpyridine* ligand recovery, which was not confirmed by the checkers: The previously retained acidic aqueous layers from a 4.46 mmol reaction are added to a 4 L Erlenmeyer flask with a 60 mm PTFE octagon stir bar. The solution is stirred and slowly brought to pH 10 by adding 119 g of NaOH pellets in 10 portions (Note 22). The solution is added to a 6 L separatory funnel and the ligand is extracted from the salty water with DCM (6 x 1.2 L). The combined organic extracts are added to a 20 L metal canister

and are dried with 20 g of MgSO₄. After filtering the drying reagent through a 5 cm funnel with a cotton plug into another 20 L canister the solvent is removed in batches by utilizing a continuous rotary evaporation setup (35 °C, 130 mm Hg, 100 rpm) (Figure 4) to afford 5.90 g (96.4%) (based on excess 9 equiv) of 2-phenylpyridine is recovered in 97.7–98.5% purity as determined by quantitative ^{1}H NMR using 1,3,5-trimethoxybenzene as the internal standard.

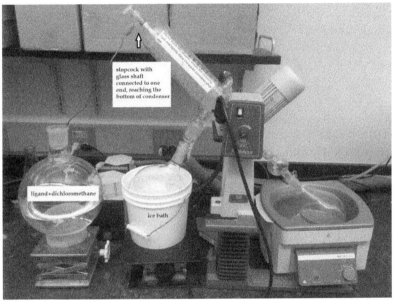

Figure 4. Diagram of a continuous rotary evaporation setup

22. One should slowly add the NaOH to avoid excessive heat generation and potential injury. We observed that extraction of ligand becomes difficult, and yields suffer when using a NaOH solution in comparison to direct addition of NaOH pellets. The use of pellets minimizes the aqueous volume into which the ligand may partition.

Working with Hazardous Chemicals

The procedures in *Organic Syntheses* are intended for use only by persons with proper training in experimental organic chemistry. All hazardous materials should be handled using the standard procedures for work with chemicals described in references such as "Prudent Practices in the Laboratory" (The National Academies Press, Washington, D.C., 2011; the full text can be accessed free of charge at http://www.nap.edu/catalog.php?record_id=12654). All chemical waste should be disposed of in accordance with local regulations. For general guidelines for the management of chemical waste, see Chapter 8 of Prudent Practices.

In some articles in *Organic Syntheses*, chemical-specific hazards are highlighted in red "Caution Notes" within a procedure. It is important to recognize that the absence of a caution note does not imply that no significant hazards are associated with the chemicals involved in that procedure. Prior to performing a reaction, a thorough risk assessment should be carried out that includes a review of the potential hazards associated with each chemical and experimental operation on the scale that is planned for the procedure. Guidelines for carrying out a risk assessment and for analyzing the hazards associated with chemicals can be found in Chapter 4 of Prudent Practices.

The procedures described in *Organic Syntheses* are provided as published and are conducted at one's own risk. *Organic Syntheses, Inc.*, its Editors, and its Board of Directors do not warrant or guarantee the safety of individuals using these procedures and hereby disclaim any liability for any injuries or damages claimed to have resulted from or related in any way to the procedures herein.

Discussion

For more than 30 years, transition metal complexes have been known to mediate single-electron transfers (SET) and have led to SET-initiated polymerization reactions,[3] dye-sensitized solar cells,[4] and light emitting diodes.[5] While the use of these complexes in synthesis dates back to the 1980's, application towards synthetic organic chemistry was sporadic.[6] Early reports of transition metal-based photocatalytic mediated organic transformations utilized Ru-based photocatalysts such as *tris*-(2,2'-bipyridine) ruthenium(II) (Ru(bpy)$_3^{2+}$), which when excited by visible light undergoes a metal to ligand charge transfer (MLCT) and facilitates one-electron oxidation or reduction of the complex.[7] In 1985, King and coworkers reported the synthesis and characterization of the neutral *fac*-tris-phenylpyridinato iridium (III), (Ir(ppy)₃) photocatalyst.[8] In general, the neutral Ir-photocatalysts tend to have a longer lived excited-state lifetime and to be substantially more reducing when compared to ruthenium based photocatalysts, and thus are a complimentary class of photocatalysts.

During the last decade photocatalysis has grown in importance to the synthetic community, and the neutral *tris*-homoleptic Ir-complexes are a vital class of photocatalyst.[9] In 2012, Stephenson[10] reported an efficient hydrodeiodination of alkyl, alkenyl, and aryl iodides using Ir(ppy)₃. In 2014, Macmillian[11] investigated several neutral complexes Ir(*p*-F-ppy)₃, Ir(dFppy)₃, Ir[*p*-F(*t*-Bu)-ppy]₃, and Ir[dF(*t*-Bu)-ppy]₃ which all out performed Ir(ppy)₃ in the decarboxylative arylation of alpha amino acids. Recently other groups,[9] including our own,[12] have utilized the *fac*-tris-homoleptic iridium (III) complexes. In addition to redox chemistry, Ir-based photocatalyts can also undergo quenching through an energy transfer process with styrenyl derivatives which leads to the contra-thermodynamic isomerization to the less conjugated alkene.[12b,13,13b,14,15,16] We have shown that while the emissive energy of the photocatalyst is an important feature in controlling this process, that unlike other sensitizers the substituents play a decisive role in determining which process will dominate.[13a] Thus, there is a need to be able to efficiently access these complexes, and others, so that a full understanding of the structure/function relationship is possible, which will elevate our collective synthetic capabilities.

Despite the rapidly increasing utility of the homoleptic iridium (III) complexes in organic syntheses, there is a significant lack of commercial availability[16] which likely stems from lack of a general synthetic method that can produce these complexes in a cost efficient manner. There are a number of reported synthetic methods that use the more expensive Ir(acac)₃[11-12, 17] ($73970/mol Ir(acac)₃ vs $18,750/mol IrCl₃), or alternatively require two synthetic steps. The first step results in the formation of a chloro-bridging dimer via coordination of two phenylpyridine ligands. Then, in a subsequent synthetic step the halogen is abstracted, often aided by silver, and a third phenylpyridine is added via cyclometalation.[18] Even in the case of the *tris*-homoleptic cyclometalated complexes, this two-step sequence is commonly employed. Ideally, one could place all three bidentate ligands in the same reaction to generate the thermodynamically favored *facial*-homoleptic iridium (III) complex. It has been known that the *meridional*-homoleptic iridium (III) complex is the kinetically favored product but the *facial* complex, obtained at temperatures greater than 200 °C, is the thermodynamically favored product.[18b] Konno[19] reported microwave synthesis of tris-cyclometalated iridium complexes, but this required a large excess of ligand (50 to 100 equiv) which limits the scope of the reaction to readily available ligands. Perhaps more importantly, the scale of the reaction is limited by the microwave equipment used. Therefore, in 2014 we developed and published a simple, reliable, and straightforward prep that would allow us to acquire the *facial*-homoleptic iridium complexes in high yields via a selective one step process starting with the least expensive source of Ir(III) salt, IrCl₃. Ideally, water was the reaction solvent (figure 5).[20]

Figure 5. Synthesis of facial-homoleptic iridium complexes

While we and others have used this method to make gram scale quantities of these complexes, there are some notable shortcomings. Namely, purification of the complexes via column chromatography was time intensive, and required copious amounts of organic solvents, and was particularly challenging because of the low solubility of these complexes. Often the isolation of the complex was more challenging than its synthesis. Since this publication, we have worked to improve the method, in particular we have focused on making the reaction more scalable. The three main objectives were to maintain high purity of complex on a gram scale, eliminate chromatography, and recover the excess ligand. Herein, we report a modified method which has been applied to Irppy$_3$ to obtain multigram quantities of the photocatalyst (up to 2.85 g) in high purity (>97%), without the use of a column, and with excellent recovery of excess ligand in good yield (up to 97.0%). While not checked, we also obtained similar results by applying this method to Ir(diFppy)$_3$ **1d** where the reaction yielded 1.82 g of catalyst (71.4% yield) with high purity (>97%) indicating that the method works with more elaborate ligands that are less accessible. Additionally, while we did not demonstrate the recovery of the organic solvent, it is conceivable that the DCM could also be recovered in high purity. We believe this method to be quite general to this family of photocatalysts. Consequently, this method will facilitate exploration of the chemistry of thssese Ir-complexes, as well as enable applications which may require more substantial quantities of Ir-photocatalysts.

References

1. 107 Physical Science, Oklahoma State University, Stillwater, OK 74078. Jimmie.Weaver@okstate.edu, We thank NIGMS (RO1 GM115697) for financial support of this research.
2. Scarpelli, F.; Ionescu, A.; Ricciardi, L.; Plastina, P.; Aiello, I.; La Deda, M.; Crispini, A.; Ghedini M.; Godbert, N. *Dalton Trans.* **2016**, 45, 17264–17273.
3. Lalevee, J.; Peter, M.; Dumur, F.; Gigmes, D.; Blanchard, N.; Tehfe, M.-A.; Morlet-Savary, F.; Fouassier, J. P. *Chem. Eur. J.* **2011**, 17, 15027–15031.
4. K. Kalyanasundaram, K.; Grätzel, M. *Coord. Chem. Rev.* **1998**, 177, 347–414.
5. Lowry, M. S.; Bernhard, S. *Chem. Eur. J.* **2006**, 12, 7970–7977.
6. (a) Pac, C.; Ihama, M.; Yasuda, M.; Miyauchi, Y.; Sakurai, H. *J. Am. Chem. Soc.* **1981**, 103, 6495–6497; (b) Cano-Yelo, H.; Deronzier, A. *J. Chem. Soc., Perkin Trans. 2*, **1984**, 1093–1098; (c) Narayanam, J. M. R.; Stephenson, C. R. J. *Chem. Soc. Rev.* **2011**, 40, 102–113.

7. McCusker, J. K. *Acc. Chem. Res.* **2003**, *36*, 876–887.

8. King, K. A.; Spellane, P. J.; Watts, R. J. *J. Am. Chem. Soc.* **1985**, *107*, 1431–1432.

9. Prier, C. K.; Rankic, D. A.; MacMillan, D. W. C. *Chem. Rev.* **2013**, *113*, 5322–5363.

10. Nguyen, J. D.; D'Amato, E. M.; Narayanam, J. M. R.; Stephenson, C. R. J. *Nat. Chem.* **2012**, *4*, 854–859.

11. Zuo, Z.; MacMillan, D. W. C. *J. Am. Chem. Soc.* **2014**, *136*, 5257–5260.

12. (a) Singh, A.; Arora, A.; Weaver, J. D. *Org. Lett.* **2013**, *15*, 5390–5393; (b) Singh, K.; Staig, S. J.; Weaver, J. D. *J. Am. Chem. Soc.* **2014**, *136*, 5275–5278; (c) Senaweera, S. M.; Singh, A.; Weaver, J. D. *J. Am. Chem. Soc.* **2014**, *136*, 3002–3005; (d) Singh, A.; Kubik, J. J.; Weaver, J. D. *Chem. Sci.* **2015**, *6*, 7206–7212.

13. (a) Singh, A.; Fennell, C. J.; Weaver, J. D. *Chem. Sci.* **2016**, *7*, 6796–6802; (b) Fabry, D. C.; Ronge, M. A.; Rueping, M. *Chem. - Eur. J.* **2015**, *21*, 5350–5354.

14. Rackl, D.; Kreitmeier, P.; Reiser, O. *Green Chem.* **2016**, *18*, 214–219.

15. Lu, Z.; Yoon, T. P. *Angew. Chem. Int. Ed.* **2012**, *51*, 10329–10332.

16. Teegardin, K.; Day, J. I.; Chank J.; Weaver, J. *Org. Process Res. Dev.* **2016**, *20*, 1156–1163.

17. (a) Arora, A.; Teegardin, K. A.; Weaver, J. D. *Org. Lett.* **2015**, *17*, 3722–3725; (b) Chu, L.; Ohta, C.; Zuo, Z.; MacMillan, D. W. C. *J. Am. Chem. Soc.* **2014**, *136*, 10886–10889; (c) Weaver, J.; Senaweera, S. *Tetrahedron* **2014**, *70*, 7413–7428; (d) Devery III, J. J.; Douglas, J. J.; Nguyen, J. D.; Cole, K. P.; Flowers II, R. A.; Stephenson, C. R. J. *Chem. Sci.*, **2015**, *6*, 537–541; (e) Terrett, J. A.; Clift, M. D.; MacMillan, D. W. C. *J. Am. Chem. Soc.* **2014**, *136*, 6858–6861; (f) Dedeian, K.; Djurovich, P. I.; Garces, F. O.; Carlson, G.; Watts, R. J. *Inorg. Chem.* **1991**, *30*, 1685–1687; (g) Singh, A.; Arora, A.; Weaver, J. D. *Org. Lett.* **2013**, *15*, 5390–5393; (h) Wallentin, C.-J.; Nguyen, J. D.; Finkbeiner, P.; Stephenson, C. R. J. *J. Am. Chem. Soc.* **2012**, *134*, 8875–8884.

18. (a) McDonald, A. R.; Lutz, M.; von Chrzanowski, L. S.; van Klink, G. P. M.; Spek, A. L.; van Koten, G. *Inorg. Chem.* **2008**, *47*, 6681–6691; (b) Tamayo, A. B.; Alleyne, B. D.; Djurovich, P. I.; Lamansky, S.; Tsyba, I.; Ho, N. N.; Bau, R.; Thompson, M. E. *J. Am. Chem. Soc.* **2003**, *125*, 7377–7387.

19. Konno. H.; Sasaki, Y. *Chem. Lett.* **2003**, *32*, 252–253.

20. Singh, A.; Teegardin, K.; Kelly, M.; Prasad, K. S.; Krishnan, S.; Weaver, J. D. *J. Organomet. Chem.* **2015**, *776*, 51–59.

Appendix
Chemical Abstracts Nomenclature (Registry Number)

IrCl$_3$: Iridium trichloride; (10025-83-9)
2-Phenylpyridine: pyridine, 2-phenyl-; (1008-89-5)
Dichloromethane (DCM): methane, dichloro-; (75-09-2)
Hydrochloric acid (HCl): hydrochloric acid; (7647-01-0)
Ceolite: diatomaceous earth; (68855-54-9)
MgSO$_4$: magnesium sulfate (7487-88-9)
Hexanes: hexanes: (110-54-3)
NaOH: sodium hydroxide: (1310-73-2)

Kip Teegardin obtained his B.Sc. in Chemistry with a concentration in Biochemistry, from the University of Arkansas-Fort Smith, in 2012 after his military service with the 82nd Airborne Division (U.S. Army) in 2008. He is currently working on his PhD in the Weaver group at Oklahoma State University where his research currently focuses on the poly-fluoroarylation of oxazolones leading to non-natural fluorinated amino acids and their derivatives.

Jimmie Weaver obtained a B.S. in Chemistry with math and physics minors from Southern Nazarene University (2004), a PhD (2010) from the University of Kansas where he worked with Prof. Jon Tunge, and then completed a post-doc position with Jon Ellman (Yale). In the fall of 2012, Jimmie took a position at Oklahoma State University where he has established an independent research program focused on the catalytic generation and exploitation of reactive intermediates for really cool applications such as C–H and C–F functionalization, cross-coupling of minimally activated partners, uphill catalysis, and developing forms of energetic currency.

Hao Wu obtained his B.S. in Chemistry from Peking University in 2009, where he studied conjugate macrocycles and oligomers under the mentorship of Prof. Dahui Zhao. He completed his Ph.D. degree in organic chemistry from Boston College in 2015 under the supervision of Prof. Amir H. Hoveyda, working on catalytic enantioselective processes of C-B, C-C and C-Si bond formations. He is currently a senior scientist in Boehringer Ingelheim Pharm. Inc. since 2015. His research interest is the development of reliable and robust catalysis for process chemistry.

Alexander Sienkiewicz obtained his M.S. in Chemistry from Kiev State University in 1977, where he studied complexes of phosphoryl ligands with metal halides under the mentorship of Prof. Victor Skopenko. He completed his Ph.D. degree in coordination chemistry from the same University in 1987 under the supervision of Prof. Victor Skopenko, working on synthesis and characterization of polynuclear cage complexes containing chelating amino alcohols, amines, and amino acids and transition metals. He is a senior scientist in Boehringer Ingelheim Pharm. Inc. since 2015. He participates in the development of reliable catalytic techniques for process chemistry.

Jon C. Lorenz received a B.A. degree in Chemistry from Whitman College, Walla Walla, WA in 1995. He then joined the United States Peace Corps and taught science in the North West Province of Cameroon. He began his graduate studies at Colorado State University, where he received a Ph.D. in organic chemistry under the guidance of Professor Yian Shi in 2002. He joined the Department of Chemical Development at Boehringer Ingelheim Pharm. Inc. in Ridgefield, CT. In 2009 Jon moved to the Scale-up support group and then the kilo lab in Ridgefield, where he is currently a Senior Research Fellow. His research interests include the development and application of catalytic asymmetric reactions, use of Process Analytic Technology in Scale-up, and continuous processing for scale-up.

Preparation of Cyclopent-2-enone Derivatives via the Aza-Piancatelli Rearrangement

Meghan F. Nichol, Luis Limon, and Javier Read de Alaniz*[1]

Department of Chemistry and Biochemistry, University of California, Santa Barbara, CA 93106-9510, USA

Checked by Feng Peng and Kevin Campos

Procedure (Note 1)

A. *Furan-2-yl(phenyl)methanol (1).* An oven-dried 500 mL three-necked, round-bottomed flask equipped with an egg-shaped, Teflon-coated magnetic stir bar (3 cm x 1.5 cm) is capped on all necks with rubber septa. Phenylmagnesium bromide (57.4 mL, 57.4 mmol, 1.0 equiv, 1M in THF) (Note 2) is charged into this flask under an atmosphere of nitrogen at 5 °C (ice-water bath). Furfural (4.76 mL, 57.4 mmol, 1.0 equiv) (Note 3) is added over the course of 30 min via syringe into the cooled reaction through a side neck while maintaining internal temperature below 25 °C. Once all furfural is added the reaction stirs for 4 h. The reaction is monitored by thin-layer chromatography (TLC) analysis on silica gel with 70% hexanes in ethyl acetate as eluent and visualized under 254 nm UV light and stained with *p*-anisaldehyde (Note 4). Upon confirmation of product formation the reaction is quenched with saturated aqueous ammonium chloride (1 x 100 mL) and transferred to a 1 L separatory funnel and extracted with ethyl acetate (3 x 100 mL). The combined organic layers are dried over MgSO₄, filtered and concentrated to produce a yellow-orange oil. The product of the crude reaction mixture is purified via column chromatography (Note 5) to afford furan-2-yl(phenyl)methanol (**1**) as a yellow oil (9.71 g, 94%) (Notes 6, 7, and 8).

B. *4-(Mesitylamino)-5-phenylcyclopent-2-en-1-one (2)*. An oven-dried 500 mL single-necked, round-bottomed flask equipped with an egg-shaped, Teflon-coated magnetic stir bar (3 cm x 1.5 cm) is capped with a rubber septum. While cooling to ambient temperature under an atmosphere of nitrogen, an oil bath is preheated to 80 °C. Once the flask is cooled, furan-2-yl(phenyl)methanol (**1**) (3.95 g, 22.7 mmol, 1.1 equiv) (Note 9) and acetonitrile (200 mL) (MeCN) (Notes 10 and 11) are added along with 2,4,6-trimethylaniline (2.78 g, 20.5 mmol, 1.0 equiv) (Note 12), resulting in a pale brown homogeneous mixture (Figure 1).

Figure 1. Homogenous mixture of furan-2-yl(phenyl)methanol and 2,4,6-trimethylaniline

Dysprosium(III) trifluoromethanesulfonate (Dy(OTf)₃, 0.628 g, 1.03 mmol, 0.05 equiv) is added (Notes 13 and 14). Immediately following addition, the flask is fitted with a water reflux condenser, placed under an atmosphere of nitrogen, and submerged in the oil bath that is preheated to 80 °C and stirred for 4 h (Note 15).

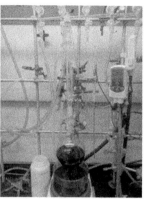

Figure 2. Reaction mixture 15 minutes into heating (left), 30 minutes into heating (center), and after 3 hours of heating (right)

The reaction mixture becomes dark brown in color upon heating (Figure 2). The reaction is followed by TLC analysis on silica gel with 85% hexanes in ethyl acetate as eluent and visualized with under 254 nm UV light and stained with *p*-anisaldehyde (Note 16). Upon confirmation that no 1,3,5-trimethylaniline remains, the stirring is stopped and the reaction mixture is allowed to cool to ambient temperature under an atmosphere of nitrogen. The cooled reaction mixture is quenched with saturated aqueous sodium bicarbonate (1 x 150 mL) and transferred to a 1 L separatory funnel and extracted with ethyl acetate (3 x 150 mL) (Figure 3).

Figure 3. Reaction mixture being quenched with sodium bicarbonate (left) and the final extraction with ethyl acetate (right)

The combined organic layers are dried over MgSO₄, filtered and concentrated (25 °C, 10 mmHg) to produce a dark brown oil. The product of the crude reaction mixture is purified via column chromatography (Note 17) to afford the cyclopentenone product (**2**) as a brown oil (4.97 g, 83%) (Notes 18, 19, 20, 21, and 22) (Figure 4).

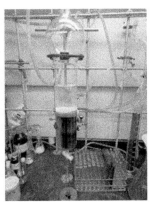

Figure 4. Column chromotagraphy on crude product (left) and the final dark brown oil cyclopentenone product (right)

Notes

1. Prior to performing each reaction, a thorough hazard analysis and risk assessment should be carried out with regard to each chemical substance and experimental operation on the scale planned and in the context of the laboratory where the procedures will be carried out. Guidelines for carrying out risk assessments and for analyzing the hazards associated with chemicals can be found in references such as Chapter 4 of "Prudent Practices in the Laboratory" (The National Academies Press, Washington, D.C., 2011; the full text can be accessed free of charge at https://www.nap.edu/catalog/12654/prudent-practices-in-the-laboratory-handling-and-management-of-chemical). See also "Identifying and Evaluating Hazards in Research Laboratories" (American Chemical Society, 2015) which is available via the associated website "Hazard Assessment in Research Laboratories" at https://www.acs.org/content/acs/en/about/governance/committees/chemicalsafety/hazard-assessment.html. In the case of this procedure,

the risk assessment should include (but not necessarily be limited to) an evaluation of the potential hazards associated with magnesium, bromobenzene, tetrahydrofuran, phenylmagnesium bromide, furfural, ethyl acetate, ammonium chloride, hexane, silica gel, acetonitrile, furan-2-yl(phenyl)methanol, dysprosium(III) trifluoromethanesulfonate, and 2,4,6-trimethylaniline.

2. The reagent solution was purchased from Sigma-Aldrich, although the same yield was obtained using phenylmagnesium bromide made from bromobenzene and magnesium turnings.

3. Furfural (99%) was purchased from Acros and distilled prior to use. The furfural can be stored in the freezer for up to 2 months.

4. When stained with *p*-anisaldehyde, furan-2-yl(phenyl)methanol stains a dark blue-brown color with a R_f of 0.46.

5. The product can be purified using column chromatography with a gradient of hexanes:ethyl acetate from 100% hexane to a 3:2 eluent. To a column (2″ in diameter) 165 g of silica (Geduran Si 60, Silicagel 60, 0.040-0.063 mm) was added and prepared with a 9:1 eluent solution. The unpurified product was dry loaded onto celite and loaded on the column. The product was eluted with 500 mL of 100% hexane eluent followed by 750 mL of 4:1 eluent, and lastly 500 mL of 3:2 eluent. The fractions containing the product were identified by TLC (Note 4), and the fractions were combined. The solvent was removed by rotatory evaporation (25 °C, 10 mmHg).

6. The product (**1**) has been characterized as follows: ^{1}H NMR (500 MHz, CDCl$_3$) δ: 2.40 (s, 1H), 5.85 (s, 1H), 6.13 (d, 1H), 6.33 (m, 1H), 7.39 (m, 6H); ^{13}C NMR (125 MHz, CDCl$_3$) δ: 70.2, 107.4, 110.2, 126.6, 128.1, 128.5, 140.8, 142.5, 155.9; IR (film) 3364, 3054, 3022, 2874, 1595, 1492, 1451, 1224, 1196, 1140, 1140, 1073, 1007, 939, 926, 884, 812, 728, 697, 625 cm^{-1}. HRMS (ESI+): [M + H] calcd for C$_{11}$H$_{11}$O$_2$: 175.0759 Found: 175.0739

7. The purity was determined to be >97% wt. by quantitative ^{1}H NMR spectroscopy in CDCl$_3$ using 174.2 mg of **1** and 168.6 mg of 1,3,5-trimethoxybenzene as an internal standard (D1 = 10 s).

8. A second reaction on 0.85X scale provided 7.61 g (94%) of the product (**1**).

9. Furan-2-yl(phenyl)methanol is freshly synthesized using a Grignard reaction between bromobenzene and furfural. If prepared freshly, the compound is a yellow oil after column purification and it becomes a dark black tar upon one week storage at room temperature. It is

recommended that the freshly prepared furan-2-yl(phenyl)methanol will be used for step 1B in less than one week upon fridge storage.

10. Acetonitrile was dried using a solvent purification system from a JC Meyer solvent dispensing system (content of water: 20-50 ppm). Reagent grade acetonitrile (Sigma Aldrich, anhydrous, 99.8%) can also be used.

11. Acetonitrile (35 mL) is added first to ensure any furan-2-yl(phenyl)methanol is washed down the sides of the flask before 4-iodoaniline addition. The final 5 mL is added after the addition of dysprosium triflate.

12. 2,4,6-Trimethylaniline (97%) was purchased from Acros Organics and used as received.

13. Dysprosium triflate (98%) was obtained from Strem Chemicals, stored in a desiccator and used as received.

14. Dysprosium triflate is added by quickly removing the rubber septum and adding the powdered catalyst to the stirring reaction mixture.

15. Reaction time may slightly vary between 3–5 h and can be determined by TLC analysis.

16. The reaction is considered complete when there is no 2,4,6-trimethylaniline is detected by TLC analysis. The R_f of the furan-2-yl(phenyl)methanol was 0.14, the R_f of the 2,4,6-trimethylaniline was 0.19, and the R_f of the product was 0.09. When stained with p-anisaldehyde, the furan-2-yl(phenyl)methanol starting material appears as a dark blue, the 2,4,6-trimethylaniline appears yellow, and the product appears as an olive brown color.

17. The product can be purified using column chromatography with a gradient of hexanes:ethyl acetate from 19:1 to 1:1. To a column (3″ diameter column) 275 g of silica (Geduran Si 60, Silicagel 60, 0.040-0.063 mm) was added and prepared with a 19:1 eluent solution. The unpurified product was dry loaded onto celite and loaded on the column. The product was eluted with 1 L of 95:5 eluent followed by 1.5 L of 85:15 eluent, then 1 L of 70:30 eluent, and lastly 0.5 L of 1:1 eluent. The fractions containing the product by TLC were combined and solvent was removed by rotatory evaporation (25 °C, 10 mmHg).

18. The product (**2**) has been characterized as follows: ^{1}H NMR (500 MHz, CDCl$_3$) δ: 2.14 (s, 6H), 2.26 (s, 3H), 3.20 (br s, 1H), 3.43 (d, 1H), 4.46 (s, 1H), 6.36 (dd, 1H), 6.84 (s, 2H), 7.06 (m, 2H), 7.31 (m, 3H), 7.67 (dd, 1H); ^{13}C NMR (125 MHz, CD$_2$Cl$_2$) δ: 18.6, 20.5, 60.4, 67.5, 127.0, 127.9, 128.7, 129.6, 132.3, 133.5, 137.9, 140.7, 162.8, 206.8; IR (film or solvent): 3346,

2914, 1704, 1585, 1481, 1452, 1373, 1335, 1230, 1155, 1109, 1040, 914, 854, 735, 697 cm^{-1}; HRMS (ESI+): calcd for C$_{20}$H$_{20}$NO [M + H] [M + H] calcd for C$_{20}$H$_{22}$NO: 292.1701 Found: 292.1704

19. The purity was determined to be >98% wt. by quantitative ^{1}H NMR spectroscopy in CDCl$_3$ using 128.1 mg of the compound 6 and 102.0 mg of trimethoxylbenzene as an internal standard (D1 = 10 s).

20. A second reaction on 0.85X scale provided 4.03 g (84%) of the product.

21. Alternatively, product was isolated as (s)-CSA salt via crystallization in MeCN: After workup of the reaction (scale of reaction based on furan-2-yl(phenyl)methanol: 10.0 g, 57.4 mmol, 1.1 equiv), the combined organic layers are dried over MgSO$_4$, filtered and concentrated to produce a dark brown oil. This dark oil was re-dissolved in 100 mL of dry MeCN. The resulting dark solution was dried until KF < 150 ppm water via constant volume distillation with dry MeCN (KF = 80 ppm). The dark mixture was then filtered to remove solid MgSO$_4$, and the resulting dark solution was heated to 50 °C. To the solution was charged (S)-CSA (10.8 g, 46.5 mmol, 0.9 equiv). The mixture was agitated at 50 °C until CSA dissolved. MeCN (50 mL) was removed by vacuum distillation (50 mmHg). During this process, the product crystallized as white solid. The slurry was cooled to room temperature and agitated for 2 h at rt. The slurry was filtered and the wet cake was washed with MeCN (2 x 10 mL). The cake was dried under vacuum (24.5 g, 80%, 98.5% purity).

22. Compound 6-CSA salt was characterized as follows: ^{1}H NMR (500 MHz, CD$_2$Cl$_2$) δ: 0.82 (s, 3H), 1.04 (s, 3H), 1.38 (m, 1H), 1.63 (m, 1H), 1.87 (d, 1H), 1.99 (m, 1H), 2.06 (t, 1H), 2.30 (s, 3H), 2.34 (s, 6H), 2.51 (ddd, 1H), 2.76 (dd, 1H), 3.27 (d 1H), 4.00 (d, 1H), 4.85 (m, 1H), 6.54 (m, 1H), 6.90 (m, 4H), 7.27 (m, 3H), 7.98 (m, 1H), 9.78 (s, 2H); ^{13}C NMR (125 MHz, CD$_2$Cl$_2$) δ: 18.8, 20.09, 20.11, 21.0, 25.3, 27.4, 43.2, 43.4, 48.3, 48.5, 53.6, 53.8, 54.0, 54.2, 54.4, 55.16, 55.24, 58.9, 69.1, 69.2, 128.1, 128.2, 128.5, 129.3, 129.4, 130.5, 131.4, 131.5, 132.7, 136.9, 137.7, 137.9, 139.8, 156.8, 156.9, 203.6, 203.7, 217.2; IR 3558, 3416, 3267, 3055, 2960, 2724, 2463, 1740, 1725, 1714, 1604, 1454, 1182, 1169, 1157, 1038 cm^{-1}; HRMS (ESI+): calcd for C$_{20}$H$_{22}$NO [M + H] 292.1701, found 292.1704.

Working with Hazardous Chemicals

The procedures in *Organic Syntheses* are intended for use only by persons with proper training in experimental organic chemistry. All hazardous materials should be handled using the standard procedures for work with chemicals described in references such as "Prudent Practices in the Laboratory" (The National Academies Press, Washington, D.C., 2011; the full text can be accessed free of charge at http://www.nap.edu/catalog.php?record_id=12654). All chemical waste should be disposed of in accordance with local regulations. For general guidelines for the management of chemical waste, see Chapter 8 of Prudent Practices.

In some articles in *Organic Syntheses*, chemical-specific hazards are highlighted in red "Caution Notes" within a procedure. It is important to recognize that the absence of a caution note does not imply that no significant hazards are associated with the chemicals involved in that procedure. Prior to performing a reaction, a thorough risk assessment should be carried out that includes a review of the potential hazards associated with each chemical and experimental operation on the scale that is planned for the procedure. Guidelines for carrying out a risk assessment and for analyzing the hazards associated with chemicals can be found in Chapter 4 of Prudent Practices.

The procedures described in *Organic Syntheses* are provided as published and are conducted at one's own risk. *Organic Syntheses, Inc.*, its Editors, and its Board of Directors do not warrant or guarantee the safety of individuals using these procedures and hereby disclaim any liability for any injuries or damages claimed to have resulted from or related in any way to the procedures herein.

Discussion

The cyclopentenone framework, often present in natural product architectures, has inspired the development of a number of elegant synthetic approaches.[2] One particularly attractive approach developed in 1976 when Piancatelli and co-workers constructed 4-hydroxycyclopentanone derivatives via an acid-catalyzed rearrangement of 2-furylcarbinols.[3] The transformation is highly diastereoselective and believed to proceed through a cascade sequence that terminates with an electrocyclic ring closure, analogous to the Nazarov cyclization (Scheme 1).[4,5]

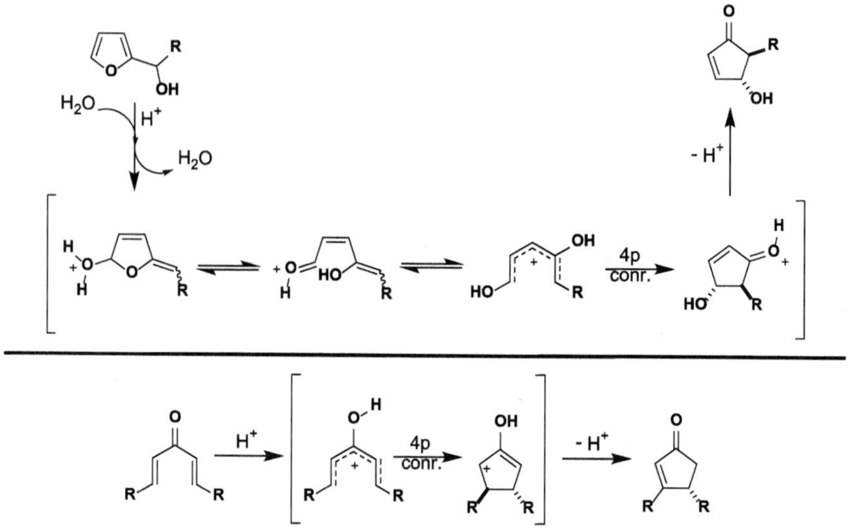

Scheme 1. Proposed Piancatelli mechanism (top) similar to a Nazarov 4π electrocyclization (bottom)

Although the Piancatelli rearrangement offers one of the most direct routes to 4-hydroxycylopentenones, its use in synthesis was largely driven by the synthesis of prostaglandins.[6] However, over the years rapid access to substituted cyclopentenones has played a critical role in a wide array of natural product synthesis.[7,8] Despite tremendous progress, a literature survey in 2009 revealed a lack of methods available for the direct synthesis of 4-aminocyclopentenone derivatives, with most relying on multistep approaches.[9] Two notable exceptions were reported independently by Denisov[10] and Batey[11] (Scheme 2) in 1993 and 2007, respectively. Inspired by these reports and Piancatelli's work we envisioned that an efficient catalytic aza-Piancatelli rearrangement could serve as a powerful general method to biologically active molecules bearing nitrogen functionality. Because furfural, the precursor to furylcarbinols, is produced from agricultural waste products like bagasse, oat hulls, and corncobs, it also provides chemists with a route to this key building block without relying on petrochemical feedstock.

Scheme 2. Research performed by various groups on acid-catalyzed rearrangements of furfural

Our investigation into an efficient aza-Piancatelli rearrangement began with the use of Dy(OTf)$_3$ in catalytic amounts (5 mol %) to facilitate the cascade rearrangement of furylcarbinols with aniline nucleophiles (Scheme 3).[12,13] While using Dy(OTf)$_3$ in our rearrangement we found the reaction was efficient in producing a trans-selective product that turned out to be adaptable with various functional groups present on either the furylcarbinol or aniline. Subsequent to our initial reports[14], a number of groups have demonstrated that other acids such as phosphomolybdic acid (PMA)[15], In(Otf)$_3$[15], La(OTf)$_3$[11], Ca(NTf$_2$)$_2$[16], In(Br)$_3$[15,17], and PPh$_3$ with DEAD[18] could also be used to catalyze the aza-Piancatelli reaction. The aza-Piancatelli reaction has also been extended to other amine nucleophiles such as hydroxylamine[14e] and tethered alkyl amines,[14a,18,19] as well as alcohol nucleophiles.[13,14b,14c] Recently the aza-Piancatelli reaction has been rendered asymmetric using chiral phosphoric acid to control the absolute stereochemistry.[2g,20,21,22]

Scheme 3. Scope of rearrangement with multiple substituted anilines
(top) as well as various 2-furylcarbinols (bottom)

References

1. Department of Chemistry and Biochemistry, University of California, Santa Barbara, CA 93106-9510, USA. Email: javier@chem.ucsb.edu.
2. (a) Tius, M. A. *Eur. J. Org. Chem.* **2005**, 2193–2206; (b) Pellissier, H. *Tetrahedron* **2005**, *61*, 6479–6517; (c) Frontier, A. J.; Collison, C. *Tetrahedron* **2005**, *61*, 7577–7606; (d) Grant, T. N.; Rieder, C. J.; West, F. G. *Chem. Commun.* **2009**, 5676–5688; (e) Blanco-Urgoiti, J.; Anorbe, L.;

Perez-Serrano,L.; Dominguez, G.; Perez-Castells, J. *Chem. Soc. Rev.* **2004**, *33*, 32–42; (f) Gibson, S. E.; Mainolfi, N. *Angew. Chem.* **2005**, *117*, 3082–3097; *Angew. Chem. Int. Ed.* **2005**, *44*, 3022–3037; (g) Simeonov, S. P.; Nunes, J. P. M.; Guerra, K.; Kurteva, V. B.; Alfonso, C. A. M. *Chem. Rev.* **2016**, *116*, 5744–5893.

3. Piancatelli, G.; Scettri, A.; Barbadoro, S. *Tetrahedron Lett.* **1976**, *17*, 3555–3558.

4. Faza, A. N.; Lopez, C. S.; Alvarez, R.; de Lera, I. R. *Chem. Eur. J.* **2004**, *10*, 4324–4329.

5. Wenz, D. R.; Read de Alaniz, J. *Eur. J. Org. Chem.* **2015**, 23–37.

6. Piancatelli, G.; Dauria, M.; Donofrio, F. *Synthesis* **1994**, 867–889.

7. Touré, B. B.; Hall, D. G. *Chem. Rev.* **2009**, *109*, 4439–4486.

8. Roche, S. P.; Aitken, D. J. *Eur. J. Org. Chem.* **2010**, 5339–5358.

9. For select examples, see: (a) Davis, F. A.; Wu, Y. Z. *Org. Lett.* **2004**, *6*, 1269–1272; (b) Dauvergne, J.; Happe, A. M.; Jadhav, V.; Justice, D.; Matos, M. C.; McCormack, P. J.; Pitts, M. R.; Roberts, S. M.; Singh, S. K.; Snape, T. J.; Whittall, J. *Tetrahedron* **2004**, *60*, 2559–2562; (c) Dauvergne, J.; Happe, A. M.; Roberts, S. M. *Tetrahedron* **2004**, *60*, 2551–2557; (d) Zaja, M.; Blechert, S. *Tetrahedron* **2004**, *60*, 9629-9634.

10. Denisov, V. R.; Shustitskaya, S.E.; Karpov, M. G. *Zh. Org. Khim.* **1993**, *29*, 249–252.

11. Li, S. W.; Batey, R. A. *Chem. Commun.* **2007**, 3759–3761.

12. Veits, G. K.; Wenz, D. R.; Read de Alaniz, J. *Angew. Chem Int. Ed.* **2010**, *49*, 9484–9487.

13. Read de Alaniz, J.; Palmer, L.; *Synlett* **2014**, *25*, 8–11.

14. (a) Palmer, L. I.; Read de Alaniz, J. *Angew. Chem. Int. Ed.* **2011**, *50*, 7167–7170; (b) Palmer, L. I.; Read de Alaniz, J. *Org. Lett.* **2013**, *15*, 476–479; (c) Fisher, D.; Palmer, L. I.; Cook, J. E.; Davis, J. E.; Read de Alaniz, J. *Tetrahedron* **2014**, *70*, 4105–4110; (d) Chung, R.; Yu, D.; Thai, V. T.; Jones, A. F.; Veits, G. K.; Read de Alaniz, J.; Hein, J. E. *ACS Catal.* **2015**, *5*, 4579–4585; (e) Veits, G. K.; Wenz, D. R.; Palmer, L. I.; St. Amant, A. H.; Hein, J. E.; Read de Alaniz, J. *Org. Biomol.Chem.* **2015**, *13*, 8465–8469.

15. Reddy, B. V. S.; Reddy, Y. V.; Lakshumma, P. S.; Narasimhulu, G.; Yadav, J. S.; Sridhar, B.; Reddy, P. P.; Kunwar, A. C. *RSC Advances* **2012**, *2*, 10661–10666.

16. Leboeuf, D.; Schulz, E.; Gandon, V. *Org. Lett.* **2014**, *16*, 6464–6467.

17. Aitken, D. J.; Eijsberg, H.; Frongia, A.; Ollivier, J.; Piras, P. P. *Synthesis* **2014**, *46*, 1–24.

18. Xu, Z. L.; Xing, P.; Jiang, B. *Org. Lett.* **2017**, *19*, 1028–1031.

19. Piutti, C.; Quartieri, F. *Molecules* **2013**, *18*, 12290–12312.
20. Cai, Y.; Tang, Y.; Atodiresei, I.; Rueping, M. *Angew. Chem. Int. Ed.* **2016**, *55*, 14126–14130.
21. Li, H.; Tong, R.; Sun, J. *Angew. Chem. Int. Ed.* **2016**, *55*, 15125–15128.
22. Gade, A. B.; Patil, N. T. *Synlett* **2017**, *28*, 1096–1100.

Appendix
Chemical Abstracts Nomenclature (Registry Number)

Furan-2-yl(phenyl)methanol: 2-Furyl(phenyl)methanol; (4484-57-5)
MeCN: acetonitrile; (75-05-8)
4-Iodoaniline: 4-Iodo-benzenamine; (540-37-4)
Dy(OTf)$_3$: Dysprosium(III) trifluoromethanesulfonate; (139177-62-1)

Meghan F. Nichol came to the University of California at Santa Barbara in 2015 and began work in the group of Dr. Read de Alaniz. She earned her B. S. degree in chemical physics from Lewis University (Romeoville, IL) in 2015 where she conducted undergraduate research under the direction of Dr. Jason J. Keleher. Meghan then moved to Santa Barbara where she is pursuing work in self-immolative polymer trigger systems.

Luis Limon is an undergraduate researcher at the University of California at Santa Barbara pursuing his B. S. degree in chemistry. He joined the Read de Alaniz group in the summer of 2016 and is currently working on development of new methodology for small molecule synthesis.

Javier Read de Alaniz joined the department of Chemistry and Biochemistry at University of California at Santa Barbara in 2009 as an Assistant Professor. He received his B. S. degree from Fort Lewis College (Durango, Colorado) in 1999 where he conducted undergraduate research under the direction of Professor William R. Bartlett. He obtained his Ph. D. in 2006 under the supervision of Professor Tomislav Rovis at Colorado State University with a research focus on asymmetric catalysis. Javier then moved to California, where he worked in the area of total synthesis with Professor Larry E. Overman at the University of California, Irvine.

Feng Peng joined the Process Research Department of Merck & Co., Inc. in 2012. His research focuses on using state-of-art organic chemistry to address critical problems in drug development. He received his B. S. degree from Beijing Normal University. He obtained his M.S. under the supervision of Professor Dennis Hall at University of Alberta with a research focus on Boron Chemistry. Feng then moved to New York City, where he obtained Ph.D. in the area of total synthesis (maoecrystal V) with Professor Samuel Danishefsky at Columbia University.

Catalytic, Metal-Free Oxidation of Primary Amines to Nitriles

Kyle M. Lambert, Sherif A. Eldirany, James M. Bobbitt, William F. Bailey*[1]

Department of Chemistry, University of Connecticut, Storrs, CT 06269

Checked by Thomas R. DeVino, Andreas R. Röthel, and Margaret Faul

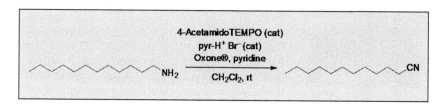

Procedure (Note 1)

A. *Dodecanenitrile.* A 1-L, three-necked, round-bottomed flask is equipped with an oval Teflon-coated magnetic stir bar, a rubber septum, a nitrogen line fitted to a glass bubbler filled with mineral oil, and a 250-mL pressure-equalizing addition funnel capped with a rubber septum pierced with a needle (Notes 2, 3, 4, and 5) (Figure 1).

While under a continuous flow of nitrogen, the flask is dried with a heat gun and then allowed to cool to room temperature. The needle on the addition funnel is removed. Then the septum on the flask is removed and the flask is charged with 200 mL of dichloromethane (Note 6), Oxone (80.0 g, 130 mmol, 2.6 equiv) (Note 7), and pyridinium bromide (365 mg, 2.25 mmol, 0.045 equiv) (Note 8) (Figure 2). The heterogeneous mixture is stirred for 15 min until a faint yellow color develops (Figure 3).

Dry pyridine (20 mL, 19.6 g, 250 mmol, 5 equiv) (Note 9) is then added *via* syringe through the rubber septum (Figure 3), upon which the yellow color of the reaction mixture intensifies (Figure 4). The reaction mixture is

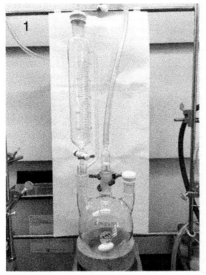
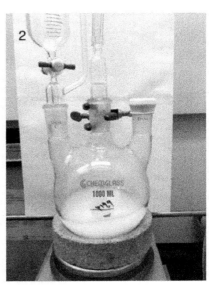

Figure 1. Initial reaction flask setup for drying (provided by checker); Figure 2. Initial mixture of Oxone and pyridinium bromide in dichloromethane (provided by checker)

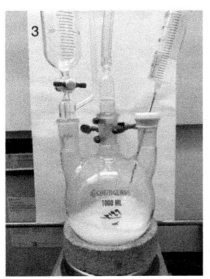
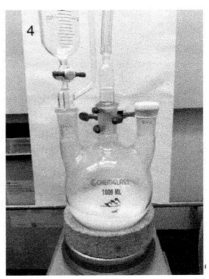

Figure 3. Reaction mixture prior to the addition of pyridine (provided by checker); Figure 4. Reaction mixture after the addition of pyridine (provided by checker)

stirred for 1–2 min, the septum is then removed and the 4-acetamidoTEMPO catalyst (533 mg, 2.50 mmol, 0.050 equiv) (Note 10) is added in one portion. The septum is placed back on the flask and the reaction mixture, which darkens upon addition of the catalyst (Figure 5), is stirred for 15–30 min.

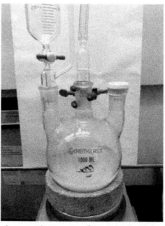

Figure 5. Reaction mixture after the addition of catalyst (provided by checker)

A clean, dry 125-mL Erlenmeyer flask is charged with dodecylamine (9.24 g, 49.9 mmol, 1 equiv) (Note 11) and 100 mL of dichloromethane. The Erlenmeyer flask is swirled to dissolve the amine. The septum on the addition funnel is removed and the homogenous solution of the amine is transferred to the addition funnel through the use of a funnel. The Erlenmeyer flask is then rinsed with additional dichloromethane (10 mL) to ensure complete transfer and to rinse any residual amine from the funnel. The funnel is then removed and the septum is placed back onto the addition funnel. An additional nitrogen line is added through the septum of the addition funnel and the addition rate is adjusted to ensure a constant delivery of the amine solution at approximately 40–50 mL/h (Figure 6). Addition of the amine takes approximately 2–2.5 h to complete, at which point additional dichloromethane (20 mL) is added *via* syringe to the funnel (Note 12) (Figure 8) allowing for any residual white solid (Figure 7) to be rinsed into the reaction flask. The light yellow reaction mixture (Figure 8) is stirred at 23 °C (Note 13) for an additional 9–15 h (Note 14).

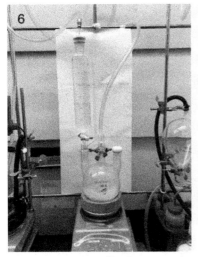
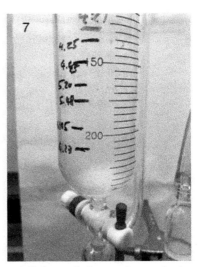

Figure 6. Reaction flask setup for the addition of the amine solution in dichloromethane (provided by checker); Figure 7. Residual solids in the addition funnel after amine addition (provided by checker)

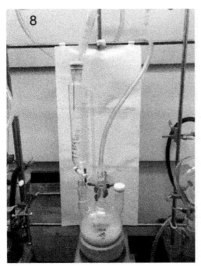
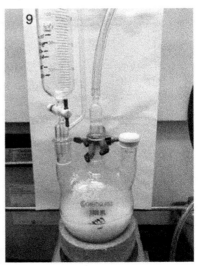

Figure 8. Rinse of residual solid with dichloromethane using a syringe (provided by checker); Figure 9. Reaction flask with final reaction mixture before extended stirring. (provided by checker)

During this time the reaction mixture turns from a yellow suspension (Figure 9) to a clear yellow solution with sticky yellow solids along the wall of the round-bottomed flask (Figure 10)

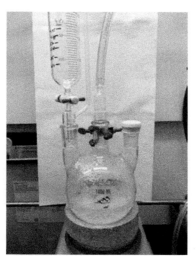

Figure 10. Reaction flask with final reaction mixture after extended stirring (provided by checker)

A 600–mL medium porosity, sintered glass Büchner funnel is filled with 150 g of silica gel (Note 15), and a small filter paper (90 mm diameter, coarse porosity) is placed on top of the silica gel. The funnel is then fitted to a 1–L round-bottomed flask, set up for vacuum filtration, and the reaction mixture is poured through the filtration setup (Figure 11). The reaction flask is rinsed with an additional 300 mL of dichloromethane and poured through the filtration setup (Note 16) (Figure 12). The yellow solids remain in the original flask (Figure 13).

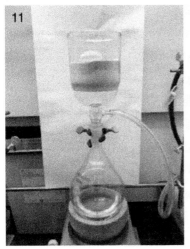 

Figure 11. Filtration of reaction mixture through silica gel. (provided by checker); **Figure 12.** Filtration setup after dichloromethane wash (provided by checker)

Figure 13. Residual solids in the reaction flask after the dichloromethane wash (provided by checker)

The filtrate is concentrated by rotary evaporation (30–40 °C, 500 mmHg then to approx. 15 mmHg for 15–30 min) (Note 17) to afford dodecanenitrile (8.80 g, 94%) (Notes 18 and 19) as a pale yellow oil (Notes 20, 21, and 22). However, the title compound is reported in the literature as a clear, colorless oil.[3] Colorless product is obtained by transferring the pale yellow crude oil (Note 23) to a 25-mL round-bottomed flask from the 1-L round-bottomed flask. The remaining oil in the 1-L round-bottomed flask is rinsed with dichloromethane (2 x 5 mL) and transferred to the 25-mL round-bottomed flask to ensure all the oil is transferred. Then the dichloromethane was removed by rotary evaporation (30–40 °C, 500 mmHg then to approx. 15 mmHg for 15–30 min). The 25-mL round-bottomed flask is then set up for bulb-to-bulb distillation (Kugelrohr, bath temp: 200 °C, 10–20 mmHg) (Figure 14) to afford analytically pure dodecanenitrile (8.37 g, 93% yield) (Notes 24, 25, 26, and 27), as a clear colorless oil.

Figure 14. Kugelrohr setup at the beginning of the distillation (provided by checker)

Notes

1. Prior to performing each reaction, a thorough hazard analysis and risk assessment should be carried out with regard to each chemical substance and experimental operation on the scale planned and in the context of the laboratory where the procedures will be carried out. Guidelines for carrying out risk assessments and for analyzing the hazards associated with chemicals can be found in references such as Chapter 4 of "Prudent Practices in the Laboratory" (The National Academies Press, Washington, D.C., 2011; the full text can be accessed free of charge at https://www.nap.edu/catalog/12654/prudent-practices-in-the-laboratory-handling-and-management-of-chemical).

See also "Identifying and Evaluating Hazards in Research Laboratories" (American Chemical Society, 2015) which is available via the associated website "Hazard Assessment in Research Laboratories" at https://www.acs.org/content/acs/en/about/governance/committees/chemicalsafety/hazard-assessment.html. In the case of this procedure, the risk assessment should include (but not necessarily be limited to) an evaluation of the potential hazards associated with dichloromethane, Oxone, pyridinium bromide, pyridine, 4-acetamidoTEMPO, dodecylamine, and silica gel, as well as the proper procedures for using a Kugelrohr apparatus.

2   The submitters used a 1-L one-necked, round-bottomed flask equipped with an oval Teflon-coated magnetic stir bar (40 mm x 20 mm) and capped with a septum that is connected *via* a nitrogen line inlet and an outlet fitted to a glass bubbler filled with mineral oil (Figure 15). The submitters flame dried the flask prior to setting up the reaction.

**Figure 15. Initial reaction flask setup for flame drying
(provided by submitter)**

3.   The submitters note that nitrogen gas is sufficient to establish an anhydrous atmosphere, and is less expensive than argon. However, if argon is available, it should be used as it results in slightly higher isolated yields ~2% greater than those obtained when nitrogen is used. The isolated yields reported in this manuscript were obtained when nitrogen gas was used.

4.   A 250-mL constant-rate addition funnel was used, but a 125-mL funnel would suffice for this scale. Due to the use of a 1-L one-necked, round-

bottomed flask the submitters had to mount a 500-mL constant-rate addition funnel available from Kontes (cat. # 186-634620-0500) after the addition of the 4-acetamidoTEMPO catalyst (Figure 16). Prior to fitting to the reaction flask, the funnel was calibrated with dichloromethane to allow for the addition of 40–50 mL of solution per hour (~ 30 drops per minute) and the graduated hash mark where the rod was located was noted. A less sophisticated pressure-equalizing addition funnel would likely suffice as long as the rate of addition does not exceed 40–50 mL/h.

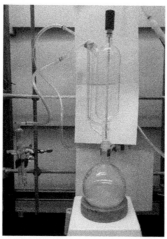

Figure 16. Reaction Flask Fitted with the Pressure-Equalizing Addition Funnel (provided by submitter)

5. The submitters report that the failure to purge the headspace of the addition funnel with nitrogen or argon gas results in the build-up of a white crystalline solid during the course of the addition of the amine substrate. This solid is likely the result of the reaction of the amine substrate with residual carbon dioxide to form a carbamic acid.[8] Build-up of the white solid can result in a slowed and problematic addition rate. However, the formation of the solid does occur to some extent even under the best attempts to remove the carbon dioxide, but its formation does not influence isolated yields of nitrile as the carbamic acid is in equilibrium with the free amine under the reaction conditions and the oxidation proceeds.

6. Dichloromethane (ACS reagent grade ≥ 99.5%, <0.01% water, 40–150 ppm amylene stabilized) was purchased from Sigma-Aldrich and was used as received. The submitters report that this solvent contained the appropriate quantity of water to solubilize a small portion of the oxidant. The reaction fails to proceed under strictly anhydrous conditions[4] and significant quantities of aldehyde byproduct result when the water content is too high. For these reasons all other sources of water should be avoided.

7. Oxone monopersulfate was purchased as a granular white powder from Alfa Aesar and used as received. Due to the hygroscopic nature of Oxone, a fresh bottle is recommended, and it should be stored in a tightly closed container in a dry location such as a desiccator. The submitters reported that additional moisture absorbed by the Oxone has a detrimental effect on the oxidation, and due to instability at high temperatures drying of the reagent is not recommended. The Oxone should be added to a rapidly stirred solvent to avoid difficulties in initiating the stirring. Oxone is a commercially available triple salt of empirical formula 2 $KHSO_5 \cdot KHSO_4 \cdot K_2SO_4$, and has a combined molecular mass of 614.7 g/mol. Each molar quantity of the triple salt contains 2 moles of the terminal oxidant ($KHSO_5$), which has a molecular mass of 152.2 g/mol. The oxidation requires at least 2.0 equivalents of $KHSO_5$ to proceed; however, it is beneficial to use an excess as it simplifies the work-up process (Note 12). The amount of Oxone employed in this procedure is 130 mmol (80.0 g, 2.6 equiv) per 50 mmol of amine substrate, and is arrived at by accounting for the fact that commercial Oxone contains only 42–44 % by mass of the active component ($KHSO_5$).[5] As a result, 33.6 g (220 mmol, 4.4 equiv relative to the starting amine) of the terminal oxidant ($KHSO_5$) is present in the reaction mixture. This was calculated as follows:

50 mmol amine x 4.4 equiv. = 220 mmol oxidant
0.220 mol $KHSO_5$ x (152.2 g $KHSO_5$/1 mol $KHSO_5$) = 33.5 g $KHSO_5$
33.5 g $KHSO_5$/0.42 = 80.0 g Oxone

8. Pyridine hydrobromide (98%) was purchased from Sigma Aldrich and used as received. Pyridinium bromide is very hygroscopic and must be kept dry and moisture-free. The submitters report that it is essential that high-quality pyridinium bromide, that does not contain any tribromide as the counterion, is used. Alternatively, a high-quality sample of

pyridinium bromide may be obtained from the reaction of freshly distilled *tert*-butyl bromide with dry pyridine.[6]

9. Pyridine (ACS reagent grade), was purchased from Sigma Aldrich and used as received. Submitters purchased from J.T. Baker and dried by magnetically stirring with calcium hydride for 24 h at room temperature in an oven-dried flask capped with a septum, then distilled under nitrogen collecting the fraction boiling at 114–116 °C into a flame-dried flask. While the oxidation requires at least 4 equivalents of pyridine to proceed,[4] the additional equivalent of pyridine ensures the amine substrate does not act as the base. When the amine substrate is protonated it fails to undergo any oxidative processes.

10. 4-AcetamidoTEMPO (4-Acetamido-2,2,6,6-tetramethylpiperidine 1-oxyl) is commercially available in 98+% from Alfa Aesar. However, this nitroxide is easily and inexpensively prepared in multi-molar quantities from known procedures.[7] The submitters report that the catalyst should be dried in an Abderhalden drying pistol at 56 °C (acetone, 10 mmHg) overnight prior to use; mp 144–146 °C (lit.[7] 145–147 °C). The checkers used commercial 4-acetamidoTEMPO as received, without any further drying.

11. Dodecylamine (98% purity) was purchased from Sigma-Aldrich and used as received. The amine turns to a light yellow color over time, but is easily purified by bulb-to-bulb distillation (Bath temp: 155 °C, 15 mmHg), if necessary prior to use. Dodecylamine was used as received by the checkers.

12. A small quantity of white solid forms upon the wall of the addition funnel (Note 4) and the additional 20 mL of dichloromethane should be added using a syringe to rinse the walls of the addition funnel so as to remove as much white solid as possible. The submitters removed the rubber septum and used a glass pipette to rinse the addition funnel (Figure 17)

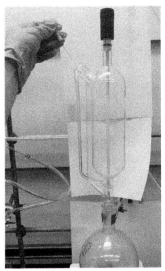

Figure 17. Residual solid is rinsed with dichloromethane with a long glass pipette (provided by submitter)

13. The heterogeneous mixture may become difficult to stir and the stir bar may stop after the addition of the amine is complete. This does not seem to affect the yield as the oxidation is complete at this time. Additionally, the reaction temperature may increase slightly during the amine addition to approximately 30 °C.

14. The oxidation of the amine substrate to the corresponding nitrile is quite rapid and is completed usually within 1 h of addition. However, due to the requirement to use an excess of an external base, pyridine, working the reaction up at this time would require an extractive work-up with acid. Removal of the excess pyridine and isolation of the product is greatly complicated, and not feasible for acid-sensitive substrates. Thus, the reaction is allowed to stir for the given period of time as the excess oxone efficiently oxidizes the majority of the excess pyridine to pyridine-*N*-oxide.[4] The pyridine-*N*-oxide is easily removed upon work-up.

15. Silica gel was purchased from Acros Organics and the particle size was 0–60 μm (60 Å). The submitters used silica gel from Silicyle or ZEOChem and the particle size was 40–63 μm (230–400 mesh).

16. The submitters report that the filtration should be stopped when the colored bands (Figure 18) have moved to within 1 cm of the glass frit.

The light yellow band is associated with the small excess of pyridine present and should not be collected. On occasion a light purple band is visible and should not be collected either. The checkers did not observe these bands right after filtration. However, they did appear upon aging of the filter cake under ambient conditions.

Figure 18. Colored bands observed during filtration (provided by submitter)

17. The submitters report that placing a small wooden boiling stick into the flask facilitates the removal of excess dichloromethane while under vacuum, and can easily be removed prior to obtaining a final weight of compound without significantly altering the isolated yield.
18. Corresponds to weight adjusted yield based on purity determined by quantitative ¹H NMR analysis (see below for ¹H QNMR spectrum)
19. The submitters found that the refractive index of the crude product was $n_D^{20} = 1.4353$, and was in agreement with the literature value of $n_D^{20} = 1.4361$.[2]
20. Additional runs of the procedure by the Checkers afforded the title compound with yields of 88% and 92%.
21. The submitters report that after storage over a period of 6 months in a refrigerator at 6 °C, in a glass vial capped with a septum, the compound maintains its initial light yellow color with no apparent decomposition. A check of the refractive index was done $n_D^{20} = 1.4368$, which is consistent with the value reported in the literature.[2]

22. The procedure described will afford the title compound with purity sufficient for most purposes. Quantitative ^{1}H NMR analysis showed this product was of 97% purity (18.9 mg of analyte with 21.2 mg benzyl benzoate of 100% purity purchased from Sigma-Aldrich as a standard; see below for ^{1}H QNMR spectrum) The submitters used dimethylfumarate as an internal standard and also assessed purity by elemental analysis. calcd for $C_{12}H_{23}N$: C 79.49, H 12.79, N 7.72, found C 79.22, H 12.97, N 7.90.

23. A small portion (*i.e.* 18.9 mg) was removed for quantitative ^{1}H NMR analysis.

24. The distillation removes any trace impurities and results in a clear, colorless oil. Quantitative ^{1}H NMR analyses showed that the distilled products were of >99.5% average purity. (*e.g.* 18.6 mg of analyte with 21.4 mg benzyl benzoate of 100% purity purchased from Sigma-Aldrich as a standard; see below for ^{1}H QNMR spectrum). Note: The submitters used dimethylfumarate as an internal standard.

25. The submitters found that the refractive index of the distilled product was $n_D^{20} = 1.4364$ and was in agreement with the literature value of $n_D^{20} = 1.4361$.[2]

26. The product was characterized as follows:[9] ^{1}H NMR (500 MHz, CDCl$_3$) δ: 0.88 (t, $J = 6.9$ Hz, 3H), 1.28 (m, 14H), 1.44 (quin, $J = 7.3$ Hz, 2H), 1.65 (quin, $J = 7.4$ Hz, 2H), 2.33 (t, $J = 7.2$ Hz, 2H). ^{13}C NMR (125 MHz, CDCl$_3$) δ: 14.3, 17.3, 22.9, 25.6, 29.0, 29.5, 29.7, 32.1, 120.0. FT–IR (neat, ATR): 2923, 2854, 2246, 1666, 1645, 1466, 1427, 1378, 1364,1329, 1246, 1163, 1122, 1070, 1054, 931, 907, 882, 846, 818, 722 cm^{-1}; GC-MS (EI): 182 ([M+1]$^+$, 7%), 181 ([M]$^+$, 0.6%), 180 ([M-1]$^+$, 3%), 138 (24%), 124 (36%), 110 (59%), 97 (85%), 82 (56%), 69 (42%), 57 (62%), 41 (100%); HRMS (DART-TOF) $C_{12}H_{23}N$ [M+H]$^+$: calc. 182.1919, found 182.1909.

27. The submitters stored the product at 6 °C in a chemical refrigerator absent of light under an atmosphere of argon for a period of several weeks and observed no apparent decomposition or discoloration. A second sample was stored at room temperature on the laboratory bench in the absence of an argon atmosphere and the product returned to the original light yellow color observed prior to distillation. A check of the refractive index resulted in an $n_D^{20} = 1.4355$, which is similar to the value obtained prior to distillation.

73   DOI: 10.15227/orgsyn.095.0060

Working with Hazardous Chemicals

The procedures in *Organic Syntheses* are intended for use only by persons with proper training in experimental organic chemistry. All hazardous materials should be handled using the standard procedures for work with chemicals described in references such as "Prudent Practices in the Laboratory" (The National Academies Press, Washington, D.C., 2011; the full text can be accessed free of charge at http://www.nap.edu/catalog.php?record_id=12654). All chemical waste should be disposed of in accordance with local regulations. For general guidelines for the management of chemical waste, see Chapter 8 of Prudent Practices.

In some articles in *Organic Syntheses*, chemical-specific hazards are highlighted in red "Caution Notes" within a procedure. It is important to recognize that the absence of a caution note does not imply that no significant hazards are associated with the chemicals involved in that procedure. Prior to performing a reaction, a thorough risk assessment should be carried out that includes a review of the potential hazards associated with each chemical and experimental operation on the scale that is planned for the procedure. Guidelines for carrying out a risk assessment and for analyzing the hazards associated with chemicals can be found in Chapter 4 of Prudent Practices.

The procedures described in *Organic Syntheses* are provided as published and are conducted at one's own risk. *Organic Syntheses, Inc.*, its Editors, and its Board of Directors do not warrant or guarantee the safety of individuals using these procedures and hereby disclaim any liability for any injuries or damages claimed to have resulted from or related in any way to the procedures herein.

Discussion

The nitrile functionality is an important motif found within a variety of pharmaceuticals,[10] natural products,[11] agrochemicals,[12] and it is often used to access other functionalities.[13] While there are many ways to prepare nitriles,[14] the vast majority of such methodologies involve a substitution reaction using a cyanide source[15] or a metal-mediated coupling reaction.[14a, 16]

There are few practical procedures for the oxidation of amines to nitriles, and these methods generally involve the use of a transition metal catalyst.[17]

The procedure described above offers a metal-free, scalable, operationally simple method for the oxidation of primary amines to nitriles in good to excellent yield. This overall conversion of an amine to a nitrile involves several intertwined catalytic cycles as described in our original report.[4] By employing catalytic quantities of a commercially available nitroxide catalyst, 4-acetamidoTEMPO, and inexpensive pyridinium bromide as a halide source in conjunction with Oxone, an environmentally benign terminal oxidant, the methodology offers a "green" approach to nitriles.

Table 1. Substrate Scope of the Oxidation[4]

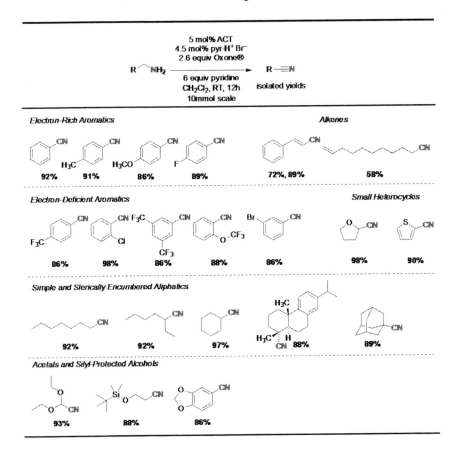

The oxidation is of wide scope as demonstrated by the results presented in Table 1.[4] Additionally, the near neutral pH of the reaction medium coupled with the simple work-up allows for the oxidation of acid-sensitive substrates.

References

1. Department of Chemistry, University of Connecticut, 55 North Eagleville Road, Storrs, Connecticut 06269, United States. willam.bailey@uconn.edu; Ph. (860) 486–2163; Fax (860) 486-2981.
2. Korosi, G.; Kovats, E. S. *J. Chem. Eng. Data* **1981**, *26*, 323–332.
3. Suzuki, Y.; Yoshino, T.; Moriyama, K.; Togo, H. *Tetrahedron* **2011**, *67*, 3809–3814.
4. Lambert, K. M.; Bobbitt, J. M.; Eldirany, S. A.; Kissane, L. E.; Sheridan, R. K.; Stempel, Z. D.; Sternberg, F. H.; Bailey, W. F. *Chem. Eur. J.* **2016**, *22*, 5156–5159.
5. *Oxone^{TM} Monopersulfate Compound General Technical Attributes*; Technical Bulletin for The Chemours Company: Wilmington, DE, November 2015.
6. Cioffi, E. A.; Bailey, W. F. *Tetrahedron Lett.* **1998**, *39*, 2679–2680.
7. (a) Bobbitt, J. M. *J. Org. Chem.* **1998**, *63*, 9367–9374; (b) Mercadante, M. A.; Kelly, C. B.; Bobbitt, J. M.; Tilley, L. J.; Leadbeater, N. E. *Nat. Protoc.* **2013**, *8*, 666–676.
8. Peterson, S. L.; Stucka, S. M.; Dinsmore, C. J. *Org. Lett.* **2010**, *12*, 1340–1343.
9. (a) Kraft, F. *Chem. Ber.* **1890**, *14*, 2363; (b) Stephenson, R. M. *J. Chem. Eng. Data*, **1993**, *38*, 625–629.
10. Fleming, F. F.; *Nat. Prod. Rep.* **1999**, *16*, 597–606.
11. Fleming, F. F.; Yao, L.; Ravikumar, P. C.; Funk, L.; Shook, B. C. *J. Med. Chem.* **2010**, *53*, 7902-7917.
12. (a) Zhou, L. Y.; Zhang, J. L.; Sun, S. L.; Ge, F.; Mao, S. Y.; Ma, Y.; Lui, Z. H.; Dai, Y. J.; Yuan, S. *J. Agric. Food Chem.* **2014**, *62*, 9957–9964; (b) Collett, M. G.; Stegelmeier, B. L.; Tapper, B. A. *J. Agric. Food Chem.* **2014**, *62*, 7370–7375.
13. (a) Larock, R. C. in *Comprehensive Organic Transformations, Vol. 1*, VCH Publishers, New York, **1989**, pp. 933; (b) Niemerier, J. K; Rothhaar, R. R.; Vicenzi, J. T.; Werner, J. A. *Org. Process Res. Dev.* **2014**, *18*, 410; (c) Lee, G.

A. *Synthesis* **1982**, *6*, 508–510; (d) Herzberger, J.; Frey, H. *Macromolecules* **2015**, *48*, 8144–8153.

14. (a) Smith, M. B. in *March's Advanced Organic Chemistry, Vol. 7*. Wiley and Sons, Hoboken, NJ, **2013**, pp. 1625–1626; (b) Fatiadi, A. J. *Preparation and Synthetic Applications of Cyano Compounds*, Wiley, New York, 1983.

15. Mowry, D. T.; *Chem. Rev.* **1948**, *42*, 189–283.

16. (a) Zhu, Y.; Zhao, M.; Lu, W.; Li, L.; Shen, Z. *Org. Lett.*, **2015**, *17*, 2602–2605; (b) Reeves, J. T.; Malapit, C. A.; Buono, F. G.; Sidhu, K. P.; Marsini, M. A.; Sader, C. A.; Fandrick, K. R.; Busaccaand, C. A.; Senanayake, C. H. *J. Am. Chem. Soc.*, **2015**, *137*, 9481–9488.

17. (a) Kim, J.; Stahl, S. S. *ACS Catal.* **2013**, *3*, 1652–1656; (b) Murahashi, S.-I.; Imada, Y. Amine Oxidation. In *Transition Metals for Organic Synthesis*; Beller, M., Bolm, C., Eds.; Wiley-VCH: Weinheim, Germany, 2008; pp 497; For a review of catalytic processes see: (c) Bartelson, A. B.; Lambert, K. M.; Bobbitt, J. M.; Bailey, W. F. *Chem. Cat. Chem.* **2016**, *8*, 3421–3430.

Appendix
Chemical Abstracts Nomenclature (Registry Number)

Dodecanenitrile: Dodecanenitrile; (2437-25-4)

Dichloromethane: Methane, dichloro-; (75-09-2)

Oxone: Potassium peroxymonosulfate sulfate; (37222-66-5 or 70693-62-8)

Pyridinium bromide: Pyridine, hydrobromide (1:1); (18820-82-1)

Pyridine: Pyridine; (110-86-1)

4-AcetamidoTEMPO: 1-Piperidinyloxy, 4-(acetylamino)-2,2,6,6-tetramethyl-; (14691-89-5)

4-Acetamido-2,2,6,6-tetramethylpiperidine 1-oxyl: 1-Piperidinyloxy, 4-(acetylamino)-2,2,6,6-tetramethyl-; (14691-89-5)

Dodecylamine: 1-Dodecanamine; (124-22-1)

Pyridine *N*-Oxide; Pyridine, 1-oxide; (694-59-7)

separatory funnel using dichloromethane (50 mL) and the organic phase is collected. The aqueous phase is extracted with dichloromethane (2 × 50 mL), and the combined organic layers are washed with deionized water (200 mL) and then concentrated with the aid of a rotary evaporator (30 °C, 80 mmHg) to afford a crude, colorless, heterogeneous mixture. This material is dissolved in dichloromethane (50 mL) and eluted through a pad of alumina (Note 6) to yield **1** as a white solid (27.5–28.0 g, 93–94%) (Figure 3) (Note 7).

Figure 3. Crude mixture after extraction and final, purified product

B. *(R)-N,N-Dibenzyl-1-phenylpropan-1-amine* (**2**). A 250-mL, two-necked, round-bottomed flask (Note 2) is equipped with a 1-cm, Teflon-coated magnetic stir bar and rubber septa, through one of which a needle connected to a manifold under a positive pressure of dry nitrogen is inserted. One septum is removed and the flask is charged sequentially with *N,N*-dibenzyl-*O*-pivaloylhydroxylamine (**1**, 7.55 g, 25.4 mmol, 1.2 equiv), copper(II) acetate (38 mg, 0.21 mmol, 0.010 equiv), (*S*)-DTBM-SEGPHOS (274 mg, 0.23 mmol, 0.011 equiv), triphenylphosphine (61 mg, 0.46 mmol, 0.011 equiv), and *trans*-β-methylstyrene (2.50 g, 2.75 mL, 21.1 mmol, 1 equiv) under nitrogen flow (Figure 4) (Note 8).

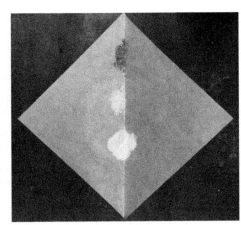

Figure 4. Compound 1, catalyst components, and substrate

The septum is reattached to the flask, and THF (21 mL) (Note 9) is added by syringe. The flask is submerged in a room-temperature (23 °C) water bath such that the solvent level is barely below the water surface. Once the mixture has become homogeneous, using a 6-mL plastic syringe, dimethoxy(methyl)silane (5.22 mL, 4.49 g, 42.3 mmol, 2 equiv) (Note 10) is added dropwise over 10 min, during which time the color of the solution gradually changes from blue to green to bright yellow to orange (Figure 5).

Figure 5. Progression of color changes upon addition of hydrosilane

At this time, the reaction flask is removed from the water bath and allowed to stir for an additional 12 h (Note 11). The septum is removed and saturated aqueous sodium bicarbonate (50 mL) is slowly added, followed by the addition of ethyl acetate (50 mL). After transferring the mixture to a 250-mL separatory funnel, the organic layer is separated and retained, and the aqueous layer is extracted with additional ethyl acetate (2 × 50 mL). The combined organic layers are concentrated with the aid of a rotary evaporator (35 °C water bath temperature, 50 mmHg) to afford a heterogeneous yellow-green mixture. This material is purified by flash column chromatography (Note 12) to yield **2** as a colorless, viscous oil (5.72 g, 86%) in 98% enantiomeric excess (Figure 6) (Note 13).

Figure 6. Reaction mixture after quenching, concentrated crude mixture, and purified product

C. *(R)-N,N-Dibenzyl-2,3,3-trimethylbutan-1-amine (3).* A 250-mL, two-necked, round-bottomed flask (Note 2) is equipped with a 1-cm, Teflon-coated magnetic stir bar and two rubber septa, through one of which a needle connected to a manifold under dry nitrogen is inserted. One septum is removed and the flask is charged sequentially with *N,N*-dibenzyl-*O*-pivaloylhydroxylamine (**1**, 9.09 g, 30.6 mmol, 1.2 equiv), copper(II) acetate (46 mg, 0.25 mmol, 0.010 equiv), *(S)*-DTBM-SEGPHOS (330 mg, 0.28 mmol, 0.011 equiv), triphenylphosphine (74 mg, 0.28 mmol, 0.011 equiv), and 2,3,3-trimethyl-1-butene (2.50 g, 3.55 mL, 25.5 mmol, 1 equiv) (Notes 8 and 14) under nitrogen flow. The flask is resealed with the septum, THF (25 mL)

(Note 9) is added by syringe, and the flask is partially submerged in an oil bath heated to 40 °C (Note 9). Once the mixture has become homogeneous, using a 6-mL plastic syringe, dimethoxy(methyl)silane (6.27 mL, 5.41 g, 50.9 mmol, 2 equiv) (Note 10) is added dropwise over 10 min, during which the color of the solution gradually changes from blue to green to bright yellow to orange. The reaction mixture is allowed to stir for additional 12 h at 40 °C. The reaction is cooled to room temperature (23 °C), the septum is removed and saturated aqueous sodium carbonate (50 mL) is slowly added, followed by the addition of ethyl acetate (50 mL). After transferring the mixture to a 250-mL separatory funnel, the organic layer is separated and retained, and the aqueous layer is extracted with additional ethyl acetate (2 × 50 mL). The combined organic layers are concentrated with the aid of a rotary evaporator (35 °C water bath temperature, 50 mmHg) and purified by flash column chromatography (Note 15) to yield **3** as a colorless, viscous oil (6.54 g, 87%) in 90% enantiomeric excess (Figure 7) (Note 16).

Figure 7. Reaction after 12 h, concentrated crude mixture, and purified product

Notes

1. Prior to performing each reaction, a thorough hazard analysis and risk assessment should be carried out with regard to each chemical substance and experimental operation on the scale planned and in the context of the laboratory where the procedures will be carried out. Guidelines for carrying out risk assessments and for analyzing the hazards associated with chemicals can be found in references such as

Chapter 4 of "Prudent Practices in the Laboratory" (The National Academies Press, Washington, D.C., 2011; the full text can be accessed free of charge at https://www.nap.edu/catalog/12654/prudent-practices-in-the-laboratory-handling-and-management-of-chemical). See also "Identifying and Evaluating Hazards in Research Laboratories" (American Chemical Society, 2015) which is available via the associated website "Hazard Assessment in Research Laboratories" at https://www.acs.org/content/acs/en/about/governance/committees/chemicalsafety/hazard-assessment.html. In the case of this procedure, the risk assessment should include (but not necessarily be limited to) an evaluation of the potential hazards associated with *N,N*-dibenzylhydroxylamine, 4-dimethylaminopyridine, dichloromethane, pivaloyl chloride, ammonium chloride, copper(II) acetate, *(S)*-DTBM-SEGPHOS, triphenylphosphine, *trans-* -methylstyrene, dimethoxy-(methyl)silane, sodium bicarbonate, ethyl acetate, 2,3,3-trimethyl-1-butene, tetrahydrofuran, sodium carbonate, silica gel, aluminum oxide, and hexanes. The reactions described in steps B and C are highly exothermic and can potentially generate a significant amount of flammable hydrogen gas. It is advisable to conduct these experiments in a large flask, adequately vented to a standard inert gas manifold or bubbler into a fume hood. In addition, the apparatus should be placed behind a weighted blast shield and inside a fume hood away from heat sources or flammable solvents. Dimethoxy(methyl)silane (DMMS) listed by various vendors as a H318 (Category 1 Causes Serious Eye Damage) or as a H319 (Category II Eye Irritant). At the end of the reaction, the work-up described in this procedure should be carried out prior to any subsequent manipulations to ensure destruction of the residual dimethoxy(methyl)silane.

2. All glassware and stir bars were dried in a conventional oven (140 °C) for at least 12 h and filled with dry nitrogen while hot. Unless otherwise stated, reactions were performed under a positive pressure of nitrogen by connection to a gas manifold.

3. *N,N*-Dibenzylhydroxylamine (>98.0%) was purchased from TCI America and used as received, except that a few colored or darker crystals, which were present in trace amounts, were discarded using standard tweezers. 4-Dimethylaminopyridine (>99%) was purchased from Sigma-Aldrich and used as received. Dichloromethane was purchased from J.T. Baker in CYCLE-TAINER solvent delivery kegs

and purified by passage under argon pressure through two packed columns of neutral alumina and copper(II) oxide.

4. Pivaloyl chloride (>98%) was purchased from Alfa Aesar and used as received.

5. The reaction was monitored by TLC analysis using glass-backed 60 Å silica gel plates purchased from SiliCycle with dichloromethane as the mobile phase. UV light (254 nm) was used as the visualization method. *N,N*-Dibenzylhydroxylamine: R_f = 0.42; **1**: R_f = 0.71.

6. Aluminum oxide (neutral, powder, reagent-grade) was purchased from J.T. Baker. The crude reaction mixture is suspended in dichloromethane (50 mL) and is loaded onto a column, with interior diameter of roughly 2 inches, packed with alumina (100 g) and wetted with hexanes. Dichloromethane is used as the eluent, and fractions are collected in Erlenmeyer flasks (50 mL each). The desired product typically elutes in fractions 2 through 25. The fractions that contain **1** are combined and the solvent is removed with the aid of a rotary evaporator (30 °C, 80 mmHg) to afford a cloudy white, viscous oil, which slowly solidifies on standing under vacuum (10 mmHg).

7. The desired product **1** has the following properties. ^{1}H NMR (400 MHz, CDCl₃) δ : 0.92 (s, 9H), 4.06 (s, 4H), 7.23 – 7.34 (m, 6H), 7.40 (d, *J* = 7.1 Hz, 4H). ^{13}C NMR (101 MHz, CDCl₃) δ : 27.1, 38.4, 62.4, 127.7, 128.3, 129.6, 136.2, 176.3. IR (neat film, NaCl) ν: 3064, 3031, 2973, 2932, 2906, 2872, 1751, 1496, 1479, 1456, 1273, 1116, 1029, 738, 698 cm⁻¹. HRMS (ESI-TOF): calculated [M+H]⁺ *m/z* 298.1802, found 298.1794. mp (capillary, uncorrected): 56–57 °C. Quantitative NMR using 1,1,2,2-tetrachloroethane (>98%, purchased from Alfa Aesar) in CDCl₃ indicates 99% purity. The compound is stable in a dry, dark environment.

8. Copper(II) acetate (anhydrous, 97%) was purchased from Strem and used as received. (*S*)-DTBM-SEGPHOS (>94%) was obtained from Takasago and used as received. Triphenylphosphine (99%) was purchased from Sigma-Aldrich and used as received. *trans*-β-Methylstyrene (97%, stabilized) was purchased from Combi-Blocks or Acros and used as received.

9. Tetrahydrofuran (THF) was purchased from J.T. Baker in CYCLE-TAINER solvent delivery kegs and purified by passage under argon pressure through two packed columns of neutral alumina and copper(II) oxide.

 DOI: 10.15227/orgsyn.095.0080

10. Dimethoxy(methyl)silane (>97%) was purchased from TCI America, stored in a freezer at −20 °C, and used without further purification.

11. The reaction was monitored by TLC analysis using glass-backed 60 Å silica gel plates purchased from SiliCycle with 2% ethyl acetate in hexanes as the mobile phase. UV light (254 nm) was used as the visualization method. Styrene reactant: R_f = 0.64; **2**: R_f = 0.36.

12. The crude reaction mixture is dissolved in a minimal quantity of benzene or toluene and is loaded onto a column, with interior diameter of roughly 2 inches, packed with silica (200 g, SiliCycle, F60/230–400 mesh) and equilibrated with hexanes. The column is eluted under air pressure with hexanes (500 mL), then 1% ethyl acetate in hexanes (1 L), then 2% ethyl acetate in hexanes (1 L). During elution, fractions are collected in test tubes (roughly 28 mL each), and the desired product **2** typically elutes around fractions 18 through 66. The fractions that contain **2** are combined and the solvent is removed with the aid of a rotary evaporator (30 °C, 80 mmHg) to afford pure **2**.

13. A second run of this experiment on 11.0 mmol scale yielded 3.04 g, (88%) of the identical product **2**, which has the following properties. ¹H NMR (400 MHz, CDCl₃) δ : 1.01 (t, *J* = 7.3 Hz, 3H), 1.89 (ddq, *J* = 14.2, 7.2, 7.1 Hz, 1H), 2.17 (ddq, *J* = 14.1, 7.2, 7.1 Hz, 1H), 3.24 (d, *J* = 13.9 Hz, 2H), 3.68 (t, *J* = 7.5 Hz, 1H), 3.91 (d, *J* = 13.8 Hz, 2H), 7.50–7.28 (m, 15H). ¹³C NMR (101 MHz, CDCl₃) δ⊠: 11.9, 24.4, 53.8, 63.8, 126.8, 127.0, 128.0, 128.3, 128.9, 129.1, 139.1, 140.6. IR (neat film, NaCl) ν: 3083, 3061, 3027, 2962, 2932, 2873, 2802, 1948, 1872, 1809, 1602, 1493, 1453, 761, 742 cm⁻¹. HRMS (ESI-TOF): calculated [M+H]⁺ *m/z* 316.2060, found 316.2049. Enantiomeric excess was determined by HPLC (Daicel Chiralpak OD-H column), eluting with 4% isopropanol in hexanes at 0.6 mL/min: 10.9 min (minor), 13.4 min (major), 98% ee for the first run and 99% ee for the second run. Specific rotation: [⊠]ᴅ = +108 (c = 1.0, chloroform). Quantitative NMR using ferrocene (98%, purchased from Sigma-Aldrich, recrystallized from pentane) in CDCl₃ indicates 99% purity. The compound is stable in a dry environment at room temperature.

14. 2,3,3-Trimethyl-1-butene (98%) was purchased from Sigma-Aldrich and used as received.

15. Silica (30 g) is added to the crude reaction mixture and the solvent removed in *vacuo*. This mixture is loaded onto a column, with interior diameter of roughly 2 inches, packed with silica (200 g, SiliCycle, F60/230–400 mesh) and equilibrated with hexanes. The column is

eluted under air pressure with 1% ethyl acetate in hexanes (2500 mL). During elution, fractions are collected in test tubes (roughly 28 mL each), and the desired product **3** typically elutes around fractions 13 through 72 (**3**: R_f = 0.34). The fractions that contain **3** are combined and the solvent is removed with the aid or a rotary evaporator (30 °C, 80 mmHg) to afford pure **3**.

16. A second run of this experiment on the same scale (25.5 mmol) yielded 6.01 g, 80% of the identical product **3**, which has the following properties. ^{1}H NMR (400 MHz, CDCl$_3$) δ⊠: 0.81 (s, 9H), 0.90 (d, J = 6.7 Hz, 3H), 1.49 (dq, J = 7.1, 3.4 Hz, 1H), 2.13 (dd, J = 12.3, 10.5 Hz, 1H), 2.39 (dd, J = 12.2, 2.8 Hz, 1H), 3.21 (d, J = 13.7 Hz, 2H), 3.83 (d, J = 13.7 Hz, 2H), 7.23 (t, J = 7.3 Hz, 2H), 7.31 (t, J = 7.4 Hz, 4H), 7.37 (d, J = 7.5 Hz, 4H). ^{13}C NMR (101 MHz, CDCl$_3$) δ⊠: 13.9, 27.6, 32.4, 41.0, 56.6, 59.1, 126.8, 128.2, 129.0, 140.2. IR (neat) v 3063, 3027, 2964, 2870, 2791, 1602, 1494, 1453, 1365, 1244, 1121, 1069, 1028, 974, 745, 698 cm^{-1}. HRMS (ESI-TOF): calculated [M+H]$^+$ m/z 296.2373, found 296.2375. Enantiomeric excess was determined by SFC (Daicel Chiralpak AD-H column, heated to 40 °C), eluting with a linear gradient over 6 min from 5% to 10% isopropanol in supercritical CO$_2$ at 2.5 mL/min.: 2.57 min (major), 2.98 min (minor), 90% ee for both runs. Specific rotation: $[\alpha]_D$ = –114 (c = 1.0, chloroform). Quantitative NMR using ferrocene (98%, purchased from Sigma-Aldrich, recrystallized from pentane) in CDCl$_3$ indicates 97% purity. The compound is stable in a dry environment at room temperature.

Working with Hazardous Chemicals

The procedures in *Organic Syntheses* are intended for use only by persons with proper training in experimental organic chemistry. All hazardous materials should be handled using the standard procedures for work with chemicals described in references such as "Prudent Practices in the Laboratory" (The National Academies Press, Washington, D.C., 2011; the full text can be accessed free of charge at http://www.nap.edu/catalog.php?record_id=12654). All chemical waste should be disposed of in accordance with local regulations. For general guidelines for the management of chemical waste, see Chapter 8 of Prudent Practices.

In some articles in *Organic Syntheses*, chemical-specific hazards are highlighted in red "Caution Notes" within a procedure. It is important to recognize that the absence of a caution note does not imply that no significant hazards are associated with the chemicals involved in that procedure. Prior to performing a reaction, a thorough risk assessment should be carried out that includes a review of the potential hazards associated with each chemical and experimental operation on the scale that is planned for the procedure. Guidelines for carrying out a risk assessment and for analyzing the hazards associated with chemicals can be found in Chapter 4 of Prudent Practices.

The procedures described in *Organic Syntheses* are provided as published and are conducted at one's own risk. *Organic Syntheses, Inc.*, its Editors, and its Board of Directors do not warrant or guarantee the safety of individuals using these procedures and hereby disclaim any liability for any injuries or damages claimed to have resulted from or related in any way to the procedures herein.

Discussion

Many methods have been developed for the synthesis of chiral aliphatic amines, primarily due to the prevalence of these fragments in organic building blocks, natural products, and synthetic compounds of biological interest. The most popular strategies include reductive amination, stereospecific substitution, asymmetric hydrogenation, the use of chiral auxiliaries, and biocatalysis.[2] Hydroamination, formally the insertion of an olefin into an N-H bond, represents an alternative approach that has been the subject of considerable academic interest.[3] The attractiveness of this approach originates from the availability and versatility of alkenes and alkynes, and the opportunity for catalyst control over chemo-, regio-, and stereoselectivity in the transformation.

Several years ago, our group[4] and the Miura group[5] contemporaneously developed an *umpolung* strategy for hydroamination of olefins, employing hydrosilanes as hydride sources and *O*-acylhydroxylamines as electrophilic amine equivalents. This approach is based on the catalytic generation of a phosphine-ligated copper–hydride intermediate, which can insert a C–C π-bond to form an alkylcopper species (Scheme 1). After trapping with the amine electrophile, σ-bond metathesis with the hydrosilane closes the catalytic cycle by regenerating the initial hydride complex.

Scheme 1. Catalytic Mechanism for CuH-Catalyzed Hydroamination

Despite diminished atom economy relative to traditional hydroamination with nucleophilic amine reagents, this new method features several practical advantages. Most importantly, the mildness of the reaction conditions preserves compatibility with useful functional groups such as alcohols, esters, amides, sulfonamides, aryl or alkyl halides, and heterocycles. Furthermore, using the same earth-abundant metal catalyst, many classes of alkenes are transformed efficiently and with excellent stereoselectivity.

Generally, the hydroamination of olefins bearing an activating substituent such as aryl, silyl, or boryl results in the regioisomer with the amine introduced adjacent to this substituent (Table 1). Using non-racemic DTBM-SEGPHOS as the supporting ligand, the Markovnikov-selective hydroamination of styrenes can be effected with high enantioselectivity, thus constructing a common α-chiral amine substructure.[4] Likewise, α-aminosilanes[6] and α-aminoboranes,[7] which are also useful fragments in organic synthesis, can be assembled using this strategy from simple olefins. Furthermore, all of the above hydroamination reactions are known to proceed with exclusive *syn*-diastereoselectivity relative to the olefin.

Table 1. Enantioselective, Markovnikov Hydroamination of Alkenes

$$R_{act}\overset{R^2}{\underset{R^1}{=}} \xrightarrow[\text{THF}]{\substack{R_2N\text{-}O_2CR' \\ \text{hydrosilane} \\ (R)\text{-DTBM-SEGPHOS} \\ Cu(OAc)_2, (Ph_3P)}} R_{act}\overset{NR_2}{\underset{H\ R^2}{\overset{|}{\diagup}}}R^1$$

(R)-DTBM-SEGPHOS
Ar = 3,5-t-Bu-4-OMe-C₆H₂

NBn₂ ... H, Cl
86% yield
92% ee

NBn₂ ... Me, ''H, OTBS
83% yield
>99% ee

Bn₂N, Me, H, Me, SiMePh₂
74% yield
99% ee

NBn₂ ... H, N-Ts
80% yield
98% ee

i-Pr, CO₂t-Bu, NH, NMe₂, H, S, N
82% yield
>20:1 dr

In comparison, the hydroamination of unactivated olefins such as terminal aliphatic alkenes[4] strongly favors the formation of the anti-Markovnikov isomer (Table 2). In the case of 1,1-disubstituted or trisubstituted substrates, highly enantioselective construction of the β-stereocenter can be achieved with a chiral ligand.[8] Recent advances have made the use of internal olefins possible, despite their relative inertness and poor binding ability toward metal centers.[9] For instance, from 2-butene, amine products bearing otherwise synthetically challenging methyl-ethyl stereocenters can be produced efficiently. Useful regioisomeric control can also be imparted, either using steric effects of the substrate, or using a directing group that electronically biases the π-bond.[10]

Table 2. Enantioselective, Anti-Markovnikov Hydroamination of Alkenes

$$R_2N\text{-}O_2CR'$$
$$\text{hydrosilane}$$
$$(S)\text{-DTBM-SEGPHOS}$$
$$R^2 \diagdown R^3 \xrightarrow[\text{THF}]{\text{Cu(OAc)}_2 \text{ or } [\text{Ph}_3\text{PCuH}]_6} R^1 \diagdown \overset{NR_2}{R^3}$$

(S)-DTBM-SEGPHOS
Ar = 3,5-t-Bu-4-OMe-C₆H₂

90% yield

91% yield
80% ee

76% yield
97% ee

(R)-ligand (S)-ligand
85% yield 86% yield
20:1 dr 1:14 dr

9:1 rr
84% yield
97% ee

To date, several classes of bond constructions remain challenging for this type of catalytic process. Highly hindered olefins and *cis*-disubstituted olefins are not efficiently converted. Moreover, very electron-deficient alkenes generally produce hydroamination products with low levels of enantioselectivity. While the design of more robust electrophilic amine reagents has allowed for the synthesis of secondary amines,[11] corresponding reagents for the synthesis of primary or aryl amines have not yet been reported. Additional limitations of copper–hydride-catalyzed hydroamination, along with many more successful examples, are described in a recent mini-review.[12]

Beyond addressing deficiencies in terms of scope, current research has also been aimed at devising synthetically valuable combinations of this hydroamination strategy with other copper-catalyzed processes,[13] or at extending this reactivity to other useful organic electrophiles, such as ketones and imines.[14]

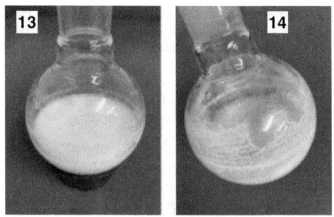

Figure 13. Reaction mixture of Step B after the reaction

Figure 14. Reaction mixture of Step B after removal of solvent and iodomethane

Figure 15. Reaction mixture of Step B washed with dichloromethane

Figure 16. Reaction mixture of Step B washed with NaCl solution

Figure 17. Product of Step B

Notes

1. Prior to performing each reaction, a thorough hazard analysis and risk assessment should be carried out with regard to each chemical substance and experimental operation on the scale planned and in the context of the laboratory where the procedures will be carried out. Guidelines for carrying out risk assessments and for analyzing the hazards associated with chemicals can be found in references such as Chapter 4 of "Prudent Practices in the Laboratory" (The National Academies Press, Washington, D.C., 2011; the full text can be accessed free of charge at https://www.nap.edu/catalog/12654/prudent-practices-in-the-laboratory-handling-and-management-of-chemical). See also "Identifying and Evaluating Hazards in Research Laboratories" (American Chemical Society, 2015) which is available via the associated website "Hazard Assessment in Research Laboratories" at https://www.acs.org/content/acs/en/about/governance/committees /chemicalsafety/hazard-assessment.html. In the case of this procedure, the risk assessment should include (but not necessarily be limited to) an evaluation of the potential hazards associated with biphenyl-2-

carboxylic acid, potassium peroxydisulfate, silver nitrate, acetonitrile, dichloromethane,ethyl acetate, sodium hydroxide, sodium chloride, sodium sulfate, iodomethane, potassium hydroxide, hexanes, and silica gel.

2. [1,1'-Biphenyl]-2-carboxylic acid (98%) was purchased from Ark Pharm and used as received.

3. Potassium peroxydisulfate (99.9%) was purchased from Alfa Aesar and used as received.

4. Silver nitrate (99.9%) was purchased from Alfa Aesar and used as received.

5. Distilled water was used from tap distilled water.

6. Acetonitrile (99.9%, HPLC grade) was purchased from Fisher Scientific and used as received.

7. It is important to run the reaction under air. Sodium hydroxide (98%) was purchased from Alfa Aesar and used as received.

8. The temperature for the oil bath was set at 55 °C to ensure the reaction mixture's temperature is at 50 °C. It is important to maintain this reaction temperature. Reaction temperature was checked by immersing the tip of thermometer into the reaction mixture through the side neck of the flask. The reaction mixture was monitored by TLC analysis on silica using ethyl acetate/hexanes (1:3). R_f of **1** = 0.15, R_f of **2** = 0.4.

Figure 18: Crude product TLC after Step A

9. Dichloromethane (99.5%) was purchased from Sigma Aldrich and used as received.

10. Ethyl acetate (99.5%) was purchased from Sigma Aldrich and used as received.

11. The silica gel pad was pre-wetted by ethyl acetate (30 mL Filter Funnel, Büchner, fine Frit height ~3.5 cm; diameter ~3.5 cm).

12. A second run on the identical scale provided 6.40 g (82%) of the product.

13. The product was dried under vacuum for 24 h to remove residue solvent. ^{1}H NMR (500 MHz, CDCl$_3$) δ: 7.28–7.33 (m, 2H), 7.44 (ddd, J = 8.4, 7.1, 1.6 Hz, 1H), 7.54 (ddd, J = 8.0, 7.3, 1.1 Hz, 1H), 7.79 (ddd, J = 8.1, 7.3, 1.5 Hz, 1H), 7.98–8.02 (m, 1H), 8.04–8.08 (m, 1H), 8.35 (ddd, J = 8.0, 1.4, 0.60 Hz, 1H). ^{13}C NMR (126 MHz, CDCl$_3$) δ: 117.8, 118.1, 121.3, 121.7, 122.8, 124.6, 128.9, 130.5, 130.6, 134.8, 134.9, 151.3, 161.2. IR (film): 3073, 1733, 1607, 1505, 1485, 1457 cm^{-1}. HRMS–APCI (m/z) calculated for C$_{13}$H$_9$O$_2$ [M + H] 197.05971, found 197.05817. Purity of **2** is 97.5%, which was determined by QNMR using ethylene carbonate as internal standard (15.6 mg ethylene carbonate and 34.7 mg **2**, NMR setting, d1 = 60 s, ns = 2).

14. Iodomethane (99%) was purchased from Spectrum Chemicals and used as received.

15. Potassium hydroxide (85%) was purchased from Alfa Aesar and used as received.

16. A needle connected to air was use to release the pressure from the flask. Inert atmosphere is not important for that step.

17. The reaction was stirred for 24 h. The submitters report that the reaction can be monitored by GC-MS.

18. Iodomethane is toxic, thus, its removal should be performed in a fume hood.

19. The crude product was diluted with dichloromethane (10 mL), loaded to a pre-wetted silica gel column, and purified by eluding with 1:3 ethyl acetate/hexanes. Column diameter ~3 cm, column height ~10 cm. Approximately 200 mL of the eluent was used, and the product was the only fraction collected.

20. A second reaction performed on the identical scale provided 6.40 (88%) of the same product.

21. The final product was dried under vacuum for 24 h to remove residual solvent. ^{1}H NMR (500 MHz, CDCl$_3$) δ: 3.66 (s, 3H), 3.72 (s, 3H), 6.91 (dd, J = 8.2, 1.0 Hz, 1H), 7.05 (td, J = 7.5, 1.1 Hz, 1H), 7.25 (dd, J = 7.4, 1.7 Hz, 1H), 7.32–7.36 (m, 2H), 7.40 (td, J = 7.6, 1.3 Hz, 1H), 7.55 (td, J = 7.6, 1.5 Hz, 1H), 7.87 (ddd, J = 7.8, 1.4, 0.5 Hz, 1H). ^{13}C NMR (126 MHz, CDCl$_3$) δ: 51.8, 55.4, 110.2, 120.9, 127.3, 129.0, 129.5, 130.1, 130.7, 131.5,

131.7, 131.8, 138.9, 156.2, 168.8. IR (film): 3064, 2998, 1725, 1597, 1291, 1249 cm^{-1}. HRMS–APCI (m/z) calculated for $C_{15}H_{15}O_3$ [M + H] 243.10157, found 243.10011. Purity of **3** is 99.5%, which was determined by QNMR, using ethylene carbonate as internal standard (61.9 mg ethylene carbonate and 45.9 mg **3**, NMR setting, d1 = 30 s, ns = 4).

Working with Hazardous Chemicals

The procedures in *Organic Syntheses* are intended for use only by persons with proper training in experimental organic chemistry. All hazardous materials should be handled using the standard procedures for work with chemicals described in references such as "Prudent Practices in the Laboratory" (The National Academies Press, Washington, D.C., 2011; the full text can be accessed free of charge at http://www.nap.edu/catalog.php?record_id=12654). All chemical waste should be disposed of in accordance with local regulations. For general guidelines for the management of chemical waste, see Chapter 8 of Prudent Practices.

In some articles in *Organic Syntheses*, chemical-specific hazards are highlighted in red "Caution Notes" within a procedure. It is important to recognize that the absence of a caution note does not imply that no significant hazards are associated with the chemicals involved in that procedure. Prior to performing a reaction, a thorough risk assessment should be carried out that includes a review of the potential hazards associated with each chemical and experimental operation on the scale that is planned for the procedure. Guidelines for carrying out a risk assessment and for analyzing the hazards associated with chemicals can be found in Chapter 4 of Prudent Practices.

The procedures described in *Organic Syntheses* are provided as published and are conducted at one's own risk. *Organic Syntheses, Inc.*, its Editors, and its Board of Directors do not warrant or guarantee the safety of individuals using these procedures and hereby disclaim any liability for any injuries or damages claimed to have resulted from or related in any way to the procedures herein.

Discussion

6*H*-Benzo[c]chromen-6-one motif presents in a variety of natural products, medicinal molecules, and functional organic compounds.[2] Due to the importance of this core, different synthetic methods were developed toward its synthesis. Existing methods include transition metal-catalyzed C–C bond formation[3] or intramolecular O–H/C–X coupling reaction.[4] Unfortunately, these methods suffer from limited availability of starting material and narrow reaction scope.

Biaryl-2-carboxylic acids are easily available materials and hence are ideal precursors for 6*H*-benzo[c]chromen-6-one synthesis. Thus, many research groups attempted to employ biaryl-2-carboxylic acids for synthesis of 6*H*-benzo[c]chromen-6-ones. The Thomson group reported a method to convert biaryl-2-carboxylic acid's silver salts into the corresponding 6*H*-benzo[c]chromen-6-ones via oxidation.[5] Similarly, Togo and Yokoyama described a *uv*-mediated procedure for synthesis of 6*H*-benzo[c]chromen-6-ones from unstable [bis(*o*-phenylphenylcarbonyloxy)iodo]benzene intermediates, which were produced from biaryl-2-carboxylic acids.[6] Another approach for the synthesis of 6*H*-benzo[c]chromen-6-ones from biaryl-2-carboxylic acids requires stoichiometric amounts of toxic Cr(VI) or Pb(IV) oxidants. Unfortunately, poor yields, narrow scope, and use of stoichiometric amounts of toxic metal reagents limit the utility of these transformations.[7] Recently, a Pd-catalyzed cyclization method was introduced by the Wang group,[8] where biaryl-2-carboxylic acids were converted into 6*H*-benzo[c]chromen-6-ones in good yields. Drawbacks of this method include employment of significant amount of expensive palladium catalyst, as well as the limitation of this method to electron-rich and -neutral substrates only. Later, the Martin group reported Cu-catalyzed remote C–H oxygenation reactions for synthesis of 6*H*-benzo[c]chromen-6-ones from biaryl-2-carboxylic acids. Independently, the Gevorgyan group reported both the Cu-catalyzed, as well as a metal-free method, for this transformation.[10] It deserves mentioning that the Cu-catalyzed method worked well with electron-rich and -neutral compounds, while less efficient for substrates bearing electron-withdrawing groups. In contrast, the metal-free O–H/C–H dehydrogenative coupling method[10] appeared to be insensitive to the electronics of the host aromatic ring.

Herein, we describe a general metal-free method for synthesis of 6*H*-benzo[c]chromen-6-ones from biaryl-2-carboxylic acids,[8] which features a general scope with respect to electronic nature of the substrate, broad functional group compatibility, and mild reaction conditions (Table 1). In addition, employment of environmental friendly and relative cheap solvent provides an economically feasible approach for the synthesis of 6*H*-benzo[c]chromen-6-ones.[10] Moreover, it was demonstrated that the 6*H*-benzo[c]chromen-6-one could be converted easily into methyl 2'-methoxy-[1,1'-biphenyl]-2-carboxylate, thus representing a remote C–H oxygenation reaction in biaryl system.[8-10]

Table 1. Selected Scope of 6H-benzo[c]chromen-6-ones

88%

87%

66%[a]

86%[a]

46%[a]

81%[a] 10:1

74%[a] 15:1

81%

45%[a]

51%

93%

53%

[a] add 10 mol% AgNO₃

References

1. Department of Chemistry, University of Illinois at Chicago, 845 W. Taylor St., Chicago, Illinois 60607 (USA). We thank National Science

Foundation (CHE-1362541) for financial support of this work. Yang Wang and Yi Shi contributed equally to this manuscript.

2. For 6*H*-benzo[c]chromen-6-one containing natural products and bioactive compounds, see: (a) Murray, R. D. H.; Mendez, J.; Brown, S. A. *The Natural Coumarins: Occurrence, Chemistry, and Biochemistry*; Wiley: New York, **1982**. (b) Omar, R.; Li, L.; Yuan, T.; Seeram, N. P. *J. Nat. Prod.* **2012**, *75*, 1505–1509. (c) Tibrewal, N.; Pahari, P.; Wang, G.; Kharel, M. K.; Morris, C.; Downey, T.; Hou, Y.; Bugni, T. S.; Rohr, J. *J. Am. Chem. Soc.* **2012**, *134*, 18181–18184. For 6*H*-benzo[c]chromen-6-one containing materials, see: (d) Yang, C.; Hsia, T.; Chen, C.; Lai, C.; Liu, R. *Org. Lett.* **2008**, *10*, 4069–4072. (e) Nakashima, M.; Clapp, R.; Sousa, J. A. *Nature Phys. Sci.* **1973**, *245*, 124–126. (f) Fletcher, S. P.; Dumur, F.; Pollard, M. M.; Feringa, B. L. *Science* **2005**, *310*, 80–82.

3. (a) Sun, C.-L.; Liu, J.; Wang, Y.; Zhou, X.; Li, B.-J.; Shi, Z.-J. *Synlett* **2011**, 883–886. (b) Thasana, N.; Worayuthakarn, P.; Kradanrat, P.; Hohn, E.; Young, L.; Ruchirawat, S. *J. Org. Chem.* **2007**, *72*, 9379–9382.

4. Hager, A.; Mazunin, D.; Mayer, P.; Trauner, D. *Org. Lett.* **2011**, *13*, 1386–1389.

5. Chalmers, D. J.; Thomson, R. H. *J. Chem. Soc.* **1968**, 848–850.

6. Togo, H.; Muraki, T.; Hoshina, Y.; Yamaguchi, K.; Yokoyama, M. *J. Chem. Soc., Perkin Trans. 1.* **1997**, 787–793.

7. For a Cr(VI)-mediated reaction, see: (a) Kenner, G. W.; Murray, M. A.; Tylor, C. M. B. *Tetrahedron* **1957**, *1*, 259-268. For a Pb(IV)-mediated reaction, see: (b) Davies, D. I.; Waring, C. *Chem. Commun.* **1965**, 263–264.

8. Li, Y.; Ding, Y.-J.; Wang, J.-Y.; Su, Y.-M.; Wang, X.-S. *Org. Lett.* **2013**, *15*, 2574–2577.

9. Gallardo-Donaire, J.; Martin, R. *J. Am. Chem. Soc.* **2013**, *135*, 9350–9353.

10. Wang, Y.; Gulevich, A.; Gevorgyan, V. *Chem. Eur. J.* **2013**, *19*, 15836–15840.

Appendix
Chemical Abstracts Nomenclature (Registry Number)

[1,1'-Biphenyl]-2-Carboxylic acid (947-84-2)
Potassium peroxydisulfate (7727-21-1)
Silver nitrate (7761-88-8)
6*H*-benzo[c]chromen-6-one (2005-10-9)
Potassium hydroxide (1310-58-3)
Iodomethane (74-88-4)

Yang Wang received his B.S. in 2008 from Wuhan University with Professor Qinghua Fan, and his M.S. in 2011 from The Chinese University of Hong Kong under the supervision of Professor Zuowei Xie. In 2011, he moved to Chicago and obtained his Ph.D. in 2017 with Professor Vladimir Gevorgyan. During his Ph.D. studies he was involved in development of novel transition metal-catalyzed, as well as metal-free, C–H functionalization methods. Currently, he is a postdoctoral fellow in the group of Prof. Tobin Marks at Northwestern University.

Dr. Yi Shi was born in Beijing in 1989. She received her B.S. degree from Peking University in 2011 under the supervision of Professor Jianbo Wang and Professor Yan Zhang. Then she received her Ph.D. degree from University of Illinois at Chicago in 2017 under the supervision of Professor Gevorgyan. Her graduate research focused on development of transition metal-catalyzed transformations of triazoles, metal carbine-involved coupling reactions and heterocycle synthesis. At present she is a postdoctoral fellow in the group of Prof. Fraser Stoddart at Northwestern University.

Vladimir Gevorgyan received his Ph.D. from the Latvian Institute of Organic Synthesis in 1984. After two years of postdoctoral research at Tohoku University (1992–1994), Japan, and a visiting professorship (1995) at CNR, Bologna, Italy, he joined the faculty at Tohoku University (Assistant Professor, 1996; Associate Professor, 1997–1999). He joined UIC as an Associate Professor in 1999 and was promoted to Professor in 2003. From 2012, he has been an LAS Distinguished Professor. His research interest is on the development of transition metal-catalyzed synthetic methodology for annulation and isomerization reactions, synthesis of carbo- and heterocycles, and C–H functionalization reactions.

Organic Syntheses

Evan R. Darzi received his B.S. in Medicinal Biochemistry from Arizona State University in Tempe, AZ, where he performed undergraduate research under Professor Edward Skibo on the synthesis of extended amidines. He received his Ph.D. from the University of Oregon in Eugene, OR, under the guidance of Professor Ramesh Jasti. There he developed syntheses of highly strained [n]Cycloparaphenylenes and other "nanohoops". He is currently a NIH postdoctoral fellow in Professor Neil K. Garg's laboratory at the University of California, Los Angeles. His postdoctoral studies are focused on the development of strained intermediates in synthetic methodology.

Synthesis of *N*-Acyl Pyridinium-*N*-Aminides and Their Conversion to 4-Aminooxazoles *via* a Gold-Catalyzed Formal (3+2)-Dipolar Cycloaddition

Matthew P. Ball-Jones and Paul W. Davies*[1]

School of Chemistry, University of Birmingham, Birmingham, B15 2TT, UK

Checked by Manuela Brütsch, Estíbaliz Merino and Cristina Nevado

Procedure (Note 1)

A. *((tert-Butoxycarbonyl)glycyl)(pyridin-1-ium-1-yl)amide* (**2**). A 500 mL single-necked, round-bottomed flask equipped with a 3 cm stirrer bar and a needle-pierced septum is charged with methyl (*tert*-butoxycarbonyl)-glycinate **1** (Note 2) (6.80 g, 36.0 mmol, 1.20 equiv) and methanol (Note 3) (225 mL). 1-Amino pyridinium iodide (Note 4) (6.67 g, 30.0 mmol) is added and the reaction is stirred for 5 min at 22 °C. Potassium carbonate (Note 5) (9.95 g, 72.0 mmol, 2.40 equiv) is added and the reaction is stirred at 22 °C for 64 h (Note 6) (the yellow heterogeneous reaction turns colorless two seconds after the addition of potassium carbonate, and forms a dark purple solution, Figure 1). After removal of the stir bar, 200 mL of the methanol is removed under reduced pressure (200 mmHg to 70 mmHg, 40 °C) to give a

112
Published on the Web 4/23/2018
© 2018 Organic Syntheses, Inc.

Figure 1. Change in reaction color: a) Reaction mixture before potassium carbonate addition; b) Reaction mixture 10 seconds after potassium carbonate addition; c) Reaction mixture after 64 h

brown-purple syrup, which is poured onto an alumina pad (Notes 7 and 8). The flask is rinsed with 10 mL of dichloromethane as well as 50 mL of eluent (dichloromethane-methanol, 9:1), and the product is eluted with 1.1 L of dichloromethane-methanol (9:1) (Note 9). The filtrate is concentrated (375 mmHg to 75 mmHg, 40 °C) and then transferred to a 500 mL single-necked round-bottomed flask and rinsed with dichloromethane (20 mL). The filtrate is concentrated further (375 mmHg to 15 mmHg, 40 °C) and then dried under vacuum (0.08 mmHg, 20 °C, 1 h) to give a brown powder (Note 10). A 3 cm Teflon coated stirrer bar is added, followed by acetone (175 mL) (Note 11). A water-cooled condenser is added to the flask and the mixture is heated to reflux until complete dissolution had occurred (15 min). The mixture is allowed to cool to room temperature over 3 h and then cooled to –22 °C in a freezer for 20 h. The resultant fine brown crystals

Figure 2. Compound 2 after recrystallization from acetone

are filtered off through a sintered S3 funnel, the flask is rinsed with diethyl ether (Note 12) (3 x 25 mL), and the contents were then added to the funnel. The powder is transferred to a 50 mL single-necked flask and dried under static vacuum (0.12 mmHg, 18 h) (6.39 g, 85%, 98% purity) (Figure 2) (Notes 13, 14, and 15).

B. *tert-Butyl ((4-((N-benzyl-4-methylphenyl)sulfonamido)-5-phenyloxazol-2-yl)methyl)carbamate (4)*. A three-necked 250 mL round-bottomed flask equipped with a septum, a glass stopper, a vacuum tap and 2 cm Teflon coated stirrer bar is flame-dried under vacuum (0.12 mmHg) for 1 min and then backfilled with nitrogen. The evacuation/nitrogen purge is repeated a total of 3 times, and then the reaction vessel is allowed to cool to room temperature (22 °C) under a positive pressure of nitrogen. The flask is sequentially charged with N-benzyl-4-methyl-N-(phenylethynyl)benzene-sulfonamide (Note 16) (4.69 g, 13.0 mmol), dichloro(2-pyridinecarboxylato)gold (Note 17) (51.02 mg, 0.13 mmol) and ((*tert*-butoxycarbonyl)glycyl)(pyridin-1-ium-1-yl)amide (3.60 g, 14.3 mmol, 1.1 equiv). Toluene (130 mL) (Note 18) is added and the flask is placed in a preheated aluminium mantle and stirred at 90 °C (controlled by an external probe) for 5.5 h (Note 19). The heterogeneous mixture becomes a homogenous orange-brown solution over the first hour (Figure 3).

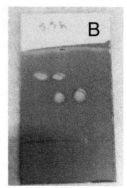

Figure 3. a) Reaction set-up for Step B; b) Reaction progress monitored by TLC. From left to right: Compound 3, control spot and reaction mixture

Upon completion, the flask is removed from the heating mantle and the reaction is allowed to cool to room temperature for 1 h. The crude reaction mixture is filtered through a silica pad (Note 20) and then the reaction flask is rinsed with ethyl acetate (2 x 50 mL) (Note 21), which is then added to the silica pad. The pad is then flushed with ethyl acetate (320 mL) into a 1 L single-necked round-bottomed flask. The solvent is removed under reduced pressure (150–75 mmHg, 40 °C). The product is transferred, with the assistance of ethyl acetate washes (2 x 10 mL), to a 250 mL single-necked, round-bottomed flask and evaporated further (160 mmHg to 30 mmHg). The resultant white-yellowish solid is then broken up with a glass rod (Figure 4a). A 3 cm Teflon coated stir bar is added followed by ethanol (Note 22) (150 mL), and the flask is equipped with a water-cooled condenser. The flask is heated to reflux in an aluminium heating mantle and stirred for 25 min until complete dissolution to give a yellow solution. The flask is allowed to cool to room temperature and then placed in a freezer at –22 °C for 19 h. The resultant crystals were filtered off and washed with hexane (3 x 25 mL) (Note 23) to give fine off-white crystals (Figure 4b).

The crystals are transferred to a 100 mL single-necked flask and then toluene (30 mL) (Note 24) is added to give a slightly grey suspension (Figure 4c). The solvents are removed under reduced pressure (75 mmHg to 35 mmHg, 40 °C) and then the product is dried under vacuum (0.14 mmHg, 1 h, room temperature) to give a fine off white powder (Figure 4d) (Note 25). The powder is transferred to a 250 mL single-necked, round-bottomed flask (using 10 mL dichloromethane to rinse the flask) containing a 3 cm Teflon coated stir bar. Dichloromethane (80 mL) is added, the mixture is stirred for 10 min and then the yellow solution (Figure 4e) is filtered through cotton wool (Note 26), which is then washed with dichloromethane (4 x 10 mL), eluting under gravity into a 250 mL round-bottomed flask (Note 27). The solvent is evaporated (450 mmHg – 35 mmHg, 40 °C) to give a white solid (Figure 4f), which is broken up with a glass rod and then dried under vacuum (60 °C, 0.12 mmHg, 96 h) (6.67 g, 96%, >99% purity) (Notes 28, 29, 30, and 31).

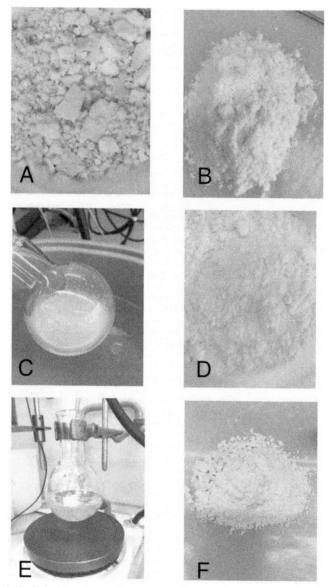

Figure 4. a) Solid after filtration through a silica pad; b) Solid after first crystallization; c) Grey suspension of the white crystals in toluene; d) The final product before filtration through cotton; e) Yellow solution of the product before filtration through cotton; f) Compound 4

Notes

1. Prior to performing each reaction, a thorough hazard analysis and risk assessment should be carried out with regard to each chemical substance and experimental operation on the scale planned and in the context of the laboratory where the procedures will be carried out. Guidelines for carrying out risk assessments and for analyzing the hazards associated with chemicals can be found in references such as Chapter 4 of "Prudent Practices in the Laboratory" (The National Academies Press, Washington, D.C., 2011; the full text can be accessed free of charge at https://www.nap.edu/catalog/12654/prudent-practices-in-the-laboratory-handling-and-management-of-chemical).
    See also "Identifying and Evaluating Hazards in Research Laboratories" (American Chemical Society, 2015) which is available via the associated website "Hazard Assessment in Research Laboratories" at https://www.acs.org/content/acs/en/about/governance/committees/chemicalsafety/hazard-assessment.html. In the case of this procedure, the risk assessment should include (but not necessarily be limited to) an evaluation of the potential hazards associated with methyl (*tert*-butoxycarbonyl)glycinate, methanol, 1amino pyridinium iodide, potassium carbonate, alumina, dichloromethane, diethyl ether, 1,3,5-trimethoxybenzene *N*-benzyl-4-methyl-*N*-(phenylethynyl)benzenesulfonamide, dichloro(2-pyridinecarboxylato)gold, toluene' silica gel, ethyl acetate, and ethanol.
2. Methyl (*tert*-butoxycarbonyl)glycinate (**1**) (97%) was purchased from Fluorochem and used as received.
3. Methanol (99.9%) was purchased from Acros Organics and used as received.
4. 1-Amino pyridinium iodide (97%) was purchased from Fluorochem. Prior to use, it was dissolved in methanol and the resulting suspension filtered to remove the non-dissolved materials. The mother liquors were evaporated and the obtained solid recrystallized with ethanol.
5. Potassium carbonate (99%) was purchased from Alfa Aesar and used as received.
6. TLC analysis is used to determine reaction completion: the R_f of the product is 0.17 (in 88:12 dichloromethane-methanol), on aluminium backed silica gel 60 F_{254} plates. The product was visualised with UV

light (254 nm) and the starting material (ester **1**, R_f 0.90) was visualized with KMnO₄ (heating 130 °C, 30 seconds).

7. Activated, neutral Brockmann type 1 aluminium oxide (60 mesh, 58 Å pore size) was purchased from Alfa Aesar and used as received.

8. Alumina was slurried with dichloromethane, and an alumina pad was used that was 9.0 cm in diameter and 7.0 cm deep.

9. Dichloromethane (99%) was purchased from Biosolve and used as received.

10. The authors point out that at this stage of the process silica gel flash column chromatography may be used to purify aminides instead of recrystallization. (Silica: 230-400 mesh size, 60 Å pore size, 40-63 μm particle size, which was purchased from Sigma-Aldrich and used as received). TLC on aluminium backed silica gel 60 F_{254} plates. R_f = 0.31 in dichloromethane-methanol [9:1]. Visualized with 254 nm UV light only. The crude powder was dissolved in CH₂Cl₂ (40 mL). Silica (20 g) was added and the solvent was then removed (500 mmHg to 30 mmHg, 40 °C). The product-adsorbed silica was then loaded onto a column (6.5 cm diameter, 15 cm depth, 180 g silica, slurry loaded with eluent) and eluted with dichloromethane-methanol [9:1] (2.65 L) with 30 mL fractions. Rotary evaporation of the collected fractions (500 mmHg to 30 mmHg, 40 °C) gave a pale yellow solid, which was dried under high vacuum (0.15 mmHg, 20 °C) for 6 h (6.69 g, 89%).

11. Acetone (99.8%) was purchased from Sigma-Aldrich and used as received.

12. Diethyl ether (99.8%) was purchased from Biosolve and used as received.

13. A reaction performed on half scale provided 3.71 g (90%) of the product.

14. These stable, hygroscopic brown crystals should be stored in a desiccator. The compound exists as a mixture of rotamers. mp 165–167 °C; IR (thin film) 3234, 2969, 1712, 1580, 1542, 1471, 1416, 1363, 1270, 1243, 1153, 1138, 1046, 934, 817, 770, 679, 612, 492 cm⁻¹; ¹H NMR (CDCl₃, 400 MHz) δ: 1.46 (s, 9H), 3.94 (d, J = 4.5 Hz, 2H), 5.34 (s, 1H), 7.66 (t, J = 7.0 Hz, 2H), 7.92 (t, J = 7.7 Hz, 1H), 8.69 (d, J = 6.2 Hz, 2H); ¹³C NMR (CDCl₃, 101 MHz) δ: 28.4 (3CH₃), 44.1 (CH₂), 78.9 (C), 126.1 (2CH), 137.2 (CH), 143.0 (2CH), 155.9 (C), 172.4 (C); HRMS (ESI) m/z calcd for C₁₂H₁₈N₃O₃ 252.13427, found 252.13413 $[M+H]^+$.

15. Purity of the product was determined to be 98% by quantitative ¹H NMR using 1,3,5-trimethoxybenzene as the internal standard.

16. *N*-Benzyl-4-methyl-*N*-(phenylethynyl)benzenesulfonamide was synthesized according to an *Org. Synth.* procedure.[6a] The data matched those reported.

17. Dichloro(2-pyridinecarboxylato)gold (no assay) was purchased from Sigma Aldrich and used as received.

18. Toluene was collected in an oven-dried (140 °C for 24 h) 250 mL Schlenk flask, which had been purged with argon atmosphere by three evacuation-backfill cycles from a dry solvent system (Innovative Technology).

19. TLC analysis is used to determine reaction completion: the R_f of the product in 3:2 hexane-ethyl acetate is 0.34, on aluminium backed silica gel 60 F_{254} plates. The product and limiting reagent (ynamide **3**, R_f 0.59) were visualised with UV light (254 nm) and $KMnO_4$ (heating 130 °C, 30 seconds). The pyridine by-product (R_f 0.14) was visualised with UV light (254 nm) only.

20. Silica was added to a sintered (S3) funnel (7.0 cm internal diameter) to a depth of 3 cm. This was wetted with ethyl acetate (50 mL).

21. Ethyl acetate (100%) was purchased from Biosolve and used as received.

22. Ethanol (99.96%) was purchased from Biosolve and used as received.

23. Hexane (98%) was purchased from Fisher Chemical and used as received.

24. Toluene (99.7%) was purchased from Biosolve and used as received.

25. The authors reported a gray/lilac powder.

26. A 2.8 cm diameter column was packed with cotton wool, to a depth of 3.5 cm.

27. The used cotton wool is a gray color.

28. The oxazole **4** contains traces of dichloromethane. Removal of the final vestiges of solvent from within the crystal structure required heating under vacuum.

29. A reaction performed on half scale provided 3.31 g (95%) of the product.

30. The product is a bench stable powder. Analysis by ^{1}H NMR indicates the presence of a mixture of carbamate rotamers. mp 164–165°C; IR (thin film) 3276, 2974, 1715, 1447, 1391, 1367, 1349, 1331, 1153, 1129, 1089, 1057, 1043, 1026, 844, 822, 752, 732, 706, 690, 668, 653, 604, 578, 555, 540 cm^{-1}; ^{1}H NMR (CDCl$_3$, 400 MHz) δ: 1.48 (s, 9H), 2.46 (s, 3H), 4.37 (d, *J* = 5.1 Hz, 2H), 4.55 (s, 2H), 4.98 (s, 1H), 7.09–7.04 (m, 3H), 7.15–7.13 (m, 2H), 7.35–7.27 (m, 5H), 7.68 (d, *J* = 7.2 Hz, 2H), 7.77 (d, *J* = 8.1 Hz, 2H); ^{13}C NMR (CDCl$_3$, 101 MHz) δ: 21.6, 28.3, 38.2, 53.9, 80.2, 125.5, 126.4,

127.9, 128.1, 128.3, 128.6, 128.8, 129.1, 129.5, 131.1, 134.6, 135.3, 144.0, 148.1, 155.3, 157.6; HRMS (ESI) m/z calcd for $C_{29}H_{32}N_3O_5S$ 534.20572, found 534.20563 $[M+H]^+$; Anal. calcd. for $C_{29}H_{31}N_3O_5S$: C; 65.27, H; 5.86, N; 7.87, found: C; 65.08, H; 5.83, N; 7.60.

31. Purity of the product was determined to be >99% by quantitative ^1H NMR using 1,3,5-trimethoxybenzene as the internal standard.

Working with Hazardous Chemicals

The procedures in *Organic Syntheses* are intended for use only by persons with proper training in experimental organic chemistry. All hazardous materials should be handled using the standard procedures for work with chemicals described in references such as "Prudent Practices in the Laboratory" (The National Academies Press, Washington, D.C., 2011; the full text can be accessed free of charge at http://www.nap.edu/catalog.php?record_id=12654). All chemical waste should be disposed of in accordance with local regulations. For general guidelines for the management of chemical waste, see Chapter 8 of Prudent Practices.

In some articles in *Organic Syntheses*, chemical-specific hazards are highlighted in red "Caution Notes" within a procedure. It is important to recognize that the absence of a caution note does not imply that no significant hazards are associated with the chemicals involved in that procedure. Prior to performing a reaction, a thorough risk assessment should be carried out that includes a review of the potential hazards associated with each chemical and experimental operation on the scale that is planned for the procedure. Guidelines for carrying out a risk assessment and for analyzing the hazards associated with chemicals can be found in Chapter 4 of Prudent Practices.

The procedures described in *Organic Syntheses* are provided as published and are conducted at one's own risk. *Organic Syntheses, Inc.*, its Editors, and its Board of Directors do not warrant or guarantee the safety of individuals using these procedures and hereby disclaim any liability for any injuries or damages claimed to have resulted from or related in any way to the procedures herein.s

Discussion

Oxazoles have been used in key steps in several total syntheses,[2] have been isolated in many natural products[3] and have been produced at large scale in the pharmaceutical and agrochemical[4] industries: convergent routes into diversely functionalized oxazoles are attractive propositions. The gold-catalyzed formal (3+2)-dipolar cycloaddition between an *N*-heteroaryl-*N*-acyl aminide and an ynamide produces 4-aminooxazole derivatives in a highly convergent and atom economical fashion.[5] As Scheme 1 depicts, the aminide furnishes C-2 of the oxazole, whilst the ynamide equips C-4 and C-5.

5 mol % PicAuCl$_2$,

0.1 M toluene, 90 °C

1.1 – 1.5 eq.

Scheme 1. The synthesis of 4-aminooxazole derivatives

The formal dipolar cycloaddition displays the exquisite chemoselectivity of gold catalysis providing a wealth of oxazoles with diverse substitution patterns (Figure 5). Functionality at C-2 includes electron donating, electron withdrawing, aromatic, heteroaromatic, alkyl and cycloalkyl groups, with either oxygen, carbon or nitrogen directly bonded to the oxazole. The reaction tolerates stereogenic centers, secondary unprotected amines and acyclic acetals and can also form disubstituted oxazoles from formyl aminides (or terminal ynamides). Using the functionality appended to the ynamide, the oxazole synthesis incorporates electron withdrawing groups, sulfur substituted systems, aromatic groups and alkyl chains at the 5-position. The 4-amino group can be protected as a cyclic carbamate, a phosphoramidate or a sulfonamide.[5]

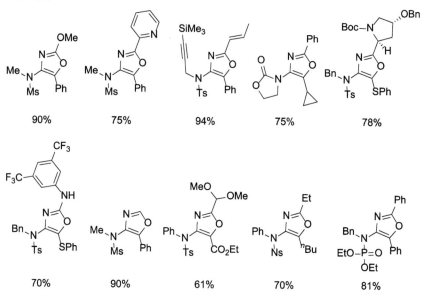

Figure 5. Selected examples of products from the formal cycloaddition

The broad applicability of this method requires access to both ynamides[6] and *N*-acyl pyridinium-*N*-aminides with substantial structural and functional group variety.

N-Acyl pyridinium-*N*-aminides have been exploited across a variety of cycloadditions and transition metal catalysed transformations to allow elaboration of the pyridine ring.[7] The anionic nitrogen atom acts as a directing group for C-H activation. Charette and co-workers showed how such pyridinium ylides undergo a range of intermolecular reactions in which the pyridyl unit is incorporated into the desired product.[8]

The potential of acyl pyridinium ylides to act as $^{1,3}N,O$ dipole equivalents is unveiled in the gold-catalysed oxazole synthesis. The aminides function as nucleophilic nitrenoids where the pyridinium moiety acts as a leaving group to reveal the acyl nitrenoid character of the aminide.

The synthesis of acyl aminides has been shown by Knaus,[9a] Akita,[9b] Charette[9c] and Alvarez-Buiilla,[9d] but formation of an acid chloride or anhydride is often prerequisite. Such features restrict the functionality tolerated.

An improved protocol was developed in order to access aminides by Davies *et al.* which incorporated a broader range of functionality at the now critical acyl position.[5] The aminide is synthesized from its respective methyl ester using 1-aminopyridinium iodide and potassium carbonate (Scheme 2), although a one-pot approach from carboxylic acids was also established. The diverse array of methyl esters at the chemist's disposal, commercial availability of the amidating reagent, and the simple reaction set-up make this an attractive route to functionalized systems.

Scheme 2. The synthesis of pyridinium aminides

References

1. School of Chemistry, University of Birmingham, Birmingham, B15 2TT, UK, p.w.davies@bham.ac.uk. We thank the University of Birmingham and the EPSRC for a studentship (M. P. B. J.).

2. (a) Ohba, M.; Natsutani, I. *Tetrahedron* **2007**, *63*, 12689–12694. (b) Ohba, M.; Izuta, R.; Shimizu, E. *Chem. Pharm. Bull.* **2006**, *54*, 63–67. (c) Jacobi, P. A.; Walker, D. G. *J. Am. Chem. Soc.* **1981**, *103*, 4611–4613. (d) Jacobi, P. A.; Kaczmarek, C. S. R.; Udodong, U. E. *Tetrahedron Lett.* **1984**, *25*, 4859–4862.

3. (a) You, S.-L.; Kelly, J. W. *J. Org. Chem.* **2003**, *68*, 9506–9509. (b) Wang, B.; Hansen, T. M.; Weyer, L.; Wu, D.; Wang, T.; Christmann, M.; Lu, Y.; Ying, L.; Engler, M. M.; Cink, R. D.; Lee, C.-S.; Ahmed, F.; Forsyth, C. J. *J. Am. Chem. Soc.* **2011**, *133*, 1506–1516. (c) Yeh, V. S. C. *Tetrahedron* **2004**, *60*, 11995–12042, (d) Jin, Z. *Nat. Prod. Rep.* **2013**, *30*, 869–915, (e) Zhang,

J.; Ciufolini, M. A. *Org. Lett.* **2011**, *13*, 390–393, (f) Bai, Y.; Chen, W.; Chen, Y.; Huang, H.; Xiao F.; Deng, G.-J. *RSC Adv.* **2015**, *5*, 8002–8005.

4. (a) Ibrar, A.; Khan, I.; Abbas, N.; Farooqa, U.; Khan, A. *RSC Adv.* **2016**, *6*, 93016–93047, (b) Ryu, C.-K.; Lee, R.-Y.; Kim, N. Y.; Kim, Y. H.; Song, A. L. *Bioorg. Med. Chem. Lett.* **2009**, *19*, 5924–5926, (c) Brown, P. Davies, D. T.; O'Hanlon, P. J.; Wilson, J. M. *J. Med. Chem.* **1996**, *39*, 446–457.

5. (a) Gillie, A. D.; Reddy, R. J.; Davies, P. W. *Adv. Synth. Catal.* **2016**, *358*, 226–239, (b) Chatzopoulou, E.; Davies, P. W. *Chem. Commun.* **2013**, *49*, 8617–8619, (c) Davies, P. W.; Cremonesi, A.; Dumitrescu, L. *Angew. Chem. Int. Ed.* **2011**, *38*, 8931–8935.

6. (a) Coste, A.; Couty, F.; Evano, G. *Org. Synth.* **2010**, *87*, 231–244. For reviews see: (b) Evano, G.; Coste, A.; Jouvin, K. *Angew. Chem. Int. Ed.* **2010**, *49*, 2840–2859, (c) DeKorver, K. A.; Li, H.; Lohse, A. G.; Hayashi, R.; Lu, Z.; Zhang, Y.; Hsung, R. P. *Chem. Rev.* **2010**, *110*, 5064–5106.

7. Representative examples: (a) Ding, S.; Yan, Y.; Jiao, N. *Chem. Commun.* **2013**, *49*, 4250–4252; (b) Xu, X.; Zavalij, P. Y.; Doyle, M. P. *Angew. Chem. Int. Ed.* **2013**, *52*, 12664–12668; (c) Zhou, Y.-Y.; Li, J.; Ling, L.; Liao, S.-H.; Sun, X.-L.; Li, Y.-X.; Wang, L.-J.; Tang, Y. *Angew. Chem. Int. Ed.* **2013**, *52*, 1452–1456; (d) Zhao, J.; Wu, C.; Li, P.; Ai, W.; Chen, H.; Wang, C.; Larock, R. C.; Shi, F. *J. Org. Chem.* **2011**, *76*, 6837–6843.

8. Representative examples: (a) Mousseau, J. J.; Bull, J. A.; Ladd, C. L.; Fortier, A.; Sustac Roman, D.; Charette, A. B. *J. Org. Chem.* **2011**, *76*, 8243–8261; (b) Mousseau, J. J., Bull, J. A; Charette, A. B. *Angew. Chem. Int. Ed.* **2010**, *49*, 1115–1118. (c) Larivée, A. ; Mousseau, J. J. ; Charette, A. B. *J. Am. Chem. Soc.* **2008**, *130*, 52–54.

9. (a) Yeung, J. M.; Corleto, L. A.; Knaus, E. E. *J. Med. Chem.* **1982**, *25*, 191–192; (b) Miyazawa, K.; Koike, T.; Akita, M. *Chem. Eur. J.* **2015**, *21*, 11677–11680; (c) Legault, C.; Charette, A. B. *J. Am. Chem. Soc.* **2003**, *125*, 6360–6361; (d) Molina, A.; de las Heras, M. A.; Martinez, Y.; Vaquero, J. J.; García Navio, J. L.; Alvarez-Builla, J.; Gomez-Sal, P.; Torres, R. *Tetrahedron* **1997**, *53*, 6411–6420.

Appendix
Chemical Abstracts Nomenclature (Registry Number)

1-Aminopyridinium iodide: Pyridinium, 1-amino, iodide (1:1); (6295-87-0)
Potassium carbonate: Carbonic acid, potassium salt (1:2); (584-0807)

N-Benzyl-4-methyl-N-(phenylethynyl)benzenesulfonamide:
Benzenesulfonamide, 4-methyl-N-(2-phenylethynyl)-N-(phenylmethyl)-;
(609769-63-3)
Dichloro(2-pyridinecarboxylato) gold: Gold, dichloro(2-
pyridinecarboxylato-κN¹,κO²)-, (SP-4-3)-; (88215-41-2)

Matthew Ball-Jones studied for a MChem degree in Chemistry at the University of Reading. In 2013 he began his Ph.D. at the University of Birmingham to work under the supervision of Dr Paul Davies. His work involves the development of cascade reactions and the synthesis of polycyclic three-dimensional heterocycles.

Paul Davies obtained his Ph.D. at the University of Bristol in 2003 with Prof. Varinder Aggarwal. After a postdoctoral stay at the Max Planck Institute fur Kohlenforschung with Prof. Alois Fürstner, he was appointed as a Lecturer and independent group leader in the School of Chemistry at the University of Birmingham in 2006, where he is now Senior Lecturer. His research interests focus on the discovery, development and application of new catalysis-based synthetic methods.

Organic
Syntheses

Manuela Brütsch completed her Master Degree in Chemistry in 2014 at the University of Zurich with Prof. Cristina Nevado. She then joined Prof. Jay S. Siegel's group at the Tianjin University in China to work as Research Assistant, followed by an appointment at the Scripps Research Institute in La Jolla with Prof. Dale Boger. In October 2016, she returned to the University of Zurich where she is working in Prof. Cristina Nevado's group.

Estíbaliz Merino obtained her Ph.D. degree from the Autónoma University (Madrid-Spain). After a postdoctoral stay with Prof. Magnus Rueping at Goethe University Frankfurt and RWTH-Aachen University in Germany, she worked with Prof. Avelino Corma in Instituto de Tecnología Química-CSIC (Valencia) and Prof. Félix Sánchez in Instituto de Química Orgánica General-CSIC (Madrid) in Spain. At present, she is research associate in Prof. Cristina Nevado's group in University of Zürich. She is interested in the synthesis of natural products using catalytic tools and in the development of new materials with application in heterogeneous catalysis.

Preparation of Solid Organozinc Pivalates and their Reaction in Pd-Catalyzed Cross-Couplings

Mario Ellwart, Yi-Hung Chen, Carl Phillip Tüllmann, Vladimir Malakhov, and Paul Knochel*[1]

Ludwig-Maximilians-Universität München, Department Chemie, Butenandtstraße 5–13, Haus F, 81377 München (Germany)

Checked by Kelsey E. Poremba and Sarah E. Reisman

Procedure (Note 1)

A. *Zinc Pivalate.* A dry, tared, 500 mL round-bottomed flask equipped with a 5x2-cm Teflon-coated magnetic stirring bar and a septum is charged with toluene (250 mL, 0.2 M) (Note 2). Pivalic acid (12.5 mL, 11.3 g, 110 mmol, 2.2 equiv) (Note 3) is added to form a colorless solution. Zinc oxide (4.07 g, 50 mmol, 1 equiv) is added in 1 g portions at 25 °C over 15 min to form a colorless suspension (Note 4). The flask is equipped with a Dean-Stark trap (10 mL) wrapped in aluminum foil and topped with a reflux condenser (20 cm) and the suspension is stirred under nitrogen at reflux in an oil bath for 16 h (Figure 1) (Note 5).

Figure 1: Step A - Dean-Stark trap and evaporation of remaining pivalic acid and water with a liquid nitrogen cold trap

A viscous colorless suspension is formed overnight. After cooling to 25 °C, the mixture is concentrated by rotary evaporation (50 °C/50 mmHg). The remaining pivalic acid and water are removed *in vacuo* from the reaction mixture using a vacuum line (0.1 mmHg) and a liquid nitrogen cold trap (1000 mL) (see Figure 1). The white solid is warmed to 100 °C in an oil bath and dried for at least 6 h (Note 6). Zinc pivalate (13.1–13.2 g,

48.9–49.7 mmol, 98–99%), is obtained as a puffy amorphous white solid (Note 7).

B. *Pyridin-3-ylzinc Pivalate.* A dry and argon flushed 1 L Schlenk-flask equipped with a 5×2-cm Teflon-coated magnetic stirring bar and a septum is filled with argon and then weighed. 3-Bromopyridine (6.32 g, 40.0 mmol, 1 equiv) (Note 8) and dry THF (50 mL, 0.8 M) are added to the flask via syringe. (Note 9). The solution is cooled in an ice-water bath under an atmosphere of argon and stirred for at least 5 min at 0 °C before *i*PrMgCl·LiCl (35.2 mL, 1.25 M, 44.0 mmol, 1.1 equiv) (Note 10) is added via a syringe pump over the period of 30 min (Note 11). The ice bath is removed and the solution is stirred for 3 h at 25 °C during which time it gradually turns from yellow to dark red (Note 12). Upon completion of the reaction, solid Zn(OPiv)$_2$ (12.3 g, 46.0 mmol, 1.15 equiv) is added in one portion under argon counterflow via a powder funnel. A slight exotherm is noticed (Note 13). The mixture is stirred at 25 °C for 30 min leading to a clear dark red solution. The solvent is removed using a vacuum line (0.1 mmHg) and a liquid nitrogen cold trap and the solid residue is dried for at least 2 h longer leading to a voluminous yellow foam (Figure 2) (Note 14). The foam is crushed with a spatula under argon counterflow to form a fine yellow powder. This powder is dried under high vacuum (0.1 mmHg) for further 2 h. The resulting pyridine-3-ylzinc pivalate (28.6–28.8 g, 1.1–1.20 mmol g^{-1}, 31.5–34.5 mmol, 79–86%) is used immediately.

After the drying process is complete, the argon-flushed flask is weighed to determine the weight of the resulting powder. To determine the actual content in zinc species and the reaction yield, a small aliquot of the powder (accurately weighed amount, ca. 1 g, see Figure 3: a) is titrated using a 1 M solution of iodine in THF (Note 15) with a color change from red (b) to bright yellow (c) until the persisting brown color of the iodine (d) indicates the completion of the titration (Note 16).

Figure 2: Step B – The color of the reaction mixture turns from yellow to dark red. Photographs of the solid foam and powder.

Figure 3: Step B – Titration of the organozinc reagent. a) solid zinc reagent. b) before the iodometric titration. c) titration before color change. d) titration after color change

C. *Ethyl 4-(Pyridin-3-yl)benzoate*. To the dry and argon-flushed 1 L flask, containing the solid pyridine-3-ylzinc pivalate (27.2 g, 32.6 mmol, 1.20 mmol g⁻¹, 1.15 equiv), a 5×2-cm Teflon-coated magnetic stirring bar and a septum, is added dry THF (65 mL, 0.44 M). Ethyl 4-bromobenzoate (4.6 mL, 6.45 g, 28.2 mmol, 1 equiv) (Note 17) is added via syringe. The septum is temporarily removed and PEPPSI-IPr (193 mg, 0.28 mmol, 1 mol%)[2] added, after which the septum is reconnected and the flask flushed with argon. The red solution is stirred for 2 h at room temperature (25 °C) under an atmosphere of argon (Note 18). Then sat. aq. NH$_4$Cl (50 mL) is added and the aqueous layer is extracted with EtOAc (3 × 70 mL). The combined organic phases are dried (12 g MgSO$_4$). After filtration and evaporation of the solvent in vacuo, purification by column chromatography (hexane:EtOAc:NEt$_3$ = 50:10:1 → 50:25:1) (Note 19) afforded ethyl 4-(pyridin-3-yl)benzoate (6.02 g, 26.5 mmol, 94%) as a yellow solid (Notes 20, 21, and 22).

Notes

1. Prior to performing each reaction, a thorough hazard analysis and risk assessment should be carried out with regard to each chemical substance and experimental operation on the scale planned and in the context of the laboratory where the procedures will be carried out. Guidelines for carrying out risk assessments and for analyzing the hazards associated with chemicals can be found in references such as Chapter 4 of "Prudent Practices in the Laboratory" (The National Academies Press, Washington, D.C., 2011; the full text can be accessed free of charge at https://www.nap.edu/catalog/12654/prudent-practices-in-the-laboratory-handling-and-management-of-chemical. See also "Identifying and Evaluating Hazards in Research Laboratories" (American Chemical Society, 2015) which is available via the associated website "Hazard Assessment in Research Laboratories" at https://www.acs.org/content/acs/en/about/governance/committees/chemicalsafety/hazard-assessment.html. In the case of this procedure, the risk assessment should include (but not necessarily be limited to) an evaluation of the potential hazards associated with pivalic acid, aluminum foil, zinc oxide, toluene, liquid nitrogen, 3-bromopyridine, tetrahydrofuran, *i*PrMgCl·LiCl, iodine, ethyl 4-bromobenzoate, PEPPSI-

IPr, ammonium chloride, ethyl acetate, magnesium sulfate, hexane, and trimethylamine.

2. Toluene was used after purification through activated alumina using a Glass Contour solvent purification system.

3. Pivalic acid (99%) was obtained from Acros Organics and warmed to 60 °C ca. 1 h prior to the addition. The molten pivalic acid (mp = 35 °C) can be easily added by quickly handling via syringe.

4. Zinc oxide was obtained from Sigma Aldrich or Panreac AppliChem and was used without any further purification

5. The trap was filled up with toluene. After 16 h, water (0.9 mL, 50 mmol) was obtained in the Dean-stark trap and residues of pivalic acid could be observed on the bottom of the condenser and the Dean-Stark trap.

6. The flask was connected to high vacuum (0.1 mmHg) with a liquid nitrogen cold trap. Vigorous stirring (800 rpm) was maintained to keep the solid from heterogenization.

7. Zinc pivalate should be stored under argon, but can be weighed on air. The product was characterized as follows: mp 305–315 °C (sublimation); ^{1}H NMR (500 MHz, DMSO-d_6) δ : 1.08 (s, 9H); ^{13}C NMR (101 MHz, DMSO-d_6) δ : 28.3, 37.8, 184.0; IR (diamond ATR, neat): 2962, 2929, 1606, 1534, 1481, 1457, 1426, 1378, 1361, 1228, 1031, 937, 899, 791, 609 cm^{-1}. Purity >97% as assessed by quantitative NMR, in which ethylene carbonate is used as the internal standard. Conditions: Zn(OPiv)$_2$ (19.8 mg); standard (17.3 mg) Solvent: DMSO.

8. 3-Bromopyridine (99%) was purchased from Apollo Scientific and used as received.

9. THF was used after purification through activated alumina using a Glass Contour solvent purification system.

10. A KDS single-syringe pump (series 100) was used with a 50 mL NORM-JECT syringe and a 1.10 × 120 mm TSK-SUPRA needle. The iPrMgCl·LiCl solution was added with a rate of 1.16 mL/min.

11. A solution of iPrMgCl·LiCl in THF was obtained from Sigma Aldrich or Albemarle (Frankfurt) and titrated against iodine prior to use. For the titration accurately weighted aliquots (e.g. 221 mg) of iodine were placed in a dry and argon flushed 20 mL Schlenk-flask with a septum and dissolved in ca. 2 mL dry THF. To the resulting solution was added the iPrMgCl·LiCl solution using a 1 mL NORM-JECT syringe from Henke Sass Wolf until the complete disappearance of the dark brown color of iodine (0.69 mL equals a concentration of 1.26 M).

12. The progress of the halogen-magnesium exchange was monitored by GC-analysis of reaction aliquots quenched with iodine or NMR analysis after NH₄Cl quench of reaction aliquot. GC analysis was performed using an Agilent Technologies 6850 Series equipped with an HP-5 column (J&W Scientific) (15m x 0.25mm x 0.25µm). Oven program for GC analysis: Starting temperature 70 °C for 0.5 min; heating to 250 °C at a rate of 50 °C/min; 5 min at 250 °C.

13. Reaction mixture warmed to 40 °C internal temperature.

14. The flask was warmed in a 20 °C water bath in order to accelerate the solvent evaporation.

15. A 1 M solution of iodine in THF was prepared in a dry and argon flushed 20 mL Schlenk-flask by dissolving 2.54 g I₂ in 9.30 mL dry THF.

16. For the titration accurately weighted aliquots (e.g. 913 mg) of the powder were placed in a dry and argon flushed 20 mL Schlenk-flask with a septum and dissolved in dry THF (ca. 3 mL). To the resulting solution was added the 1 M iodine solution using a 1 mL NORM-JECT syringe from Henke Sass Wolf until the persistence of the dark brown color of iodine (e.g. 1.19 mL equals a concentration of 1.30 mmol/g).

17. The following reagents in this section were purchased from commercial sources and used without further purification: ethyl 4-bromobenzoate (99+%, Apollo Scientific), PEPPSI-IPr (98%, Sigma-Aldrich).

18. The cross-coupling was monitored by NMR analysis of reaction aliquots.

19. Five grams of ISOLUTE HM-N adsorbed with the crude product was dry-loaded onto a column (diameter: 6.0 cm, height: 25.0 cm) packed with silica gel (250 g) slurry in 50:10:1 hexane:EtOAc:NEt₃ (R$_f$(product): 0.12; visualized with UV light) and 100 mL fractions were collected. The desired product was obtained in fractions 20-80 (50:25:1 hexane:EtOAc:NEt₃), which are concentrated by rotary evaporation (40 °C, 250 mmHg). The purified product is stored under an inert atmosphere in the dark for long-term storage.

20. The product has been characterized as follows: mp: 44–46 °C; ¹H NMR (400 MHz, CDCl₃) δ : 1.41 (t, *J* = 7.1 Hz, 3H), 4.40 (q, *J* = 7.2 Hz, 2H), 7.38 (ddd, *J* = 7.9, 4.8, 0.9 Hz, 1H), 7.61–7.68 (m, 2H), 7.89 (ddd, *J* = 7.9, 2.4, 1.6 Hz, 1H), 8.10–8.17 (m, 2H), 8.63 (dd, *J* = 4.8, 1.6 Hz, 1H), 8.87 (dd, *J* = 2.4, 0.9 Hz, 1H); ¹³C NMR (101 MHz, CDCl₃) δ : 14.5, 61.2, 123.7, 127.1, 130.2, 130.4, 134.6, 135.7, 142.2, 148.5, 149.4, 166.3; IR (diamond ATR, neat): 2985, 2909, 1700, 1608, 1471, 1425, 1366, 1275, 1192, 1181,

1124, 1102, 1021, 1000, 856, 815, 765, 711, 700 cm^{-1}. HRMS (ESI-TOF): m/z calc. for [C$_{14}$H$_{13}$NO$_2$]: 227.0946; found: 227.0950 (M$^+$).
21. A second reaction on similar scale provided 5.17 g (91%) of the product.
22. Purity of the product was determined by the Checkers as >97% by quantitative NMR, in which ethylene carbonate is used as the internal standard. Conditions: product (27.2 mg); standard (17.0 mg) Solvent: CDCl$_3$. Purity of the product was determined by the authors to be >98% by GC analysis. The Submitters performed GC analysis using an Agilent Technologies 6850 Series equipped with an HP-5 column (J&W Scientific) (15m x 0.25mm x 0.25µm). Oven program for GC analysis: Starting temperature 70 °C for 0.5 min; heating to 250 °C at a rate of 50 °C/min; 5 min at 250 °C.

Working with Hazardous Chemicals

The procedures in *Organic Syntheses* are intended for use only by persons with proper training in experimental organic chemistry. All hazardous materials should be handled using the standard procedures for work with chemicals described in references such as "Prudent Practices in the Laboratory" (The National Academies Press, Washington, D.C., 2011; the full text can be accessed free of charge at http://www.nap.edu/catalog.php?record_id=12654). All chemical waste should be disposed of in accordance with local regulations. For general guidelines for the management of chemical waste, see Chapter 8 of Prudent Practices.

In some articles in *Organic Syntheses*, chemical-specific hazards are highlighted in red "Caution Notes" within a procedure. It is important to recognize that the absence of a caution note does not imply that no significant hazards are associated with the chemicals involved in that procedure. Prior to performing a reaction, a thorough risk assessment should be carried out that includes a review of the potential hazards associated with each chemical and experimental operation on the scale that is planned for the procedure. Guidelines for carrying out a risk assessment and for analyzing the hazards associated with chemicals can be found in Chapter 4 of Prudent Practices.

The procedures described in *Organic Syntheses* are provided as published and are conducted at one's own risk. *Organic Syntheses, Inc.,* its

Editors, and its Board of Directors do not warrant or guarantee the safety of individuals using these procedures and hereby disclaim any liability for any injuries or damages claimed to have resulted from or related in any way to the procedures herein.

Discussion

The performance of cross-couplings between Csp²-centers is a major synthetic concern due to the importance of the resulting products as potential pharmaceuticals, agrochemicals, or new organic materials. Organozincs are excellent nucleophilic candidates for such cross-couplings since the labile carbon-zinc bond undergoes fast transmetalations with numerous transition metals under mild conditions due to the presence of empty low-lying p-orbitals at the zinc center. Also, zinc (II) salts are ecologically friendly salts of low inherent toxicity. The only drawbacks are the moisture and air sensitivity of organozinc derivatives which precludes their handling in air. In 2011, our research group developed a range of aryl- and heteroaryl zinc derivatives with significantly enhanced air and moisture stability.[3] we have reported that the treatment of an aryl bromide such as **1a** with magnesium powder in the presence of Zn(OPiv)₂·2LiCl provides the corresponding zinc organometallic species (**2a**) conveniently abbreviated as **3a** and called organozinc pivalates knowing that the improved water and air stability may be due to the presence of magnesium pivalate[4] (Scheme 1).

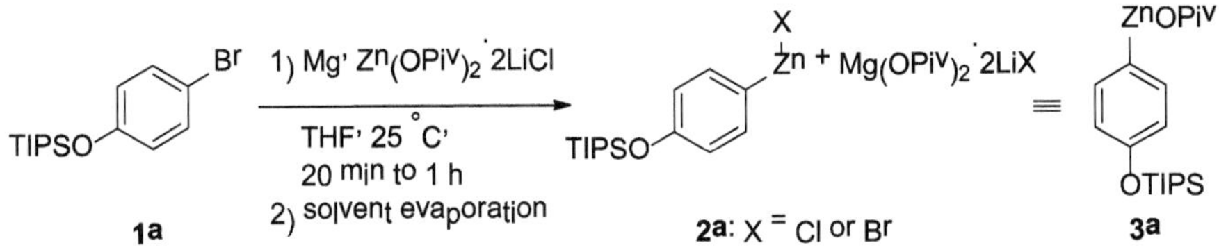

Scheme 1: Preparation of a solid arylzinc pivalate via the direct magnesium insertion in the presence of Zn(OPiv)₂·2LiCl

Alternatively, these organozinc pivalates may be prepared by directed metalation using either TMPMgCl·LiCl[5,6] followed by the addition of Zn(OPiv)₂ or TMPZnOPiv[7] prepared in situ. Thus, the treatment of ethyl 3-

fluorobenzoate (**1b**) with TMPMgCl·LiCl at 0 °C in THF followed by the addition of Zn(OPiv)$_2$ and subsequent solvent evaporation produces the functionalized solid arylzinc pivalate **3b** in 92% yield.[5] Heterocyclic zinc reagents are prepared similarly and the use of TMPZnOPiv may be advantageous. Thus, the reaction of 4,6-dichloropyrimidine (**1c**) with TMPZnOPiv·LiCl in THF at ambient temperatures furnishes after solvent evaporation the corresponding heteroarylzinc pivalate (**3c**) in 78%.[7] This particular heteroarylzinc pivalate is air-stable with almost no activity loss after 4 h in air (Scheme 2).

Scheme 2: Preparation of solid aryl and heteroarylzinc pivalates by directed metalation using TMPMgCl·LiCl and TMPZnOPiv·LiCl

This method has a broad scope and a range of polyfunctional zinc reagents like **3d-s** have been prepared in satisfactory yields (Scheme 3).

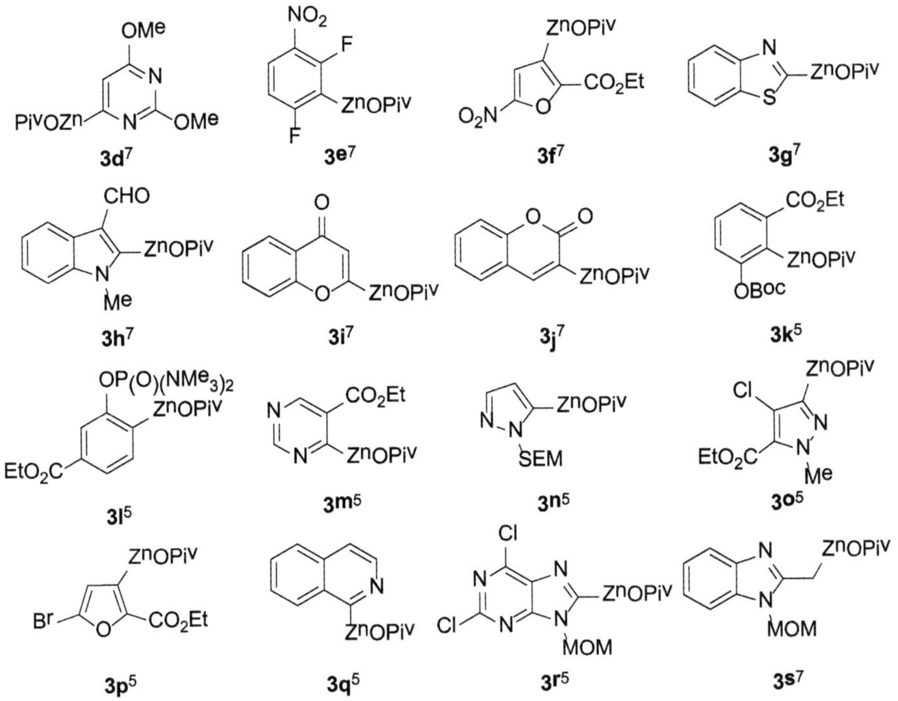

Scheme 3: Various organozinc pivalates prepared by directed metalation

All these aryl- and heteroarylzinc reagents have improved air and moisture stability.[8] They readily undergo palladium- or cobalt catalyzed cross-coupling reactions (Scheme 4).[5,9]

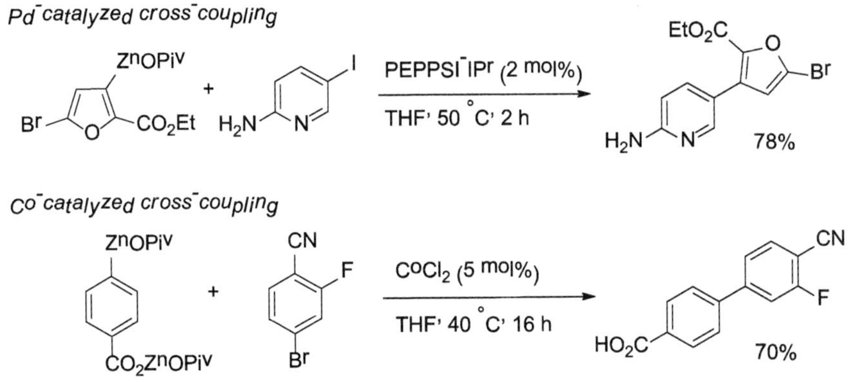

Pd-catalyzed cross-coupling

Co-catalyzed cross-coupling

Scheme 4: Pd- and Co-catalyzed cross-couplings of organozinc pivalates

The cross-coupling conditions are usually mild and comparable to those of regular organozinc halides. Also these organozinc pivalates undergo readily acylation reactions and copper-catalyzed allylations.[10] Finally, benzylic zinc pivalates can be prepared according to the same methods as well as allylic zinc pivalates which are also stable in a solid state (in the latter case under argon).[11]

In conclusion, aryl and heteroarylzinc pivalates are convenient zinc reagents suitable for a wide range of carbon-carbon bond forming reactions. They are easy to prepare using either a direct magnesium insertion in the presence of zinc pivalate or can be prepared by a directed metalation using either TMPMgCl·LiCl and Zn(OPiv)$_2$ or TMPZnOPiv·LiCl. Their low toxicity and compatibility with a wide range of functional groups makes them valuable organometallic reagents for both academic and industrial applications.

References

1. Ludwig-Maximilians-Universität München, Department Chemie, Butenandtstrasse 5–13, Haus F, 81377 München (Germany), E-mail: paul.knochel@cup.uni-muenchen.de.
2. O'Brien, C. J.; Kantchev, E. A. B.; Valente, C.; Hadei, N.; Chass, G. A.; Lough, A.; Hopkinson, A. C.; Organ, M. G. *Chem. Eur. J.* **2006**, *12*, 4743–4748.
3. Bernhardt, S.; Manolikakes, G.; Kunz, T.; Knochel, P. *Angew. Chem.* **2011**, *123*, 9372; *Angew. Chem. Int. Ed.* **2011**, *50*, 9205–9209.
4. Hernán-Gómez, A.; Herd, E.; Hevia, E.; Kennedy, A. R.; Knochel, P.; Koszinowski, K.; Manolikakes, S. M.; Mulvey, R. E.; Schnegelsberg, C. *Angew. Chem. Int. Ed.* **2014**, *53*, 2706–2710.
5. Stathakis, C. I.; Bernhardt, S.; Quint, V.; Knochel, P. *Angew. Chem.* **2012**, *124*, 9563; *Angew. Chem. Int. Ed.* **2012**, *51*, 9428–9432.
6. Haag, B.; Mosrin, M.; Ila, H.; Malakhov, V.; Knochel, P. *Angew. Chem.* **2011**, *123*, 9968; *Angew. Chem. Int. Ed.* **2011**, *50*, 9794–9824.
7. Stathakis, C. I.; Manolikakes, S. M.; Knochel, P. *Org. Lett.* **2013**, *15*, 1302–1305.
8. Colombe, J. R.; Bernhardt, S.; Stathakis, C.; Buchwald, S. L.; Knochel, P. *Org. Lett.* **2013**, *15*, 5754–5757.

9. Hammann, J. M.; Lutter, F. H.; Haas, D.; Knochel, P. *Angew. Chem.* **2017**, *129*, 1102; *Angew. Chem. Int. Ed.* **2017**, *56*, 1082–1086.
10. Manolikakes, S. M.; Ellwart, M.; Stathakis, C. I.; Knochel, P. *Chem. Eur. J.* **2014**, *20*, 12289–12297.
11. Ellwart, M.; Knochel, P. *Angew. Chem. Int. Ed.* **2015**, *54*, 10662–10665; *Angew. Chem.* **2015**, *127*, 10808.

Appendix
Chemical Abstracts Nomenclature (Registry Number)

Zinc Pivalate: Propanoic acid, 2,2-dimethyl-, zinc salt (2:1); (15827-10-8)
Pivalic acid: Propanoic acid, 2,2-dimethyl-; (75-98-9)
Zinc oxide: Oxozinc; (1314-13-2)
Pyridin-3-ylzinc Pivalate: Zinc, (2,2-dimethylpropanoato-κO)-3-pyridinyl-; (1344727-29-2)
3-Bromopyridine: Pyridine, 3-bromo-; (626-55-1)
*i*PrMgCl·LiCl: Magnesate(1-), dichloro(1-methylethyl)-, lithium (1:1); (745038-86-2)
Ethyl 4-(Pyridin-3-yl)benzoate: Benzoic acid, 4-(3-pyridinyl)-, ethyl ester; (4385-71-1)
Ethyl 4-bromobenzoate: Benzoic acid, 4-bromo-, ethyl ester; (5798-75-4)
PEPPSI-IPr: Palladium, [1,3-bis[2,6-bis(1-methylethyl)phenyl]-1,3-dihydro-2*H*-imidazol-2-ylidene]dichloro(3-chloropyridine-κN)-, (*SP*-4-1)-; (905459-27-0)

Mario Ellwart was born in Munich (Germany) in 1987. He studied chemistry at the Ludwig-Maximilians-Universität München and joined the research group of Prof. Paul Knochel in 2012. His research focuses on the synthesis of novel organozinc reagents and their applications in organic synthesis.

Yi-Hung Chen was born in Taipei (Taiwan) in 1977. He studied chemistry at the National Tsing Hua university in Taiwan and started his Ph.D. in the research group of Prof. Frank E. McDonald at Emory university in 2002. His research focused on the polyketide based natural product synthesis. In 2007, he completed his Ph.D. and joined the group of Prof. Paul Knochel as a Humboldt fellow and continued to stay in the same group as research assistant. His research focused on the preparation organometallic reagents and their applications in organic synthesis.

Carl Phillip Tüllmann was born in Düsseldorf (Germany) in 1992. He studied chemistry at the Albert-Ludwigs-University in Freiburg and the Ludwig-Maximilians-University in Munich. He joined the research group of Prof. Paul Knochel in 2017. His research focuses on the synthesis of novel organozinc reagents and their applications in organic synthesis.

Vladimir Malakhov was born in Moscow (Russia) in 1965. He completed his undergraduate studies in pharmacy at the I. M. Sechenov Moscow Medical Academy (1985–1990). In 1997, he joined Prof. P. Knochel's group at the Philipps-Universität Marburg (Germany) and moved in 1999 with him to the Ludwig-Maximilians-Universität Munich (Germany). He completed his Ph.D. under supervision of Prof. P. Knochel, which focused on polyfunctional organozinc reagents.

Organic
Syntheses

Paul Knochel was born 1955 in Strasbourg (France). He carried out his undergraduate studies at the University of Strasbourg (France) and his Ph.D. at the ETH-Zürich with Prof. Dieter Seebach. He spent four years at the CNRS at the University Pierre and Marie Curie in Paris with Prof. Jean F. Normant and one year of postdoctoral studies at Princeton University in the laboratory of Prof. Martin F. Semmelhack. In 1987, he accepted a position as Assistant Professor at the University of Michigan at Ann Arbor, USA. In 1991, he became Full Professor, then in 1992, he moved to the Philipps University at Marburg (Germany) as C4-Professor in organic chemistry. In 1999, he moved again to the Chemistry Department of the Ludwig-Maximilians-University in Munich (Germany). His research interests include the development of novel organometallic reagents and methods for their use in organic synthesis, asymmetric catalysis and natural product synthesis.

Kelsey Poremba received her B.A. from the College of the Holy Cross in Worcester, Massachusetts in 2014, where she conducted research in the lab of Professor Bianca R. Sculimbrene. She is currently pursuing her Ph.D. in the lab of Professor Sarah E. Reisman. Her graduate research is focused on the development of nickel-catalyzed asymmetric reductive cross-coupling reactions.

Enantioselective Synthesis of (S)-Ethyl 2-((*tert*-butoxycarbonyl)((*tert*-butyldimethylsilyl)oxy)amino)-4-oxobutanoate

Thibault J. Harmand, Claudia E. Murar, Hikaru Takano, and Jeffrey W. Bode[1]*

Eidgenössische Technische Hochschule Zürich, HCI F317, 8093 Zürich, Switzerland

Checked by Jacob C. Timmerman, Yu-Wen Huang, and John L. Wood

A.

$$\text{Boc}\underset{H}{\overset{}{N}}\text{-OH} \xrightarrow[\substack{CH_2Cl_2 \\ 4\ ^\circ C\ \text{to rt}}]{TBSCl,\ Et_3N} \text{Boc}\underset{H}{\overset{}{N}}\text{-OTBS}\quad \mathbf{1}$$

B.

$$\text{(pyrrolidine-Ph,Ph-OH)} \xrightarrow[\substack{THF \\ 4\ ^\circ C\ \text{to rt}}]{TMSCl,\ Imidazole} \text{(pyrrolidine-Ph,Ph-OTMS)}\quad \mathbf{2}$$

C.

$$\mathbf{1} + \mathbf{2}\ (50\ \text{mol }\%) + EtO_2C\text{-CH=CH-CHO} \xrightarrow[\substack{CHCl_3 \\ 4\ ^\circ C\ \text{to rt}}]{} \text{Boc-N(OTBS)-CH(CO_2Et)CH_2CHO}\quad \mathbf{3}$$

Procedure (Note 1)

A. *tert-Butyl (tert-butyldimethylsilyl)oxycarbamate (**1**)*. *N*-Boc hydroxyl-amine (44.0 g, 330 mmol, 1.0 equiv) (Note 2) is introduced into a 2-L round-bottomed flask equipped with a 4-cm oval Teflon-coated stir-bar and is dissolved with CH₂Cl₂ (1.10 L). The reaction mixture is cooled to 4 °C in an ice-water bath after which a 250-mL addition funnel is attached. Triethylamine (49.0 mL, 363 mmol, 1.10 equiv) is transferred into the addition funnel *via* a graduated cylinder, and added dropwise over 15 min. Dichloromethane (20 mL) is used to ensure that no reagents are left on the side. *tert*-Butyldimethylsilylchloride (49.7 g, 330 mmol, 1.0 equiv) is dissolved in CH_2Cl_2 (150 mL) and added dropwise over 60 min *via* the addition funnel at 4 °C. Dichloromethane (50 mL) is used to ensure that no reagent is left on the side of the addition funnel (final concentration of substrate is 0.25 M). The addition funnel is removed, the flask is equipped with a nitrogen inlet, and the reaction is stirred for 16 h at 23 °C under nitrogen (Note 3). Water (250 mL) is added and the mixture is poured into a 2-L separatory funnel. The reaction flask is rinsed with CH_2Cl_2 (50 mL) and the combined organic layer is separated and washed with saturated aqueous NaCl (250 mL). The organic layer is dried over $MgSO_4$ (20 g) and filtered by suction using a fritted funnel. Dichloromethane (50 mL) is used to wash the $MgSO_4$ and the filtrate is concentrated by rotary evaporation in a 2-L round-bottomed flask (40 °C bath, 425–30 mmHg). The resulting oil, which contains *tert*-butyl (*tert*-butyldimethylsilyl)oxycarbamate, is transferred to a 500-mL round-bottomed flask using CH_2Cl_2, which is then evaporated (40 °C bath, 425–30 mmHg). The product is dried on the vacuum pump (0.5 mmHg) for 48 h affording the desired compound (81.3 g, 329 mmol, 99.5% yield, 97.5% purity) as a white solid (Notes 4 and 5).

B. *(S)-(–)-α,α-Diphenyl-2-pyrrolidinemethanol trimethylsilyl ether* (**2**). (S)-(-)-α,α-Diphenyl-2-pyrrolidinylmethanol (25.0 g, 98.8 mmol, 1.0 equiv) (Note 6) is introduced in a 1-L three-necked round-bottomed flask (equipped with a 4-cm oval Teflon-coated stir-bar, an internal thermometer, and a glass stopper) and is dissolved with THF (220 mL). To the resulting solution, imidazole (20.0 g, 294 mmmol, 3.0 equiv) is added in one portion. After complete dissolution of the imidazole, the reaction mixture is cooled to 4 °C in an ice-water bath. A 250-mL addition funnel is attached and then charged with trimethylchlorosilane (31.3 mL, 247 mmol, 2.50 equiv) *via* a 50 mL syringe. The TMSCl is added dropwise via the addition funnel over 20 min (Figure 1). Tetrahydrofuran (50 mL) is used to rinse the addition funnel and ensure that no reagent is left on the side of the addition funnel. The addition funnel is removed, the flask is equipped with a nitrogen inlet, and the reaction is stirred for 15 h at 23 °C under nitrogen. Methyl *tert*-butyl ether (MTBE) (150 mL) is added and the reaction stirred for an additional 15 min. The resultant heterogeneous mixture is filtered through a improvement. Despite all of our attempts to optimize it we were not able to make this process catalytic.

The present synthesis of *(S)-ethyl 2-((tert-butoxycarbonyl)((tert-butyldimethylsilyl)oxy)amino)-4-oxobutanoate* provides a convenient route to the synthesis of unnatural amino acids and hydroxylamine building blocks for the KAHA ligation. This synthesis will be the starting point for the design and synthesis of a large number of cyclic hydroxylamines as new monomers for KAHA ligation.

References

1. Eidgenössiche Technische Hochschule Zürich, HCI F317, 8093 Zürich, Switzerland. Email: bode@org.chem.ethz.ch.
2. Boeckman and co-workers, *Org. Synth.* **2015**, *92*, 309-319.
3. (a) Chen, Y. K.; Yoshida, M.; MacMillan, D. W. C. *J. Am. Chem. Soc.* **2006**, *128*, 9328–9329. (b) Vesely, J.; Ibrahem, I.; Rios, R.; Zhao, G. L.; Xu, Y.; Córdova, A. *Tetrahedron Lett.* **2007**, *48*, 2193–2198. (c) Ibrahem, I.; Rios, R.; Vesely, J.; Zhao, G. L.; Córdova, A., *Chem. Commun.* **2007**, 849–851.
4. (a) Pattabiraman, V. R.; Ogunkoya, A. O.; Bode, J. W. *Angew. Chem., Int. Ed.* **2012**, *51*, 5114–5118. (b) Ogunkoya, A. O.; Pattabiraman, V. R.; Bode, J. W. *Angew. Chem., Int. Ed.* **2012**, *51*, 9693–9697.
5. Murar, C. E.; Thuaud, F.; Bode, J. W. *J. Am. Chem. Soc.* **2014**, *136*, 18140–18148.
6. (a) Fleischer, I.; Pfaltz, A. *Chem. Eur. J.* **2010**, *16*, 95–99. (b) Hayashi, Y.; Gotoh, H.; Hayashi, T. Shoji, M. *Angew. Chem., Int. Ed.* **2005**, *44*, 4212–4215. (c) Jensen, K.L.; Dickmeiss, G.; Hao, J.; Albrecht, L.; Jørgensen, K. A. *Acc. Chem. Res.* **2012**, *45*, 248–264.

Appendix
Chemical Abstracts Nomenclature (Registry Number)

N-Boc hydroxylamine; (36016-38-3)
tert-Butyldimethylsilyl chloride; (18162-48-6)
Triethylamine; (121-44-8)
(*S*)-(–)-α,α-Diphenyl-2-pyrrolidinemethanol; (112068-01-6)
Ethyl (2*E*)-4-oxo-2-butenoate; (2960-66-9)

Organic
Syntheses

Thibault Harmand studied Chemistry at the University of Nantes and Lyon (France). After his master thesis with Prof. George Fleet at the University of Oxford, he joined the group of Prof. Jeffrey Bode at ETH Zurich. His research focuses on the total chemical synthesis of proteins such as the hormone protein betatrophin and the antiviral membrane protein IFITM3 as well as the development of amino acids building blocks for chemical ligation.

Claudia Murar was born in Deva, Romania in 1988. She moved for her studies to France where she studied chemistry at the National Institute of Applied Sciences (INSA), Rouen (France). She worked for one year at GSK, King of Prussia (USA) after which she completed her master thesis at Novartis, Basel (Switzerland). She joined the group of Prof. Jeffrey Bode at ETH Zurich in 2012 for her Ph.D. Her research focuses on the total chemical synthesis of hormone proteins and therapeutic proteins, development of new building blocks for chemical ligation and synthesis of unnatural amino acids.

Hikaru Takano was born in Nagasaki, Japan in 1989. He received his B.Eng. degree in 2012 from Saitama University (Prof. Katsukiyo Miura). He then moved to Tokyo Medical and Dental University and is currently pursuing his Ph.D. degree under the supervision of Prof. Hirokazu Tamamura (2012-present). His current research focuses on the development of chemical probes, especially fluorescent dyes, photolabile protecting groups and caged compounds.

Jeffrey Bode is Professor of Synthetic Organic Chemistry at ETH Zürich. In addition, he serves as an Executive Editor for the *Encyclopedia of Reagents for Organic Synthesis*, co-Editor in Chief of *Helvetica Chimica Acta*, and a Principal Investigator at the *Institute of Transformative bio-Molecules (ITbM)* at Nagoya University in Japan. His research group focuses on the development of new reactions, including methods for *N*-heterocycles, chemical protein synthesis, bioconjugation, and chemical biology.

Jacob C. Timmerman graduated from the University of North Carolina at Chapel Hill in 2012 with a B.A. in Chemistry. Later in 2012, Jacob began his graduate studies at Duke University under the advisement of Professor Ross A. Widenhoefer where his research focused on the development and mechanistic studies of gold(I)-catalyzed hydrofunctionalization reactions of alkenes. After completing his Ph.D. in 2017, Jacob joined the laboratories of Professor John L. Wood as a postdoctoral research associate where his research focuses on the total synthesis of natural products.

Yu-Wen Huang was born in Hsinchu, Taiwan (R.O.C.) in 1982. He received his bachelor's degree from National Cheng Kung University in 2005. He then joined the M.S. program at the National Tsing Hua University working under the supervision of Professor Shang-Cheng Hung on carbohydrate synthesis. In 2016, he received his Ph.D. degree from the University of Rochester where he worked with Professor Alison J. Frontier on 1,6-conjugate addition initiated Nazarov reactions and sequential 1,5-hydride transfer chemistry. He is currently a post-doctoral fellow in the CPRIT lab (Baylor University) with Professor John L. Wood working on the total synthesis of natural products.

Preparation of (S)-N-Boc-5-oxaproline

Claudia E. Murar, Thibault J. Harmand, Hikaru Takano, and Jeffrey W. Bode*[1]

Eidgenössiche Technische Hochschule Zürich, HCI F317, 8093 Zürich, Switzerland

Checked by Jacob C. Timmerman, Yu-Wen Huang, and John L. Wood

Procedure (Note 1)

A. *2-(tert-Butyl) 3-ethyl (S)-isoxazolidine-2,3-dicarboxylate* (**4**). The aldehyde **1** (15.0 g, 39.9 mmol, 1.0 equiv, 94% ee) (Note 2) is dissolved in MeOH (200 mL, concentration of substrate is 0.20 M) (Note 3) in a 500-mL, three-necked, round-bottomed flask equipped with a 4-cm Teflon-coated magnetic stir-bar, a plastic stopper, a low-temperature thermometer and a rubber septum through which a positive nitrogen atmosphere is ensured

(Note 3) (Figure 1). The reaction mixture is cooled to –20 °C using an CH₃CN-dry ice bath. Sodium borohydride (3.02 g, 79.8 mmol, 2.0 equiv) is added in ten portions (~300 mg every three min) via the neck with the

Figure 1. Glassware assembly for the reduction step (picture obtained from submitters)

plastic stopper, and the internal temperature is maintained at –20 °C. After complete addition, the reaction is stirred for 45 min in the CH₃CN-dry ice bath at –20 °C, after which the reaction is allowed to warm up to 0 °C. The reaction is monitored by TLC in 25% EtOAc in hexanes using ninhydrin to stain (Note 4). After stirring for 35 min at 0 °C, TLC analysis shows disappearance of starting material. To the completed reaction, a mixture of ice-water (170 mL) is added to the solution with vigorous stirring and the solution is stirred for 10 min at 0 °C (Note 5). To this mixture EtOAc (900 mL) and H₂O (50 mL) are added and the resultant mixture is poured into a 2-L separatory funnel. Ethyl acetate (50 mL) and H₂O (50 mL) are used to rinse the flask. The aqueous layer is extracted with EtOAc (3 x 150 mL). The combined organic layers are washed with saturated aqueous NH₄Cl (100 mL) and saturated aqueous NaCl (100 mL). The organic layer is dried over Na₂SO₄ (20 g) and filtered by suction using a fritted funnel (9 cm diameter, medium porosity). Additional EtOAc (50 mL) is used to wash the Na₂SO₄ and the filtrate is concentrated by rotary evaporation into a 2-L round-bottomed flask (40 °C bath, 140–30 mmHg). The resulting pale yellow oil, containing ethyl *N-(tert-*butoxycarbonyl)-*N-((tert-*

butyldimethylsilyl)oxy)-L-homoserinate (i.e., **2**) is transferred to a 500-mL, single-necked round-bottomed flask using CH$_2$Cl$_2$, which is then evaporated (40 °C bath, 440–30 mmHg) on a rotary evaporator. The flask is equipped with a 4-cm oval Teflon-coated magnetic stir-bar, and the viscous oil stirred while being dried on the vacuum pump (0.15 mmHg, 24 °C) for 5 h to obtain a pale yellow oil **2** (14.7 g). The material is used without further purification (Notes 6 and 7).

The oil **2** (14.7 g, 38.9 mmol, 1.0 equiv) is diluted with CH$_2$Cl$_2$ (175 mL) and transferred, using a long-stemmed plastic funnel, into a 500-mL, three-necked, round-bottomed flask equipped with a 4-cm Teflon-coated magnetic stir-bar, a 125-mL addition funnel, a thermometer fitted with a glass adaptor, and a rubber septum through which an active nitrogen atmosphere is ensured (Figure 2). The flask from which **2** is transferred is

Figure 2. Glassware assembly for the mesylation step (picture obtained from submitters)

washed with CH$_2$Cl$_2$ (10 mL) to ensure that no product is left. The receiving flask is cooled to 4 °C using an ice-water bath. The addition funnel is charged with Et$_3$N (16.6 mL, 119 mmol, 3.0 equiv, via 20-mL disposable syringe), which is then added dropwise over 15 min maintaining the internal temperature to 3–4 °C. The addition funnel is washed with CH$_2$Cl$_2$ (5 mL) to ensure that no reagents are left. The same addition funnel is next charged with methanesulfonyl chloride (7.40 mL, 95.6 mmol, 2.40 equiv, via 10-mL disposable syringe) which is then added dropwise over approximately 20 min. The internal temperature is maintained at 4 °C through the entire course of the addition. The addition funnel is washed using CH₂Cl₂ (5 mL) to ensure that no reagents are left (final concentration of substrate is 0.2 M). The addition funnel is removed and the flask is equipped with a glass stopper. The reaction mixture is stirred for 15 min at 4 °C after which the ice-water bath is removed and the reaction is stirred at 24 °C for 1 h. The reaction is monitored by TLC in 40% EtOAc in hexanes using ninhydrin stain (Note 8). After stirring for 1 h at 24 °C, TLC analysis shows disappearance of starting material. To the completed reaction, saturated aqueous NH₄Cl (150 mL) is added and the mixture is poured into a 2-L separatory funnel. The flask is washed with CH₂Cl₂ (10 mL) to ensure that no reagents are left. The aqueous layer is separated and extracted with CH₂Cl₂ (4 x 150 mL). The combined organic layers are washed sequentially with saturated aqueous NH₄Cl (200 mL), saturated aqueous NaHCO₃ (200 mL) and saturated aqueous NaCl (200 mL). The organic layer is dried over Na₂SO₄ (20 g) and filtered by suction using a fritted funnel (9 cm diameter, medium porosity). Dichloromethane (25 mL) is used to wash the Na₂SO₄ and the filtrate is concentrated by rotary evaporation into a 2-L round-bottomed flask (40 °C bath, 440–30 mmHg). The resulting dark-orange oil, containing ethyl *N*-(*tert*-butoxycarbonyl)-*N*-((*tert*-butyldimethylsilyl)oxy)-*O*-(methylsulfonyl)-L-homoserinate (i.e., **3**) is transferred to a 1-L round-bottomed flask using CH₂Cl₂, which is then evaporated (40 °C bath, 440–30 mmHg) on a rotary evaporator. The residue is transferred to a pre-weighed 250-mL single-necked round-bottomed flask equipped with pre-weighed 4-cm oval Teflon-coated magnetic stir-bar, and dried while stirring on the vacuum pump (0.15 mmHg, 24 °C) for 6 h to afford a dark-orange oil **3** that is used in the next step with no further purification (17.7 g) (Note 9).

The compound **3** (17.7 g, 38.8 mmol, 1.0 equiv) is diluted with THF (750 mL) and transferred, using a long-stemmed plastic funnel, into a 1-L, three-necked, round-bottomed flask equipped with a 4-cm Teflon-coated magnetic stir-bar, a 125-mL addition funnel, a thermometer fitted with a glass adaptor, and a rubber septum through which an active nitrogen atmosphere is ensured. THF (15 mL) is used to rinse the flask and the remaining solution transferred into the reaction flask using a 10-mL pipette. The flask is cooled down to 4 °C using an ice-water bath (Figure 3). The addition funnel is charged with tetrabutylammonium fluoride (1 M in THF,

58.0 mL, 58.0 mmol, 1.50 equiv, via a 100-mL graduated cylinder), which is then added to the reaction flask dropwise over 90 min, maintaining the internal temperature to 3–4 °C. Tetrahydrofuran (10 mL) is used to ensure that no reagents are left on the side of the addition funnel (final concentration of substrate 0.05 M). The addition funnel and the

Figure 3. Glassware assembly for the cyclization step (picture obtained from submitters)

thermometer are removed and the flask is equipped with a glass stopper and a rubber septum. The reaction mixture is stirred for 1 h at 4 °C. The reaction is monitored by TLC in 40% EtOAc in hexanes using ninhydrin to stain (Notes 10 and 11). Upon completion of the reaction (as noted by disappearance of starting material (Notes 10 and 11), saturated aqueous NaHCO$_3$ (120 mL) is added to the reaction and stirred for 10 min. The resultant biphasic mixture is diluted with Et$_2$O (300 mL) and the mixture poured into a 2-L separatory funnel. The aqueous layer is separated and washed with Et$_2$O (3 x 150 mL). The combined organic layers are washed sequentially with saturated aqueous NaHCO$_3$ (150 mL) and saturated aqueous NaCl (150 mL). The organic layer is dried over Na$_2$SO$_4$ (30 g) and filtered by suction using a fritted funnel (9 cm diameter, medium porosity). Diethyl ether (200 mL) is used to wash the Na$_2$SO$_4$ and the filtrate is concentrated by rotary evaporation (40 °C bath, 500–30 mmHg). The resulting brown oil, containing (S)-2-*tert*-butyl 3-ethyl isoxazolidine-2,3-dicarboxylate (i.e., **4**) is transferred to a pre-weighed 250-mL round-bottomed flask using CH₂Cl₂, which is then evaporated (40 °C bath, 400–30 mmHg). The flask is equipped with a pre-weighed 4-cm oval Teflon-coated magnetic stirbar and dried while stirring on the vacuum pump (0.15 mmHg, 24 °C) for 10 h to provide a yellow oil (10.3 g). Column chromatography with 20% EtOAc in hexanes (Note 12) furnished **4** (6.35 g, 97.0 % purity, 64.9% based on compound **1**) as a clear yellow oil (Notes 13, 14, and 15).

B.   *(S)-2-(tert-Butoxycarbonyl)isoxazolidine-3-carboxylic acid* (**(S)-N-Boc-5-Oxaproline**) **(5)**. The oil **4** (6.35 g, 25.9 mmol, 1.0 equiv) is diluted with THF (95 mL) and transferred, using a long-stemmed plastic funnel, into a 500-mL, three-necked, round-bottomed flask equipped with a 4-cm Teflon-coated magnetic stir-bar, a 125-mL addition funnel, a thermometer fitted with a glass adaptor, and a nitrogen inlet. Tetrahydrofuran (10 mL) is used to rinse the flask and the remaining solution transferred into the reaction flask using a 10-mL pipette. The flask is cooled to 4 °C using an ice-water bath. Using a 250-mL graduated cylinder, a solution of aqueous 1 M LiOH (103 mL, 103 mmol, 4.0 equiv) (Note 16), which had been chilled to 4 °C in a refrigerator, is added to the addition funnel and is subsequently added to

Figure 4. Glassware assembly for the hydrolysis reaction- Step B (picture obtained from submitters)

the flask dropwise over 45 min while maintaining an internal temperature of 3–4 °C (Figure 4). After the addition is complete, the funnel is removed and the flask is equipped with a glass stopper. The reaction mixture is stirred for 1.5 h at 24 °C and is monitored by TLC in 5% MeOH in CH_2Cl_2 using ninhydrin as stain (Note 17). After 1.5 h at 24 °C, TLC analysis indicates disappearance of starting material, and chloroform (250 mL) is added. The solution is acidified to pH 3 with an aqueous solution of 2 M $KHSO_4$ (100 mL). The mixture is poured into a 1-L separatory funnel. The aqueous layer is separated and washed with $CHCl_3$ (5 x 100 mL). The combined organic layers are dried over Na_2SO_4 (20 g) and filtered by suction using a fritted funnel (9 cm diameter, medium porosity). Chloroform (100 mL) is used to wash the Na_2SO_4, and the filtrate is concentrated by rotary evaporation into a 1-L round-bottomed flask (40 °C bath, 210–30 mmHg). The resulting pale yellow oil, containing (S)-2-(*tert*-butoxycarbonyl)isoxazolidine-3-carboxylic acid (i.e., **5**) is transferred to a 250-mL round-bottomed flask using CH_2Cl_2 which is evaporated (40 °C bath, 450–30 mmHg). A pre-weighed 4 cm Teflon-coated oval stir-bar is added, and the oily reside is stirred and dried on the vacuum pump (0.15 mmHg, 24 °C) for 18 h to afford a clear yellow oil **5** that solidifies upon standing (5.53 g, 25.4 mmol, 98% yield, 94.1% ee) (Notes 18, 19, 20, and 21).

Notes

1. Prior to performing each reaction, a thorough hazard analysis and risk assessment should be carried out with regard to each chemical substance and experimental operation on the scale planned and in the context of the laboratory where the procedures will be carried out. Guidelines for carrying out risk assessments and for analyzing the hazards associated with chemicals can be found in references such as Chapter 4 of "Prudent Practices in the Laboratory" (The National Academies Press, Washington, D.C., 2011; the full text can be accessed free of charge at https://www.nap.edu/catalog/12654/prudent-practices-in-the-laboratory-handling-and-management-of-chemical).
    See also "Identifying and Evaluating Hazards in Research Laboratories" (American Chemical Society, 2015) which is available via the associated website "Hazard Assessment in Research Laboratories" at https://www.acs.org/content/acs/en/about/governance/committees

/chemicalsafety/hazard-assessment.html. In the case of this procedure, the risk assessment should include (but not necessarily be limited to) an evaluation of the potential hazards associated with methanol, sodium borohydride, acetonitrile, dry ice, ethyl acetate, hexanes, ninhydrin, ammonium chloride, sodium chloride, sodium sulfate, methylene chloride, triethylamine, sodium bicarbonate, methanesulfonyl chloride, tetrahydrofuran, tetrabutylammonium fluoride, diethyl ether, silica gel, lithium hydroxide, chloroform, and potassium bisulfate.

2. The protocol for preparing *(S)-ethyl 2-((tert-butoxycarbonyl)((tert-butyldimethylsilyl)oxy)amino)-4-oxobutanoate (1)* is described in *Org. Synth.* **2018**, *95*, 142-156.

3. The following reagents and solvents are used as received: The submitters purchased methanol from Sigma-Aldrich (chromasolv, ≥99.9%), while the checkers purchased methanol (99.9%) from Fisher Scientific. The submitters and checkers purchased the following chemicals: chloroform from Sigma-Aldrich, (chromasolv, ≥99.8%), diethyl ether (Sigma-Aldrich, ≥99.8%), dimethylformamide (Sigma-Aldrich, ≥99.8%), isopropyl alcohol (Sigma-Aldrich, ≥99.5%), triethylamine (Sigma-Aldrich, ≥99.5%), tetrabutylammonium fluoride (Sigma-Aldrich, 1.0 M in THF), sodium borohydride (Sigma-Aldrich, 98%), *N,N'*-diisopropylcarbodiimide (Sigma-Aldrich, 99%), 1-hydroxybenzotriazole hydrate (Sigma-Aldrich, ≥97%), sodium sulfate (Sigma-Aldrich, ≥99%), and lithium hydroxide monohydrate (Sigma-Aldrich, ≥98.5%). The submitters purchased tetrahydrofuran from Merck (for analysis, EMSURE), and the checkers purchased tetrahydrofuran from Fisher Scientific, (HPLC, >99.9%). The submitters purchased dichloromethane from Sigma-Aldrich (chromasolv, ≥99.5%) and the checkers purchased dichloromethane from Fisher Scientific (HPLC <99.9%). The submitters purchased methanesulfonyl chloride from Acros Organics (99.5%), and the checkers purchased from methanesulfonyl chloride Sigma-Aldrich (>99.7%). Potassium bisulfate (Fluka, 98%) was purchased by the submitters, while the checkers purchased the material from Oakwood Chemical (99%). The submitters used silica gel purchased from Fluka (high purity grade, pore size 60 Å, 230-400 mesh particle size) and the checkers purchased silica from EMD Millipore (60 Å, 230-400 mesh). Glass-backed extra hard layer TLC plates (60 Å (250 μm thickness containing F-254 indicator) were purchased by the submitters from Silicycle, and the checkers purchased TLC plates from EMD Millipore. The submitters purchased

Chloroform-d from Arma (99.8 atom% D), and the checkers purchased the material from Sigma-Aldrich (98.8% atom% D).1,3,5-Trimethoxybenzene (ABCR, 99%) was purchased by the submitters, while the checkers purchased the material from Sigma-Aldrich (>99 %). Deionized water is used throughout. The following salts are used as saturated aqueous solutions made by dissolving the salt in H_2O until saturation is reached: $NaHCO_3$ (Sigma-Aldrich, -40 +140 mesh, Na_2CO_3 2-5%), NaCl (ABCR, 99%), and NH_4Cl (Panreac Applichem, 99.5%).

4. TLC of the crude alcohol **2** is monitored in 25% EtOAc in hexanes (stain with ninhydrin). (Product **2** has R_f = 0.33 and stains violet; starting material **1** has R_f = 0.65 and stains yellow). TLC data obtained from submitters.

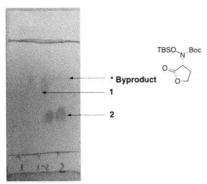

Byproduct (*) corresponds to the lactone. The ratio of byproduct to desired alcohol will be higher when the reaction is allowed to warm to room temperature for a long time or too much sodium borohydride is added at once.

5. Do not initiate the work-up procedure by evaporation of the MeOH from the reaction, since this reduces the yield and purity of the final compound. The work-up procedure described above should be followed.

6. Reactions 1, 2 and 3 in Step A should be done in a timely fashion, as intermediates are not very stable and will start to decompose over a couple days. Checkers consistently obtained a yield of <45 % when Step A was done over the period of 5 days. In particular, crude mesylate **3** should be taken to the next step immediately upon concentration. The checkers observed significant darkening of the mesylate (**3**) on stirring under high vacuum over the course of 3 h.

7. The identify of product **2** was determined, as follows. ^{1}H NMR
    (400 MHz, CDCl$_3$) δ: 0.17–0.20 (m, 6 H), 0.92 (s, 9 H), 1.27 (t, J = 7.1 Hz,
    3 H), 1.48 (s, 9 H), 2.09–2.28 (m, 2 H), 2.32 (t, J = 6.1 Hz, 1 H), 3.72–3.78
    (m, 2 H), 4.11–4.30 (m, 2 H), 4.45 (dd, J = 8.9, 5.7 Hz, 1 H); ^{13}C NMR
    (101 MHz, CDCl$_3$) δ: −4.9, −4.7, 14.1, 17.9, 25.8, 28.1, 31.4, 59.7, 61.2, 63.2,
    82.3, 157.8, 170.1. HRMS (ESI) calcd. for C$_{17}$H$_{36}$NO$_6$Si [M+Na]$^+$ 400.2131,
    found 400.2136.

8. The mesylation reaction is monitored by TLC with 40% EtOAc in
    hexanes (stain with ninhydrin). Product **3** has R_f = 0.65 and starting
    material **2** has R_f = 0.57. TLC data obtained from submitters.

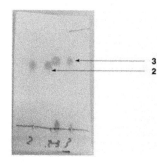

9. The identity of product **3** is confirmed by the following characterization
    data. ^{1}H NMR (400 MHz, CDCl$_3$) δ: 0.13–0.20 (m, 6 H), 0.92 (s, 9 H),
    1.27 (t, J = 7.1, 1.0 Hz, 3 H), 1.48 (s, 9 H), 2.17–2.35 (m, 1H), 2.35–2.51 (m,
    1 H), 3.01 (s, 3 H), 4.11–4.30 (m, 1H), 4.30–4.48 (m, 2 H), 4.48–4.56 (m,
    1 H). ^{13}C NMR (101 MHz, CDCl$_3$) δ: 14.9, 28.3, 33.1, 59.6, 61.8, 68.5, 82.7,
    156.0, 170.8. HRMS (ESI) calcd. for C$_{18}$H$_{37}$NO$_8$SSi [M+Na]$^+$ 478.1907,
    found 478.1915.

10. The cyclization reaction is monitored by TLC with 40% EtOAc in
    hexanes (stain with ninhydrin). Product **4** has R_f = 0.45 and stains
    yellow and starting material **3** has R_f = 0.65 and stains brown). TLC data
    obtained from submitters.

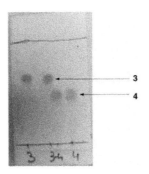

<table>
<tr><td></td><td>← 3</td></tr>
<tr><td></td><td>← 4</td></tr>
</table>

11. The checkers observed incomplete cyclization (ca. 50% conversion of **3**) after the addition of 1.5 equivalents TBAF solution. Dropwise addition of an additional 0.25 equivalents TBAF (1.0 M solution in THF, 10 mL) over the course of 5 minutes led to complete conversion **3** within 10 min; the reaction was quenched and worked up at this point.

12. The column (8 x 30 cm) was packed with 450 g of silica gel. The silica gel was loaded in 20% EtOAc in hexanes (~1 L). The crude material was dissolved in 10 mL of the eluent and loaded onto the silica gel. The flask is washed with 10 mL eluent in order to ensure that no product remains on the side of the flask, and the eluent added to the column. Sand (300 g) (~1.5 cm) is carefully added to the top of the column. No pressure is applied. Elution is performed with 20% EtOAc in hexanes and fractions collected in 50-mL tubes after one column volume had eluted. The desired product was obtained in fractions 30–53. The fractions containing the desired product were concentrated by rotary evaporation (35 °C bath, 340–8 mmHg).

13. ^{1}H-NMR, ^{13}C-NMR and HRMS confirm the purity of product **4** and match literature values.[1] ^{1}H NMR (400 MHz, Chloroform-d): δ: 4.68 (dd, J = 9.4, 4.8 Hz, 1 H), 4.23 (q, J = 7.1 Hz, 2 H), 4.18 – 4.07 (m, 1 H), 3.91 – 3.75 (m, 1 H), 2.71 – 2.53 (m, 1 H), 2.55 – 2.39 (m, 1 H), 1.50 (s, 9 H), 1.30 (t, J = 7.1 Hz, 3 H). ^{13}C NMR (151 MHz, Chloroform-d): δ: 170.8, 155.9, 82.7, 68.4, 61.8, 59.6, 33.1, 28.3, 14.2. HRMS (ESI) calcd. for $C_{11}H_{20}NO_5$ [M+Na]$^+$ 268.1161, found 238.1159.

14. The purity of the compound was calculated by qNMR with a delay of relaxation of 30 seconds, using 17.1 mg of 1,3,5-trimethoxybenzene (purity ≥99%) and 21.9 mg of the compound **4**.

15. A second run performed at the same scale provided 6.60 g (67%) of the identical product (**4**).

16. The LiOH solution (1 M) is prepared by dissolving 42 g LiOH monohydrate in 1000 mL deionized water.

17. TLC of the hydrolysis reaction is monitored in 5% MeOH in CH₂Cl₂ (stain with ninhydrin). TLC data obtained from submitters.

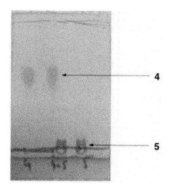

18. The identify of product **5** was characterized with the following data, which matched the literature values.[2] ^{1}H NMR (600 MHz, CDCl₃) δ: 1.44 (s, 9H), 2.43 – 2.54 (m, 1H), 2.54–2.66 (m, 1H), 3.67–3.91 (m, 1H), 3.98 – 4.20 (m, 1H), 4.67 (dd, *J* = 9.5, 4.9 Hz, 1H), 10.53 (s, 1H). ^{13}C NMR (151 MHz, CDCl₃) δ: 28.2, 32.9, 59.6, 68.6, 83.5, 156.2, 174.9. HRMS (ESI) calcd. for C₉H₁₅NO₅ [M+Na]⁺ 240.0848, found 240.0849. The purity of the compound was calculated by qNMR with a delay of relaxation of 30 sec, using 10.40 mg of 1,3,5-trimethoxybenzene (purity ≥99%) and 14.41 mg of the compound **5**.

19. A second run performed at the same scale provided 5.26 g (94%) of the identical product (**5**).

20.

p-bromo aniline
DIC, HOBt
DMF

5   **5'**

Derivatization of (S)-N-Boc-5-Oxaproline

For the determination of the enantiomeric excess of the final compound a derivatization is required for a more accurate result. To *(S)*-N-Boc-5-oxaproline **5** (10.0 mg, 46 μmol, 1.0 equiv) in DMF (200 μL) are added

N,N'-diisopropylcarbodiimide (7.2 µL, 47 µmol, 1.0 equiv) and 1-hydroxybenzotriazole hydrate (7.1 mg, 46 µmol, 1.0 equiv) and the solution is stirred for 2 min. para-Bromoaniline (8.4 mg, 49 µmol, 1.1 equiv) is added to this mixture in one portion, and the reaction is stirred for 2 h at room temperature. The mixture is diluted with CH_2Cl_2 (3 mL) and washed with H_2O (3 mL), brine (3 mL), dried over Na_2SO_4, filtered and concentrated. The solution is loaded on silica and the product is isolated by column chromatography using a gradient of 10–40% EtOAc in hexanes (8.7 mg, 51% yield, 94.1% ee). ^1H NMR (600 MHz, $CDCl_3$) δ: 1.54 (s, 9H), 2.59–2.75 (m, 2H), 3.73–3.91 (m, 1H), 4.07–4.23 (m, 1H), 4.77–4.89 (m, 1H), 7.39–7.51 (m, 4H), 8.49 (s, 1H); ^{13}C **NMR** (151 MHz, $CDCl_3$) δ: 28.2, 32.5, 62.8, 69.5, 84.2, 117.3, 121.5, 132.1, 136.4, 157.8, 168.5; **HRMS (ESI)** calcd. for $C_{15}H_{20}BrNO_4$ [M+Na]$^+$ 393.0426, found 393.0429. In order to prepare a racemic sample of **5′**, racemic **1** was prepared as described in *Org. Synth.* **2018**, *95*, 142-156 and taken through an analogous procedure for that used to prepare (*S*)-**5′** from (*S*)-**1**.

21. Enantiomeric excess of the enantiomeric amides (94%) was determined by chiral HPLC. Separation was performed by HPLC on a Chiralcel IA column using hexanes/isopropyl alcohol (8:2), 25 °C, with a flow rate of 1.0 mL/min, while monitoring at 210 nm. Retention time (t_R) of the minor enantiomer = 8.58 min, and retention time (t_R) of the major enantiomer = 15.39 min.

Working with Hazardous Chemicals

The procedures in *Organic Syntheses* are intended for use only by persons with proper training in experimental organic chemistry. All hazardous materials should be handled using the standard procedures for work with chemicals described in references such as "Prudent Practices in the Laboratory" (The National Academies Press, Washington, D.C., 2011; the full text can be accessed free of charge at http://www.nap.edu/catalog.php?record_id=12654). All chemical waste should be disposed of in accordance with local regulations. For general guidelines for the management of chemical waste, see Chapter 8 of Prudent Practices.

In some articles in *Organic Syntheses*, chemical-specific hazards are highlighted in red "Caution Notes" within a procedure. It is important to recognize that the absence of a caution note does not imply that no significant hazards are associated with the chemicals involved in that procedure. Prior to performing a reaction, a thorough risk assessment should be carried out that includes a review of the potential hazards associated with each chemical and experimental operation on the scale that is planned for the procedure. Guidelines for carrying out a risk assessment and for analyzing the hazards associated with chemicals can be found in Chapter 4 of Prudent Practices.

The procedures described in *Organic Syntheses* are provided as published and are conducted at one's own risk. *Organic Syntheses, Inc.,* its Editors, and its Board of Directors do not warrant or guarantee the safety of individuals using these procedures and hereby disclaim any liability for any injuries or damages claimed to have resulted from or related in any way to the procedures herein.

Discussion

As part of an effort to contribute to the field of chemical ligation[2] we reported in 2012 the α–ketoacid–hydroxylamine ligation (KAHA ligation) with (S)-N-Boc-5-oxaproline. This reaction makes possible the chemical synthesis of proteins from unprotected peptide segments.[3] KAHA ligation with 5-oxaproline leads to a depsipeptide ester which rearranges to the corresponding amide in basic buffers generating a homoserine residue at the ligation site (Figure 1a).[4] The utility of the KAHA ligation has been illustrated in the synthesis of Pup, CspA, UFM1, SUMO2, SUMO3, betatrophin, irisin and IFITM3 proteins[3,5] and can be considered as a complementary method to native chemical ligation (NCL).[6]

Unlike NCL, the KAHA ligation does not use an N-terminal cysteine residue and C-terminal thioester or thioester surrogate.[5] (S)-N-Boc-5-oxaproline presents high stability to acidic cleavage conditions of the peptide from resin and reactivity to the α–ketoacid functionality. It is manually coupled to the resin containing the peptide fragment using standard coupling conditions and does not affect any aspect of the peptide synthesis or purification (Figure 1b).

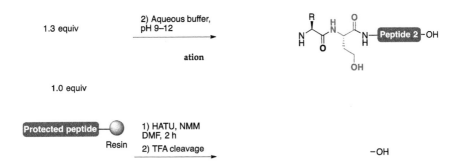

Scheme 1. (a) General description of the KAHA ligation. (b) Coupling of (S)-N-Boc-5-oxaproline on protected peptide segment followed by cleavage of peptide fragment from resin.

It is important to note that this building block can be readily converted to orthogonal (S)-N-protected-5-oxaprolines through the free-hydroxylamine intermediate **6** (Figure 2). This is significant for the use of sequential KAHA ligations for the synthesis of small and medium-size proteins.[3,5]

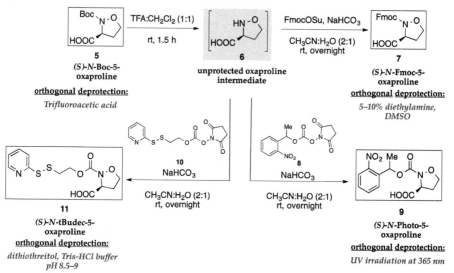

Scheme 2. Conversion from (S)-N-Boc-5-oxaproline to orthogonal protected oxaprolines.

Our original approach to the synthesis of (S)-N-Boc-5-oxaproline was based on a modified procedure of Vasella et al[8] and utilized a [3+2] cycloaddition with ethylene in a pressurized reactor. Although the protocol is well established it requires an expensive chiral auxiliary and affords a 6:4 diastereoselectivity ratio after the cycloaddition reaction. The two diastereoisomers can be easily separated by two recrystallizations but with relatively poor recovery.[3] Importantly, the requirement of using a pressurized reactor created a major bottleneck for preparing this building block and using KAHA ligation as an alternative for the chemical synthesis of peptides or proteins.

This encouraged us to enable an efficient, economical route to enantiopure (S)-N-Boc-5-oxaproline. We describe here a scalable and practical route to the synthesis of (S)-N-Boc-5-oxaproline. The route begins with the sodium borohydride reduction of (S)-ethyl 2-((tert-butoxycarbonyl)((tert-butyldimethylsilyl)oxy)amino)-4-oxobutanoate[9] **1** to obtain alcohol **2**. Following the mesylation of the alcohol **2** and the one-pot TBS deprotection-cyclization we subsequently formed the cyclic hydroxylamine **4**. Hydrolysis of the ethyl ester afforded the (S)-N-Boc-5-oxaproline building block **5** with 95% ee and good overall yield.

The applicability of the newly developed route to the (S)-N-Boc-5-oxaproline has been showcased in the synthesis of SUMO2, SUMO3, betatrophin and irisin proteins.[5] In summary we have developed a scalable and practical multistep synthesis of the oxaproline building block.

References

1. Eidgenössiche Technische Hochschule Zürich, HCI F317, 8093 Zürich, Switzerland. Email: bode@org.chem.ethz.ch

2. Harmand, T. J.; Murar, C. E.; Bode, J. W. *Curr. Opin. Chem. Biol.* **2014**, *22*, 115–121.

3. (a) Pattabiraman, V. R.; Ogunkoya, A. O.; Bode, J. W. *Angew. Chem., Int. Ed.* **2012**, *51*, 5114–5118. (b) Ogunkoya, A. O.; Pattabiraman, V. R.; Bode, J. W. *Angew. Chem., Int. Ed.* **2012**, *51*, 9693–9697.

4. Wucherpfennig, T. G.; Rohrbacher, F.; Pattabiraman, V. R.; Bode, J. W. *Angew. Chem., Int. Ed.* **2014**, *53*, 12244–12247.

5. (a) Wucherpfennig, T. G.; Pattabiraman, V. R.; Limberg, F. R.; Ruiz-Rodriguez, J.; Bode, J. W. *Angew. Chem., Int. Ed.* **2014**, *53*, 12248–52. (b) Harmand, T. J.; Murar, C. E.; Bode, J. W. *Nature Protocols* **2016**, *11*, 1130–1147. (c) Wucherpfennig, T. G.; Müller, S; Wolfrum, C.; Bode, J. W.; *Helv. Chim. Acta* **2016**, *99*, 897–907. (d) Harmand, T. J.; Pattabiraman, V. R.; Bode, J. W.; *Angew. Chem., Int. Ed.* **2017**, *56*, 12639–12643.

6. (a) Dawson, P. E.; Muir, T. W.; Clark-Lewis, I.; Kent, S. B. *Science* **1994**, *266*, 776–779. (b) Blanco-Canosa, J. B.; Dawson, P. E. *Angew. Chem., Int. Ed.* **2008**, *47*, 6851–6855.

7. Thuaud, F.; Rohrbacher, F.; Zwicky, A.; Bode, J. W. *Helv. Chim. Acta* **2016**, *99*, 868–894.

8. Vasella, A.; Voeffray, R. *J. Chem. Soc. Chem. Commun.* **1981**, 97–98.

9. Thibault, J. H.; Murar, C. E.; Takano, H.; Bode, J.W. *Org. Synth.* **2018**, *95*, 142-156.

Appendix
Chemical Abstracts Nomenclature (Registry Number)

Sodium borohydride; (16940-66-2)
Methanesulfonyl chloride; (124-63-0)

Organic
Syntheses

Yu-Wen Huang was born in Hsinchu, Taiwan (R.O.C.) in 1982. He received his bachelor's degree from National Cheng Kung University in 2005. He then joined the M.S. program at the National Tsing Hua University working under the supervision of Professor Shang-Cheng Hung on carbohydrate synthesis. In 2016, he received his Ph.D. degree from the University of Rochester where he worked with Professor Alison J. Frontier on 1,6-conjugate addition initiated Nazarov reactions and sequential 1,5-hydride transfer chemistry. He is currently a post-doctoral fellow in the CPRIT lab (Baylor University) with Professor John L. Wood working on the total synthesis of natural products.

Syntheses of Substituted 2-Cyano-benzothiazoles

Hendryk Würfel* and Dörthe Jakobi

Institute of Organic Chemistry and Macromolekular Chemistry, Friedrich-Schiller University, 07743 Jena, Germany

Caitlin Lacker, Travis J. DeLano, and Sarah Reisman

Procedure (Note 1)

A. *2-Thioxo-2-(p-tolylamino)acetamide* (**3**). A 500-mL, three-neck, round-bottomed flask is equipped with a 4-cm egg-shaped teflon-coated magnetic stir bar, a rubber septum with a thermometer inserted through it, and a stopper, leaving the central neck open (Figure 1). To the flask are added dimethylformamide (DMF) (250 mL) (Note 2), *p*-toluidine (30.3 g, 280 mmol, 1 equiv), sulfur (18.0 g, 560 mmol, 2 equiv) and triethylamine (NEt₃) (120 mL, 861 mmol, 3 equiv) (Note 3), and the contents of the flask are stirred at 700 rpm at 23 °C. Approximately 1 g portions of

Published on the Web 7/13/2018
© 2018 Organic Syntheses, Inc.

2-chloroacetamide (31.5 g, 340 mmol, 1.2 equiv) (Note 4) are added every two minutes over the course of an hour (Note 5) (Figure 2). The temperature of the reaction slowly increases upon addition of 2-chloroacetamide, and after approximately 16 min the internal temperature reaches 25 °C. The flask is then cooled using a 10 °C water bath, and the internal temperature of the reaction is maintained between 16–19 °C by occasional addition of ice to the water bath. Halfway through the addition, the stirring rate is increased to 850 rpm to account for the increasing viscosity of the reaction mixture. After complete addition, the vessel is protected against moisture with a drying tube filled with calcium chloride and is left to stir at 23 °C for an additional 12 h (Note 6).

Figure 1. Reaction setup for step A

After that time, the reaction mixture is poured into 2.5 L deionized water acidified with 30 mL of concentrated HCl. The resulting precipitate is vacuum filtered using a medium-porosity 15 cm diameter glass frit. The crude material (**3**) is dissolved in 2.8 M NaOH in deionized water (150 mL) and then vacuum filtered using a 6 cm diameter Büchner funnel lined with filter paper (grade 413) to obtain a yellow solution of the thioamide (**3**). The

Figure 2. a: Before 2-chloroacetamide addition. b-e: Reaction darkens as 2-chloroacetamide is added. f: Drying tube attached to stir 12 h

basic solution is added dropwise to 1 L of deionized water acidified with concentrated HCl (30 mL). The product that precipitates is collected by vacuum filtration with a medium-porosity 7 cm diameter glass frit and washed with deionized water (2 x 550 mL). The product is dissolved in 300 mL isopropyl alcohol heated to 75–80 °C using a heat gun (Note 7), then allowed to recrystallize at 23 °C for 18 h. The resulting yellow crystals are collected via vacuum filtration through a medium-porosity 7 cm diameter glass frit and dried at 60 °C at 0.15 mmHg for 24 h to provide 17.5 g (32%) of compound **3** (Notes 8, 9, and 10) (Figure 3).

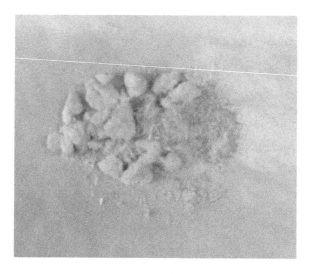

Figure 3. 2-Thioxo-2-(*p*-tolylamino)acetamide (3) recrystallized from isopropyl alcohol

B. *6-Methylbenzo[d]thiazole-2-carboxamide (4)*. Potassium ferricyanide (141 g, 429 mmol, 4.82 equiv) is dissolved in 1.2 L deionized water in a 2 L Erlenmeyer flask equipped with a 7.5 cm Teflon-coated stir bar and allowed to stir at 23 °C. Compound (**3**) (17.3 g, 89 mmol, 1 equiv) (Note 11) is dissolved in 3 M NaOH (100 mL, prepared with deionized water) by heating to 30–40 °C with a heat gun, and the warm solution is transferred to a 125 mL addition funnel. The solution is then added dropwise to the stirring solution of potassium ferricyanide (500 rpm) over 1 h (Note 12). The addition should be performed under a good working fume hood (Note 13). After a small amount of compound (**3**) is added the product (**4**) begins to precipitate. Over the course of the addition the reaction changes color from orange to green to yellow (Figure 4). After the complete addition of the

Figure 4. Reaction setup (a,b) and progress (c,d) for step B

thioamide, the stirring is continued for 30 min. The precipitate is collected by vacuum filtration on a medium-porosity 7 cm diameter glass frit, and the solid is washed with distilled water until the filtrate is clear and colorless (2 L). The product (**4**) is then dried for 24 h at 60 °C at 0.15 mmHg. The dried material is dissolved in DMF (Note 2) (30 mL DMF per 5 g), heating to 110–120 °C with a heat gun (Note 7). After cooling to 23 °C, the flask is placed in a refrigerator (5 °C) and allowed to recrystallize for 15 h. The product is collected by vacuum filtration through a medium-porosity 2 cm diameter glass filter frit, washed with diethyl ether (3 x 20 mL), and dried for 24 h at 80 °C at 0.15 mmHg to provide 10.3 g (60 %) of a crystalline solid (Notes 14, 15, and 16) (Figure 5).

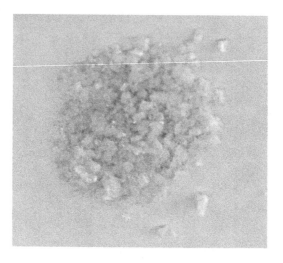

Figure 5. 6-Methylbenzo[d]thiazole-2-carboxamide recrystallized from DMF

C. *6-Methylbenzo[d]thiazole-2-carbonitrile (5).* In a 250-mL, three-necked, round-bottomed flask equipped with a 2-cm Teflon-coated magnetic stir bar, a septum, and a thermometer, dry DMF (100 mL) (Note 17) and compound **4** (6.0 g, 31.2 mmol, 1 equiv) are added, and the mixture is cooled to 0 °C in an ice-water bath. Over the course of 12–14 min phosphorous (V) oxychloride (4.8 mL, 51 mmol, 1.65 equiv) (Note 18) is added dropwise via syringe, while maintaining an internal temperature no greater than 5 °C. After the addition is complete, the mixture is removed from the cooling bath and allowed to warm to 23 °C (Figure 6). During this period the starting material reacts leading to a clear yellow solution from which TLC samples are taken to monitor reaction progress (Note 19). After about 1 h the reaction is complete as judged by disappearance of starting material by TLC. The reaction is poured into deionized water (500 mL). The resulting white precipitate is filtered through a 4.5 cm diameter medium porosity fritted filter and washed with deionized water until the filtrate is neutral (500–750 mL). The product is dried in a vacuum desiccator for 24 h

Figure 6. a: compound 4 dissolved in DMF. b: ice bath for phosphorous (V) oxychloride addition. c,d: color change as reaction progresses

at 0.15 mmHg over calcium chloride. The crude material is dissolved in *n*-heptane (80 mL) heated to 70–80 °C using a heat gun (Note 7). A small amount of brown residue remains undissolved, and is removed by vacuum filtration using a 6 cm diameter Büchner funnel lined with filter paper (grade 413). The filtrate is re-heated to dissolve any material that prematurely precipitated, and then the product is allowed to recrystallize for 18 h at 23 °C. The product is collected via vacuum filtration through a

4.5 cm diameter medium porosity fritted filter to obtain off-white crystals (3.6 g, 66%) (Notes 20, 21, and 22). The overall yield starting from 30 g of compound **1** ranges from 13 - 15%.

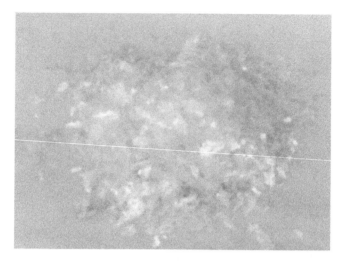

Figure 7. 6-Methylbenzo[d]thiazole-2-carbonitrile

Notes

1. Prior to performing each reaction, a thorough hazard analysis and risk assessment should be carried out with regard to each chemical substance and experimental operation on the scale planned and in the context of the laboratory where the procedures will be carried out. Guidelines for carrying out risk assessments and for analyzing the hazards associated with chemicals can be found in references such as Chapter 4 of "Prudent Practices in the Laboratory" (The National Academies Press, Washington, D.C., 2011; the full text can be accessed free of charge at https://www.nap.edu/catalog/12654/prudent-practices-in-the-laboratory-handling-and-management-of-chemical).
    See also "Identifying and Evaluating Hazards in Research Laboratories" (American Chemical Society, 2015) which is available via the associated website "Hazard Assessment in Research Laboratories" at https://www.acs.org/content/acs/en/about/governance/committees

/chemicalsafety/hazard-assessment.html. In the case of this procedure, the risk assessment should include (but not necessarily be limited to) an evaluation of the potential hazards associated with, as well as the proper procedures for dimethylformamide, *p*-toluidine, triethylamine, sulfur, 2-chloroacetamide, calcium chloride, concentrated hydrochloric acid, sodium hydroxide, isopropyl alcohol, potassium ferricyanide, phosphorous (V) oxychloride, and *n*-heptane.

2. Dimethylformamide (99.5 %) was obtained from TCI Deutschland GmbH and used as received. The checkers obtained 99.5% DMF from TCI America.

3. *p*-Toluidine (99 %), sulfur (99 %) and triethylamine (99 %) were obtained from ABCR-GmbH Germany and used as received. The checkers obtained *p*-toluidine (99 %) and triethylamine (99 %) from Aldrich and sulfur from Strem (≥99 %).

4. 2-Chloroacetamide (98.5 %) was purchased from Sigma Aldrich Germany and used as received. The checkers obtained 2-chloroacetamide from Aldrich (≥98 %).

5. During each addition the brown slurry becomes partially red; an exothermic reaction occurs over time. If the addition is carried out too rapidly, temperatures of > 60 °C could be observed, leading to a remarkable decrease in yield.

6. The drying tube should not produce a closed system.

7. Care should be taken to remove any extraneous flammable solvent from proximity of the heat gun.

8. Melting point: 170–173 °C (isopropyl alcohol); IR (thin film): 3385, 3229, 3166, 1700, 1548, 1532, 1403, 1386, 1299, 1176, 1104, 1045, 822, 782, 768, 738, 652 cm^{-1}. ^{1}H NMR (400 MHz, DMSO-d6) δ: 2.31 (s, 3H), 7.19–7.28 (m, 2H), 7.79–7.89 (m, 2H), 8.13 (d, J = 19.6 Hz, 2H), 12.04 (s, 1H). ^{13}C NMR (101 MHz, DMSO-d6) δ: 20.8, 38.9, 39.1, 39.3, 39.5, 39.7, 39.9, 40.2, 123.2, 129.0, 136.0, 136.2, 162.4, 185.7. HRMS [M + H] calcd for $C_9H_{11}N_2OS$: 195.0592. Found: 195.0590. TLC: R_f = 0.63 in 50% *n*-heptane-ethylacetate (1:1 *n*-heptane:ethylacetate) solvent system.

9. Purity of the product was assessed as >98% by Q NMR using ethylene carbonate as the internal standard.

10. A second reaction performed on equivalent scale provided 18.0 g (33%) of the identical product.

11. Potassium hexacyanoferrate (III) (98 %) was purchased from Applichem GmbH and used without further purification. The checkers obtained potassium hexacyanoferrate (III) from Alfa Aesar (≥98 %).

12. The solution in the addition funnel cools to 23 °C over the course of the addition. Gentle heating of the joint of the addition funnel may be necessary to break up clogs.

13. During the addition, toxic vapors evolve.

14. Melting point 252–254 °C (DMF); IR (ATR): 3305, 3167, 1689, 1651, 1615, 1498, 1393, 1141, 1111, 1079, 1049, 810, 699 cm^{-1}. ^{1}H NMR (400 MHz, DMSO-d6) δ: 2.47 (s, 3H), 7.35–7.51 (m, 1H) 7.92–8.11 (m, 3H), 8.44 (s, 1H). ^{13}C NMR (101 MHz, DMSO-d6) δ: 21.2, 38.9, 39.1, 39.3, 39.5, 39.7, 39.9, 40.2, 122.4, 123.6, 128.66, 136.6, 136.9, 151.0, 161.4, 163.8. [M + H] calcd for $C_9H_9N_2OS$: 193.0436. Found: 193.0427. TLC R$_f$ = 0.54 in 50% n-heptane-ethyl acetate (1:1 n-heptane:ethyl acetate) solvent system.

15. Purity of the product was assessed as >97% by Q NMR using ethylene carbonate as the internal standard.

16. A second reaction performed on equivalent scale provided 11.5 g (67%) of the identical product. The submitters reported yields of 7.1–7.4 g (38–40%) of material with mp = 254–255 °C.

17. Dry dimethylformamide (99.8 %) extra dry over molecular sieves were purchased from Acros Organics and used as received.

18. Phosphorous (V) oxychloride (99 %) was purchased from Fisher Scientific GmbH and used as received. The checkers obtained phosphorous (V) oxychloride from Alfa Aesar (99 %).

19. The TLC is run with n-heptane-ethyl acetate (1:1) on silica gel 60 coated (0.2 mm) aluminum plates with fluorescence indicator UV$_{254}$. R$_f$ of compound **5** = 0.78

20. Melting point: 86–88 °C (n-heptane); IR (thin film): 3417, 2943, 2228, 1642, 1453, 1316, 1244, 1133, 816 cm^{-1}. ^{1}H NMR (400 MHz, CDCl$_3$) δ: 2.52–2.59 (m, 3H), 7.46 (ddd, J = 8.4, 1.6, 0.5 Hz, 1H), 7.75 (dq, J = 1.6, 0.8 Hz, 1H), 8.09 (d, J = 8.5 Hz, 1H). ^{13}C NMR (101 MHz, CDCl$_3$) δ: 22.0, 76.8, 77.2, 77.5, 113.3, 121.4, 124.8, 129.9, 135.4, 135.8, 139.7, 150.6. HRMS [M + H] calcd for $C_9H_7N_2O$: 175.0330. Found: 175.0335.

21. Purity of the product was assessed as >98% by Q NMR using ethylene carbonate as the internal standard.

22. A second reaction performed on equivalent scale provided 3.6 g (66%) of the identical product.

Working with Hazardous Chemicals

The procedures in *Organic Syntheses* are intended for use only by persons with proper training in experimental organic chemistry. All hazardous materials should be handled using the standard procedures for work with chemicals described in references such as "Prudent Practices in the Laboratory" (The National Academies Press, Washington, D.C., 2011; the full text can be accessed free of charge at http://www.nap.edu/catalog.php?record_id=12654). All chemical waste should be disposed of in accordance with local regulations. For general guidelines for the management of chemical waste, see Chapter 8 of Prudent Practices.

In some articles in *Organic Syntheses*, chemical-specific hazards are highlighted in red "Caution Notes" within a procedure. It is important to recognize that the absence of a caution note does not imply that no significant hazards are associated with the chemicals involved in that procedure. Prior to performing a reaction, a thorough risk assessment should be carried out that includes a review of the potential hazards associated with each chemical and experimental operation on the scale that is planned for the procedure. Guidelines for carrying out a risk assessment and for analyzing the hazards associated with chemicals can be found in Chapter 4 of Prudent Practices.

The procedures described in *Organic Syntheses* are provided as published and are conducted at one's own risk. *Organic Syntheses, Inc.*, its Editors, and its Board of Directors do not warrant or guarantee the safety of individuals using these procedures and hereby disclaim any liability for any injuries or damages claimed to have resulted from or related in any way to the procedures herein.

Discussion

Thioamides are a very distinguished group of chemical compounds with a broad spectrum of applications,[2] including in tuberculosis medication.[3] Furthermore the compounds are used for synthesis of heterocyclic compounds – especially thiazoles and benzothiazoles[4a,b] – and have been applied as building blocks for Firefly Luciferin precursors.[5a,b]

A drawback of this compound class is its typical preparation route, which makes use of hazardous compounds such as phosphorus pentasulfide and hydrogen sulfide.[6] An efficient and selective method for the preparation of thioamides is the aminolysis of dithiocarboxylates, which can be synthesized *in situ* by oxidation of CH-acidic compounds such as 2-chloroacetamide or α–halo acetophenones at 23 °C.[7] This "mild thiolation" method utilizes elemental sulfur in an aprotic polar solvent such as DMF and a tertiary amine base, thereby avoiding harsher conditions.[8] Into a mixture of sulfur, base and solvent, the CH-acidic compound is added slowly, forming the dithiocarboxylate, which can be further transformed into dithiocarboxylic acid esters or thioamides (Scheme 1).

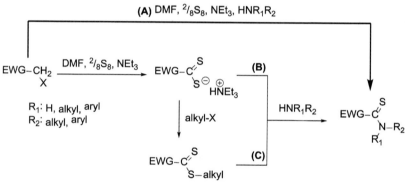

Scheme 1. Mechanism of the thioamide formation. (A) one-pot synthesis as used in the described procedure above. (B) Step wise protocol. (C) Step wise protocol for the thioamide formation employing amines with low nucleophilicity.

The amine employed for the thioamide synthesis can already be present at the beginning of the oxidation reaction or be added later (Scheme 1, A and B). The simple isolation of the reaction product is also a benefit in the "mild thiolation" strategy. The reaction mixture is poured into an excess of water and the product is filtered off. Due to the crystalline nature of the thioamide, recrystallization provides an efficient and simple method to purify the product. The synthetic scope of this reaction procedure was reviewed earlier.[9] The second step, a Jacobsen cyclization, employs a basic aqueous solution of the thioamide and potassium hexacyanoferrate(III).[10] The resulting benzothiazole precipitates as the thioamide is added. This cyclization method can be applied for *p*-alkyl and *p*-alkoxy-substituted thioamide derivatives; however, *p*-hydroxy or *p*-nitro aniline derivatives are incompatible substrates for this cyclization due to the single electron transfer processes involved in the cyclization step.[11] The colorless benzothiazole carboxamide can be easily dehydrated. An early method employs boiling phosphorus oxychloride as the dehydration reagent.[12] More recently, it was shown that a small excess (1.05 to 1.2 equiv.) of this reagent in DMF at low temperatures is sufficient to form the corresponding nitrile.[13] In the procedure presented here 1.65 equiv. of the dehydrating reagent had to be applied. It is believed that a Vilsmeier-Haack complex is responsible for this mild dehydration.

The "mild thiolation" concept illustrated in this paper offers an excellent synthetic route to dithiocarboxylic acid derivatives. The transformation into a benzothiazole and subsequent dehydration shows how precursor structures for Firefly luciferin derivatives can be obtained with only moderate synthetic effort.

References

1. Institute of Organic Chemistry and Macromolekular Chemistry, Friedrich-Schiller University, 07743 Jena, Germany. E-mail: hendryk.wuerfel@uni-jena.de. The authors are grateful for the financial support of Prof. Dr. Rainer Beckert and Prof. Dr. Thomas Heinze.
2. (a) Batjargal, S.; Huang, Y.; Wang, Y. J.; Petersson, E. J. *J. Pept. Sci.* **2014**, *20*, 87–91. (b) Lynen, F.; Reichert, E. *Angew. Chem.* **1951**, *63*, 47–48.
3. Thee, S.; Garcia-Prats, A. J.; Donald, P. R.; Hesseling, A. C.; Schaaf, H. S. *Tuberculosis* **2016**, *97*, 126–136.
4. (a) Jagodzinski, T. S. *Chem. Rev.* **2003**, *103*, 197–227. (b) Belskaya, N. P.; Dehaen, W.; Bakulev, V. A. *ARKIVOC*, **2010**, 275–332.
5. (a) White, E. H.; McCapra, F.; Field, G. F.; McElroy W. D. *J. Am. Chem. Soc.* **1961**, *83*, 2402–2403. (b) Würfel, H.; Weiss, D.; Beckert, R.; Güther, A. *J. Sulfur Chem.* **2012**, *33*, 9–16.
6. Ozturk, T.; Ertas, E.; Mert, O. *Chem Rev.* **2010**, *110*, 3419–3478.
7. Mayer, R.; Viola, H.; Hopf, B. *Z. Chem.* **1978**, *18*, 90.
8. Mayer, R. *Z. Chem.* **1976**, *16*, 260–267.
9. Würfel, H.; Weiss, D.; Beckert, R. *J. Sulfur Chem.* **2012**, *33*, 619–638.

Preparation of *N*-(1,7,7-Trimethylbicyclo[2.2.1]heptan-2-ylidene)nitramide

Emerson Teixeira da Silva, Adriana Marques Moraes, Adriele da Silva Araújo, and Marcus Vinícius Nora de Souza[1*]

Instituto de Tecnologia em Fármacos, Farmanguinhos – Fundação Oswaldo Cruz, Rua Sizenando Nabucco 100, Manguinhos, Rio de Janeiro, Brazil

Checked by Philipp C. Roosen, Christopher J. Borths, and Margaret M. Faul

Procedure (Note 1)

A. *1,7,7-Trimethylbicyclo[2.2.1]heptan-2-one oxime* (**1**). A 250 mL three-necked, round-bottomed flask containing a stir bar is equipped with a condenser, internal thermometer and glass stopper under nitrogen atmosphere. The flask is charged with D-camphor (11.0 g, 99.0 wt%, 71.6 mmol) (Note 2) and ethanol (36 mL) (Note 3), and the solution is stirred at 24 °C until the solid dissolves. Deionized water (55 mL) is added to the stirring solution (Note 4) followed by hydroxylamine hydrochloride (7.83 g, 112.7 mmol, 1.6 equiv) (Note 5) and sodium acetate (7.46 g, 90.9 mmol, 1.3 equiv) (Note 6). The flask is placed into an aluminum heating block set to 60 °C. The reaction is stirred overnight until TLC indicates complete consumption of the starting material (Note 7) (Figure 1).

Org. Synth. **2018**, *95*, 192-204
DOI: 10.15227/orgsyn.095.0192

Published on the Web 7/17/2018
© 2018 Organic Syntheses, Inc.

Figure 1. Reaction in progress

The solution is cooled to room temperature and concentrated in vacuo by rotary evaporator (30 mL volatiles are collected at 25 °C and 20 mmHg). The contents are extracted with diethyl ether (3 x 50 mL) in a 250 mL separatory funnel. The organic phases are combined, dried over anhydrous sodium sulfate (15 g) and filtered into a 250 mL one-necked round-bottomed flask. Volatiles are removed in vacuo (25 °C, 10 mmHg) to furnish a white solid that is dried under vacuum (10 mmHg) for 1 h, furnishing 12.6 g of crude material. The solid is dissolved in hot ethanol (15 mL, 75 °C) and vacuum filtered through a 30 mL fine-fritted funnel into a 50 mL one-necked round-bottomed flask. The filtrate is cooled to room temperature (24 °C) and then stored at 4 °C overnight. The resulting crystals are collected by suction filtration on a Büchner funnel, washed with cold ethanol (10 mL, 0 °C), and then transferred to a tared 100 mL round-bottomed flask and dried at 10 mmHg overnight to provide 6.90 g (56%) of white crystals (Notes 8 and 9) (Figure 2).

Figure 2. Recrystallized oxime 1

B. *N-(1,7,7-Trimethylbicyclo[2.2.1]heptan-2-ylidene)nitramide (2).* A 50 mL, two-necked round-bottomed flask, equipped with a 1-cm Teflon-coated magnetic stir bar, a pressure-equalizing addition funnel, and an internal thermometer is charged with oxime (2.0 g, 11.98 mmol) (Note 10) and acetic acid (10 mL). The mixture is stirred until homogenous (Note 11) under room temperature (24 °C) conditions. Aqueous sodium nitrite (11 mL

**Figures 3 and 4. Initial addition of nitrite sodium
solution and a color change**

15 wt%, 24.0 mmol) (Note 12) is added drop-wise by addition funnel over 10 min (Note 13). During and after the addition the internal temperature does not exceed 27 °C, and the color of the medium changes from translucent to pale yellow (Figures 3 and 4).

The temperature of the solution is maintained at room temperature, (24–27 °C) for 16 h at which time TLC analysis indicates complete consumption of the starting material (Note 14). During the reaction the color of the solution changes from yellow to pale white (Figures 5 and 6). The solution is then cooled in an ice bath. To this solution is added 12.5 M NaOH solution (13 mL) (Note 15) over the course of 1–2 min until the solution becomes basic (pH 12) (Note 16), after which distilled water (5 mL) is added. The solution is extracted with ethyl ether (3 x 20 mL) using a 125 mL separatory funnel. The organic phases are combined, dried over anhydrous sodium sulfate (10 g) and filtered into a 125 mL tared, one-necked round-bottomed flask. The solution is concentrated in vacuo (25 °C, 75 mmHg). The product is dried under vacuum (0.01 mmHg) (Figure 7) for 16 h to deliver 2.1 g (87%) of camphor nitro-imine as a yellow solid (Notes 17 and 18).

Figures 5 and 6. Reaction under stirring with a color change

Figure 7. Product after work-up under high vacuum

Notes

1. Prior to performing each reaction, a thorough hazard analysis and risk assessment should be carried out with regard to each chemical substance and experimental operation on the scale planned and in the context of the laboratory where the procedures will be carried out. Guidelines for carrying out risk assessments and for analyzing the hazards associated with chemicals can be found in references such as Chapter 4 of "Prudent Practices in the Laboratory" (The National Academies Press, Washington, D.C., 2011; the full text can be accessed free of charge at https://www.nap.edu/catalog/12654/prudent-practices-in-the-laboratory-handling-and-management-of-chemical. See also "Identifying and Evaluating Hazards in Research Laboratories" (American Chemical Society, 2015) which is available via the associated website "Hazard Assessment in Research Laboratories" at https://www.acs.org/content/acs/en/about/governance/committees/chemicalsafety/hazard-assessment.html. In the case of this procedure, the risk assessment should include (but not necessarily be limited to) an evaluation of the potential hazards associated with D-camphor, ethanol, hydroxylamine hydrochloride, sodium acetate, diethyl ether, acetic acid, sodium nitrite, sodium hydroxide, and sodium sulfate.

2. The checkers purchased D-camphor (≥97%, FG) from Aldrich.

3. The authors used ethanol (96%) from Synth (Brazil) as received. The checkers used 200 proof ethanol from Decon labs.

4. Camphor precipitates out upon addition of water, but it redissolves upon warming.

5. Hydroxylamine hydrochloride (98%) from Aldrich was used as received. The use of 2.5 equiv is important for total consumption of starting material.

6. Sodium acetate from JT Baker (ACS reagent) was used as received.

7. The reaction may be heterogenous. The reaction was monitored by TLC analysis on Merck silica gel 60 F254 plates developing with hexanes/ethyl acetate (10:1). The R_f values of oxime **1** and camphor are 0.29 and 0.64, respectively. Visualized with a 10 % ethanol solution of phosphomolybdic acid (PMA). After dipping the TLC plate to the PMA solution, the plate is developed by heating.

8. A second reaction on the same scale provided 7.31 g (60%) of the identical product with > 99% purity.

9. Oxime **1**: Purity of 99% was determined by qNMR using guaiacol as an internal standard. mp 116–117 °C (lit[2] 119–121 °C). $[\alpha]^{20}_D$ = - 40.3 (c=1, EtOH). ^{1}H NMR (CDCl$_3$, 400 MHz) δ: 0.80 (s, 3H), 0.92 (s, 3H), 1.01 (s, 3H), 1.16–1.28 (m, 1H), 1.42–1.52 (m, 1H), 1.70 (td, J = 12.2, 4.2 Hz, 1H), 1.79–1.90 (m, 1H), 1.92 (t, J = 4.3 Hz, 1H), 2.06 (d, J = 17.8 Hz, 1H), 2.55 (dt, J = 17.8, 3.8 Hz, 1H), 8.93 (s, 1H). ^{13}C NMR (100.6 MHz, CDCl$_3$) δ: 11.1, 18.5, 19.4, 27.2, 32.6, 33.0, 43.7, 48.3, 51.8, 169.9. IR 3287, 2958, 1686 cm^{-1}.

10. Oxime was prepared as described above, but the oxime can also be purchased from commercial sources.

11. Acetic acid was obtained from Aldrich, with purity of 98% and was used as received.

12. Sodium nitrite (99%) was obtained from Sigma-Aldrich Co. (USA) and used as received.

13. The release of nitrous gas occurred without any problem, however the reaction was performed in the fume hood. The use of 2.0 equiv of NaNO$_2$ is important for total consumption of starting material in this time and temperature.

14. The reaction progress was monitored by TLC analysis on Merck silica gel 60 F254 plates developing with hexanes/ethyl acetate (10:1). The R_f values of oxime **1** and nitro-imine **2** are 0.29 and 0.67, respectively. Visualized with a 10 % ethanol solution of phosphomolybdic acid (PMA). After dipping the TLC plate to the PMA solution, the

chromatogram is stained by heating. Only nitro-imine is visible with UV 254 nm lamp.

15. Sodium hydroxide (98%) was obtained from Aldrich and used as received in the preparation of the solution, as follows. To a graduated Erlenmeyer flask of 500 mL was added NaOH (250 mL). Distilled water was added slowly, with magnetic stirring, until the targeted volume (500 mL) was achieved. Cooling was necessary because the dissolution is exothermic. After total dissolution, the solution was allowed to warm to room temperature prior to use. The checkers added water (15 mL) rapidly to NaOH (12.5 g, 313 mmol) with magnetic stirring and diluted the solution to the target volume of 25 mL after the solution cooled to room temperature.

16. The pH of the solution between 12 and 14 is sufficient. Other bases, such as carbonate, can be used, but care must be taken due to the release of gases.

17. A second reaction on the same scale provided 2.26 g (90%) of the identical product with > 97% purity.

18. Nitro-imine **2**: Purity of 97% was determined by qNMR using guaiacol as the internal standard; mp 34–37 °C, although the submitters report formation of a light yellow oil at rt. $[\alpha]^{20}_D$ –26.9 (c=1, EtOH). ^{1}H NMR (CDCl$_3$, 400 MHz, mixture of isomers E:Z, 17:1) δ: 0.89 (s, 3H), 0.99 (s, 3H), 1.05 (s, 3H), 1.27–1.37 (m, 1H), 1.51–1.62 (m, 1H), 1.80–1.99 (m, 2H), 2.04 (t, J = 4.4 Hz, 1H), 2.13 (d, J = 18.7 Hz, 1H), 2.69 (ddd, J = 18.6, 4.7, 2.4 Hz, 1H). ^{13}C NMR (100 MHz, CDCl$_3$) δ: 10.3, 18.6, 19.4, 26.7, 31.5, 35.1, 43.4, 48.8, 54.1, 189.4, IR: 2965, 1643, 1554, 1332 cm^{-1}. The product is stable at room temperature, but it is recommended to store the material at refrigerated temperature. Further purification is not necessary.

Working with Hazardous Chemicals

The procedures in *Organic Syntheses* are intended for use only by persons with proper training in experimental organic chemistry. All hazardous materials should be handled using the standard procedures for work with chemicals described in references such as "Prudent Practices in the Laboratory" (The National Academies Press, Washington, D.C., 2011; the full text can be accessed free of charge at http://www.nap.edu/catalog.php?record_id=12654). All chemical waste should be disposed of in accordance with local regulations. For general guidelines for the management of chemical waste, see Chapter 8 of Prudent Practices.

In some articles in *Organic Syntheses*, chemical-specific hazards are highlighted in red "Caution Notes" within a procedure. It is important to recognize that the absence of a caution note does not imply that no significant hazards are associated with the chemicals involved in that procedure. Prior to performing a reaction, a thorough risk assessment should be carried out that includes a review of the potential hazards associated with each chemical and experimental operation on the scale that is planned for the procedure. Guidelines for carrying out a risk assessment and for analyzing the hazards associated with chemicals can be found in Chapter 4 of Prudent Practices.

The procedures described in *Organic Syntheses* are provided as published and are conducted at one's own risk. *Organic Syntheses, Inc.*, its Editors, and its Board of Directors do not warrant or guarantee the safety of individuals using these procedures and hereby disclaim any liability for any injuries or damages claimed to have resulted from or related in any way to the procedures herein.

Discussion

The nitro-imine **2** is a versatile precursor for a preparation of various compounds of interest in organic synthesis, as auxiliary chiral,[2] bioactive compounds[4] and others.[5,6] In our research group it has been utilized for obtaining diverse bioactive compounds[7] with appreciable biological activities. Literature reports that **2** has been obtained from oxime **1** in reasonable scale in long reaction times, utilizing large quantities of solvents and reagents, with only modest yields (60-70%). Nitro-imine **2** has been prepared from the oxime using diethyl ether, H_2SO_4, and $NaNO_2$ in a procedure[4] that includes a complicated isolation procedure and difficult purification. In other work, compound **2** was prepared from the oxime using NaOCl,[6] which resulted in a product with poor purity and low yield.

Page, et al,[2]. described the synthesis using $NaNO_2$ and AcOH in large excess. In this procedure, the authors used excess of $NaNO_2$ (1.8 equiv; 5% solution) and a large volume of AcOH (900 mL for 30 g of the oxime), making the medium very dilute, and causing a longer reaction time and lower yield. Following this protocol, we obtained the product only after 20h of reaction at room temperature.

Examples of Use of the Nitro-imine 2

Page, et al.[2] described the first stable chiral oxaziridine **3** from nitro-imine **2** (Scheme 1). Oxaziridines are known to induce amination of nitrogen, sulfur and carbon nucleophiles, as well as aziridination of alkenes and amination of enolates. The oxaziridine **3** was converted to compounds **4a-4e**, with oxidant properties, by derivatization of the nitrogen atom. The known asymmetric oxidant N-phenylsulfonyloxaziridine **4b** was prepared as a single diastereoisomer.

Scheme 1. Synthesis of stable oxaziridines from 2

Squire, et al.[4b] has reported on the synthesis of chiral, non-racemic *vic*-amino-alcohols derived from D-camphor as a potential chiral auxiliary for asymmetric synthesis (Scheme 2). In initial experimentation the authors

Scheme 2. Formation of *vic*-amino-alcohols

attempted to condense D-camphor with the amino alcohol (1R,2S)-norephedrine under harsh conditions to obtain the respective imine derivative in poor yield. To circumvent this issue nitro-imine **2** was used to promote the formation of desired imines that were reduced to the amino-alcohols **6**.

Nickerson, et al.[5] has identified nitro-imines as impressive starting points for the syntheses of otherwise inaccessible, sterically encumbered enamines (Scheme 3). In this work authors prepared various nitro-imines, among these, the nitro-imine **2** derived from camphor. The activation of nitro-imines with urea catalysts for reaction with a variety of amines enables the formation of highly substituted enamines. For example, formation of compounds **7a-7c** were described in this paper (Scheme 3).

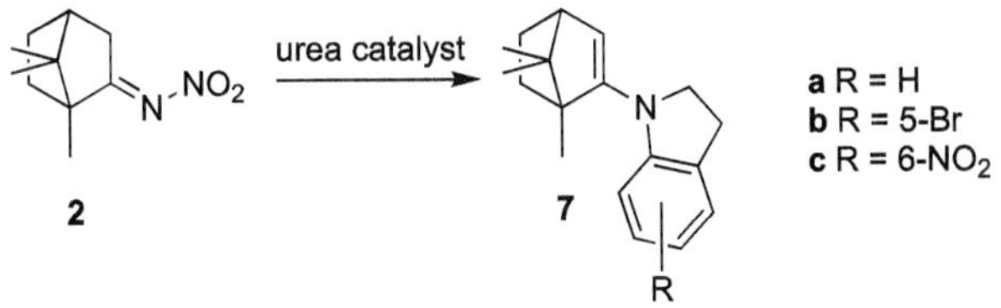

Scheme 3. Synthesis of enamines by activation of 2

Starting from **2**, our research group has synthesized and evaluated various camphor imine **8** and hydrazone **10** derivatives as potential antitubercular, anticancer and antiviral agents (Scheme 4).[7] Many of these compounds has reasonable activity against *Mycobacterium tuberculosis* and the compound **8** (R= 2-HO-Ph) showed the highest activity with MIC=3.12 (minimal inhibitory concentration), comparable to ethambutol, a classical anti-TB agent.

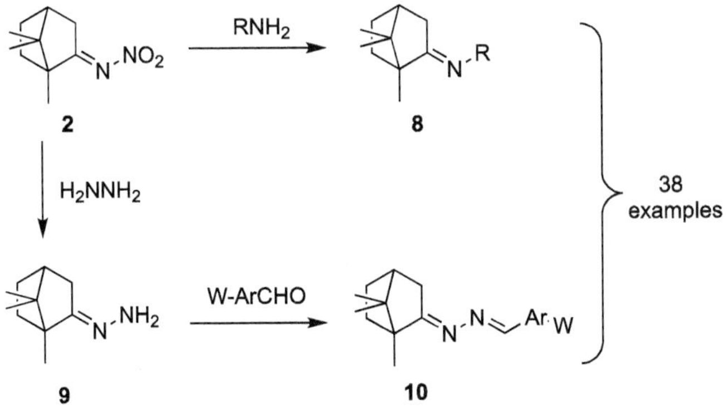

Scheme 4. Synthesis of biological active camphor derivatives

References

1. E.T. da Silva, esilva@far.fiocruz.br; A.M. Moraes, m_drik@yahoo.com.br; A. da S. Araújo, adriele.cp2@gmail.com; M. V. N. de Souza, marcos_souza@far.fiocruz.br. Instituto de Tecnologia em Fármacos, Farmanguinhos – Fundação Oswaldo Cruz - Fiocruz, Rua Sizenando

Nabucco 100, Manguinhos, Rio de Janeiro, Brazil. The authors thank Fiocruz and CNPq for financial support.

2. Page, P. C. B.; Murrell, V. L.; Limousin, C.; Laffan, D. D. P.; Bethell, D.; Slawin, A. M. Z.; Smith, T. A. D. *J. Org. Chem.* **2000**, *65*, 4204–4207.

3. Nath, U.; Das, S. S.; Deb, D.; Das, P. J. *New J. Chem.* **2004**, *28*, 1423–1425.

4. (a) Aboul-Enein, M. N.; EL-Azzouny, A. A.; Maklad, Y. A.; Sokeirik, Y. S.; Safwat, H. *J. Iran. Chem. Soc.* **2006**, *3*, 191–208. (b) Squire, M. D.; Burwell, A.; Ferrence, G. M.; Hitchcock, S. R. *Tetrahedron: Asymmetry.* **2002**, *13*, 1849–1854.

5. Nickerson, D. M.; Angeles, V. V.; Mattson, A. E. *Org. Lett.* **2013**, *15*, 5000–5003.

6. Guziec, F. S. Jr.; Russo, J. M. *Synthesis* **1984**, 479–481.

7. da Silva, E. T.; Araújo, A. da S.; Moraes, A. M.; de Souza, L. A.; Lourenço, M. C. S.; de Souza, M. V. N.; Wardell, J. L.; Wardell, S. M. S. *Sci. Pharm.* **2016**, *84*, 467–483.

8. Palomo, C.; Oiarbide, M.; Garcia, J. M.; Banuelos, P.; Odriozola, J. M.; Razkin, J.; Linden, A. *Org. Lett.* **2008**, *10*, 2637–2640.

Appendix
Chemical Abstracts Nomenclature (Registry Number)

Sodium hydroxide: sodium, hydroxide (1310-73-2)
Sodium nitrite: sodium, nitrite (7632-000-0)
Acetic acid: acetic acid (64-19-7)
D-Camphor: 1,7,7-Trimethylbicyclo[2.2.1]heptan-2-one (464-49-3)
Oxime **1**: 1,7,7-Trimethylbicyclo[2.2.1]heptan-2-one oxime (37850-13-8)
Nitro-imine **2**: N-(1,7,7-Trimethylbicyclo[2.2.1]heptan-2-ylidene)nitramide (114717-12-3)

Organic
Syntheses

Marcus Vinícius Nora de Souza is a senior technologist at Fiocruz, and leads a research group in organic chemistry, with emphasis in organic synthesis and medicinal chemistry, acting on neglected diseases (tuberculosis and malaria), cancer and synthetic methodologies. He has authored more than 250 manuscripts, 5 patents and 3 books, and their formation includes a Ph.D. in organic and bioorganic chemistry (Universite de Paris XI (Paris-Sud) – France) with stage at the University of Barcelona – Spain, and three postdocs in organic synthesis and medicinal chemistry (University of Florida – USA, Federal University of Juiz de Fora – MG, Pharmaceutical industry Genzyme - Boston – USA).

Emerson Teixeira da Silva received his PhD in UFRJ in 2004. He is a Technologist in Research at Fiocruz, and works with organic chemistry, with emphasis in organic synthesis and medicinal chemistry, acting on neglected diseases, cancer and asthma. He has authored more than 10 patents deposited in various countries and various published manuscripts. He worked with process chemistry in chemical industry.

Adriana Marques Moraes is a postgraduate student at UFRJ in Chemistry and develops the experimental part of her work at Fiocruz. She has worked in the area of organic synthesis since her graduation and has co-authored publications in the field. She is interested in bioactive synthetic compounds and natural products.

Adriele da Silva Araújo is an undergraduate student at UFRJ in Chemistry and develops scientific research at Fiocruz since the secondary education. She has worked with organic synthesis and has authored publication. She is interested in bioactive synthetic compounds and organic synthesis.

Philipp C. Roosen obtained a B.A. in biology from Albion College and an M.S. in chemistry exploring ortho-directed Ir-catalyzed C-H borylations of arenes with Professor Milton Smith and Professor Robert Maleczka at Michigan State. He studied natural product synthesis at the University of California Irvine with Professor Christopher Vanderwal, focusing his Ph.D. dissertation on the synthesis of 7,20-diisocyanoadociane. In 2016 he joined the Pivotal Drug Substance Technology organization at Amgen where he is involved in the development of pharmaceutically relevant chemical processes.

Christopher J. Borths earned a Ph.D. in synthetic organic chemistry from the California Institute of Technology in 2004 for developing novel organocatalytic methods with Prof. David MacMillan. After completing his graduate studies, he joined the Chemical Process Research and Development Group at Amgen. He is currently a Principal Scientist in the Synthetic Technologies and Engineering group within the Pivotal Drug Substance Technology department where he works on the development of robust and safe manufacturing processes to produce active pharmaceutical ingredients, including traditional synthetic small molecule drugs, antibody-drug conjugates, and oligonucleotide therapeutics.

Preparation of Alkyl Boronic Esters Using Radical-Polar Crossover Reactions of Vinylboron Ate Complexes

Marvin Kischkewitz and Armido Studer*[1]

Organisch-Chemisches Institut, Westfälische Wilhelms-Universität, Corrensstraße 40, 48149, Münster, Germany

Checked by Helene Wolleb and Erick M. Carreira

Procedure (Note 1)

4,4,5,5-Tetramethyl-2-(7,7,8,8,9,9,10,10,10-nonafluoro-5-methyldecan-5-yl)-1,3,2-dioxaborolane (2). An oven-dried 400 mL Schlenk tube equipped with a 3 cm Teflon-coated magnetic stir bar and a glass stopper is evacuated and backfilled with argon three times (Note 2). The glass stopper is substituted by a rubber septum and the tube is charged with diethyl ether (65 mL) (Note 3) and isopropenyl boronic acid pinacol ester (95%, 3.76 mL, 20.0 mmol, 1.00 equiv) (Note 4) (Figure 1). The vigorously stirred mixture is cooled to 0 °C by a water/ice bath and *n*-butyllithium (1.6 M in hexanes, 13.8 mL, 22.0 mmol, 1.10 equiv) (Note 5) is added dropwise over 10 min using a syringe pump (Note 6) (Figure 2).

Org. Synth. **2018**, *95*, 205-217
DOI: 10.15227/orgsyn.095.0205

Published on the Web 7/26/2018
© 2018 Organic Syntheses, Inc.

Figure 1. Reaction set-up

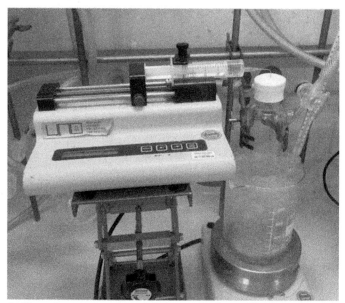

Figure 2. Reaction setup for addition of *n*-butyllithium (formation of boron ate complex)

After the addition is completed the pale yellow mixture is stirred for 5 min at 0 °C, the cooling bath is removed and stirring is continued at room temperature for 30 min (Note 7) (Figure 3). The septum of the flask is switched to a vacuum connection with a separate evacuated cooling trap in liquid nitrogen (Note 8) (Figure 4).

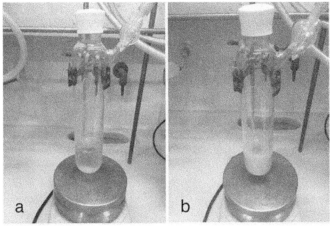

Figure 3. (a) Pale yellow solution after completed addition of *n*-butyllithium and removal of the cooling bath (b) Yellow suspension after stirring at room temperature for 30 min

Figure 4. Setup for solvent removal under reduced pressure using a separate cooling trap

The solvents are carefully removed under reduced pressure (0.4 mmHg, 23 °C) and the resulting boron ate complex is further dried for 30 min (0.06 mmHg, 23 °C) and appears as a yellow solid (Figure 5).

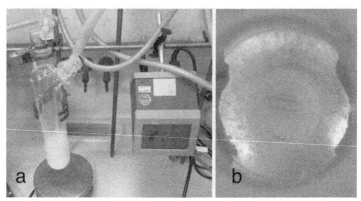

Figure 5. (a) Yellow solid after solvent removal (b) Dried boron ate complex after 30 min

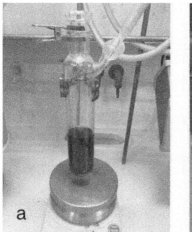

Figure 6. (a) Reaction mixture after addition of nonafluoro-1-iodobutane (b) Reaction mixture in light chamber after 18 h

The vacuum connection is closed, the Schlenk tube is backfilled with argon, the vacuum connection is substituted by a rubber septum and the residue is dissolved with the addition of acetonitrile (130 mL) (Note 9). To the resulting flaky suspension, nonafluoro-1-iodobutane (5.2 mL, 30 mmol,

1.5 equiv) (Note 10) is added under stirring in one portion and the mixture turns brown. The Schlenk tube is sealed with a glass stopper, which is fixed with a metal clamp. The reaction is transferred to a light chamber and stirred for 18 h under irradiation with visible light (Note 11) (Figure 6).

The resulting dark brown solution is transferred to a 500 mL round-bottomed flask [The Schlenk tube is rinsed with diethyl ether (3 x 15 mL)] and then concentrated under reduced pressure by rotary evaporation (7.5 mmHg, 40 °C) (Note 12). The concentrated mixture is transferred with diethyl ether (3 x 60 mL) to a 500 mL separatory funnel and washed with saturated aqueous solution of sodium thiosulfate (90 mL) (Note 13). The separated organic phase is washed with demineralized water (90 mL) and dried over sodium sulfate (30 g). The drying agent is removed by a glass sinter (porosity 4) and flushed with diethyl ether (3 x 25 mL). The filtrate is concentrated under reduced pressure by rotary evaporation (7.5 mmHg, 40 °C).

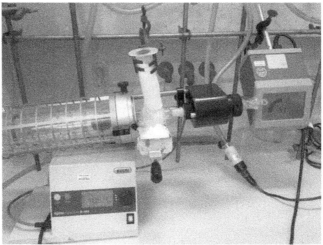

Figure 7. Kugelrohr distillation with dry ice cooling and manometer

The brown oily residue is purified by Kugelrohr distillation (Figure 7) (Note 14). The air bath temperature is slowly increased from room temperature at a pressure of 0.35 mmHg. Pure product is collected (air bath temperature 110 – 115 °C, 0.35 mmHg) as the first fraction in a bulb that is continuously cooled with dry ice over a period of approximately 2 h. The desired 4,4,5,5-tetramethyl-2-(7,7,8,8,9,9,10,10,10-nonafluoro-5-methyldecan-

5-yl)-1,3,2-dioxaborolane (**2**) is obtained as a pale yellow oil (7.76 g, 87%) (Notes 15, 16 and 17) (Figure 8).

Figure 8. Pure product after Kugelrohr distillation

Notes

1. Prior to performing each reaction, a thorough hazard analysis and risk assessment should be carried out with regard to each chemical substance and experimental operation on the scale planned and in the context of the laboratory where the procedures will be carried out. Guidelines for carrying out risk assessments and for analyzing the hazards associated with chemicals can be found in references such as Chapter 4 of "Prudent Practices in the Laboratory" (The National Academies Press, Washington, D.C., 2011; the full text can be accessed free of charge at https://www.nap.edu/catalog/12654/prudent-practices-in-the-laboratory-handling-and-management-of-chemical. See also "Identifying and Evaluating Hazards in Research Laboratories" (American Chemical Society, 2015) which is available via the associated website "Hazard Assessment in Research Laboratories" at https://www.acs.org/content/acs/en/about/governance/committees/chemicalsafety/hazard-assessment.html. In the case of this procedure, the risk assessment should include (but not necessarily be limited to) an evaluation of the potential hazards associated with isopropenyl boronic acid pinacol ester, diethyl ether, *n*-butyllithium, acetonitrile, nonafluoro-1-iodobutane, sodium thiosulfate and sodium sulfate.

2. All glassware was oven-dried, quickly assembled, and evacuated under vacuum (0.075 mmHg) before backfilling with argon. All reaction steps are performed under a partial positive argon atmosphere using an argon gas line connected to an external paraffin oil bubbler.

3. Diethyl ether (99%) was purchased from Acros Organics, refluxed over K and was freshly distilled from K-Na-alloy before use. The checkers used diethyl ether from a LC Technology Solutions SP-1 solvent purification system.

4. Isopropenyl boronic acid pinacol ester (95%, contains phenothiazine as stabilizer) was purchased from Sigma Aldrich and used as received.

5. *n*-Butyllithium (1.6 M solution in hexanes, AcroSeal) was purchased from Acros Organics and used as received.

6. A syringe pump from UNO B.V. was used with the following settings: diameter: 20 mm; flow rate: 89 mL/h; total time: 10 min.

7. After addition of *n*-butyllithium the mixture turns yellow (sometimes dark yellow to orange). After completed addition, the cooling bath was removed and stirring was continued at room temperature for 30 min while the mixture becomes cloudy.

8. The boron ate complex is sensitive to water and oxygen. The switch from the septum to the evacuated cooling trap was performed under positive argon atmosphere of the Schlenk tube. The side arm of the Schlenk tube was closed, the vacuum connection was carefully opened and the solvents were condensed in a separate cooling trap (250 mL Schlenk round-bottomed flask in a liquid nitrogen filled dewar).

9. Acetonitrile (99.9%), Extra Dry over Molecular Sieve, AcroSeal was purchased from Acros Organics and used as received.

10. Nonafluoro-1-iodobutane (98%) was purchased from Sigma Aldrich and used as received.

11. The light chamber is equipped with a Philips Master HPI-T Plus (400W) light source. The distance between bulb and the reaction tube is approximately 20 cm. Although the light chamber is equipped with a ventilator a temperature of approximately 40 °C is reached in the chamber due the heat of the bulb. The checkers used a fume hood shielded with aluminum foil as a substitute for the light chamber.

12. First diethyl ether was removed (600 mmHg, 40 °C) before the pressure was gradually reduced (7.5 mmHg, 40 °C) and held for approximately 30 min. Due to the volatility of the fluorinated product the pressure should not be further decreased.

13. The saturated aqueous solution of sodium thiosulfate (90 mL) was also used to transfer the Et₂O insoluble residue from the 500 mL flask to the 500 mL separatory funnel. As a precaution the first 10 mL of this solution were added dropwise over 5 min at room temperature to the separatory funnel. As a precaution the first 10 mL of this solution were added dropwise over 5 min at room temperature to the separatory funnel and the remaining solution can then be added in one portion.

14. A Kugelrohr Distillation Apparatus Büchi Glass Oven B-585 from Büchi Labortechnik GmbH was used with the following settings: temperature: 110 °C (±5 °C); rotation: 15 rpm; pressure: 0.35 mmHg.

15. If the distilled product shows turbidity after distillation, it can be passed through a disposable syringe filter (PTFE membrane, pore size 0.2 μm).

16. 4,4,5,5-Tetramethyl-2-(7,7,8,8,9,9,10,10,10-nonafluoro-5-methyldecan-5-yl)-1,3,2-dioxaborolane (2) has the following spectroscopic properties: FTIR (neat): ν (cm⁻¹) 2980, 2963, 2934, 2876, 1472, 1380, 1374, 1324, 1233, 1216, 1133, 1020, 878, 849, 738, 729. ^{1}H NMR (500 MHz, CDCl₃) δ: 0.89 (t, J = 7.2 Hz, 3H), 1.08 (d, J = 2.1 Hz, 3H), 1.15 – 1.24 (m, 13H), 1.25 – 1.36 (m, 4H), 1.42 – 1.51 (m, 1H), 1.88 (dddt, J = 32.3, 14.9, 10.0, 2.2 Hz, 1H), 2.27–2.42 (m, 1H). ^{13}C NMR (126 MHz, CDCl₃) δ: 14.0 (s, 1C), 20.8 (s, 1C), 23.4 (s, 1C), 24.6 (s, 2C), 24.7 (s, 2C), 27.1 (s, 1C), 38.4 (t, J = 20.7 Hz, 1C), 39.3 (s, 1C), 83.5 (s, 2C), 104.9 –122.9 (m, 4C). The signal of the α-B-carbon was not observed. ^{19}F NMR (470 MHz, CDCl₃) δ: –81.1 (tt, J = 9.7, 3.4 Hz, 3F), –106.9 – –107.8 (m, 1F), –110.8 – –111.7 (m, 1F), –124.9 – –125.1 (m, 2F), –125.7 – –125.9 (m, 2F). ^{11}B NMR (160 MHz, CDCl₃) δ (ppm) 34.1. HRMS (EI) m/z = 443.17984 calcd. for $C_{17}H_{25}BF_9O_2$ [M–H]⁺; found: 443.17982. CHN Anal. calcd for $C_{17}H_{26}BF_9O_2$: C, 45.97; H, 5.90; found: C, 46.03; H, 6.03.

17. A second reaction on equivalent scale provided 7.58 g (85%). When using 5.0 equiv of nonafluoro-1-iodobutane on the same scale the yield can be improved to 94%.

Working with Hazardous Chemicals

The procedures in *Organic Syntheses* are intended for use only by persons with proper training in experimental organic chemistry. All hazardous materials should be handled using the standard procedures for work with chemicals described in references such as "Prudent Practices in the

Laboratory" (The National Academies Press, Washington, D.C., 2011; the full text can be accessed free of charge at http://www.nap.edu/catalog.php?record_id=12654). All chemical waste should be disposed of in accordance with local regulations. For general guidelines for the management of chemical waste, see Chapter 8 of Prudent Practices.

In some articles in *Organic Syntheses*, chemical-specific hazards are highlighted in red "Caution Notes" within a procedure. It is important to recognize that the absence of a caution note does not imply that no significant hazards are associated with the chemicals involved in that procedure. Prior to performing a reaction, a thorough risk assessment should be carried out that includes a review of the potential hazards associated with each chemical and experimental operation on the scale that is planned for the procedure. Guidelines for carrying out a risk assessment and for analyzing the hazards associated with chemicals can be found in Chapter 4 of Prudent Practices.

The procedures described in *Organic Syntheses* are provided as published and are conducted at one's own risk. *Organic Syntheses, Inc.*, its Editors, and its Board of Directors do not warrant or guarantee the safety of individuals using these procedures and hereby disclaim any liability for any injuries or damages claimed to have resulted from or related in any way to the procedures herein.

Discussion

Vinyl boronic esters are highly important building blocks for C–C bond formations in the *Suzuki-Miyaura* coupling.[2] Furthermore, vinyl boronic esters can be used in three component coupling reactions, in which two new bonds are formed and the valuable boronic ester moiety remains in the product. The strategy is based on the ability of vinyl boronic esters to form boron ate complexes with carbon nucleophiles. A subsequent 1,2-metalate rearrangement can be triggered by different mechanisms (Scheme 1).

(a) Electrophile-induced 1,2-R-migration:

Zweifel, 1967

(X = halogen)

E^+

(b) Transition metal induced 1,2-R-migration:

Morken, 2016

ML_n

(c) Radical-induced 1,2-R-migration:

Studer, 2017

$\cdot R_f$

Scheme 1. Reactivity of vinyl boron ate complexes in three component couplings

In the *Zweifel* reaction the 1,2-R-shift is induced by initial electrophilic halogenation of the vinyl group.[3] In 2016 *Morken et al.* utilized *in situ* generated vinyl boron ate complexes in transition metal-catalyzed enantioselective conjunctive cross-couplings with aryl iodides.[4] Along these lines, we recently demonstrated that *in situ* generated vinyl boron ate complexes react efficiently with electrophilic alkyl radicals, generated from alkyl iodides. The resulting radical anions undergo a radical-polar crossover reaction and a 1,2-alkyl/aryl shift from boron to the α-carbon sp[2] center eventually provides valuable secondary and tertiary alkyl boronic esters.[5] Notably, the cascade proceeds without the help of any transition metal and uses commercial starting materials, allowing a rapid construction of molecular complexity. The reaction sequence tolerates α- and β-substituted vinyl boronic esters and the scope of the radical precursor includes perfluoralkyl iodides, α-iodo esters, iodoacetonitrile and α-iodo phosphonates. Initiation of the chain reaction can be achieved either by addition of catalytic amounts of BEt₃ (Scheme 2),[5] by photo-[6] or photoredox initiation.[7]

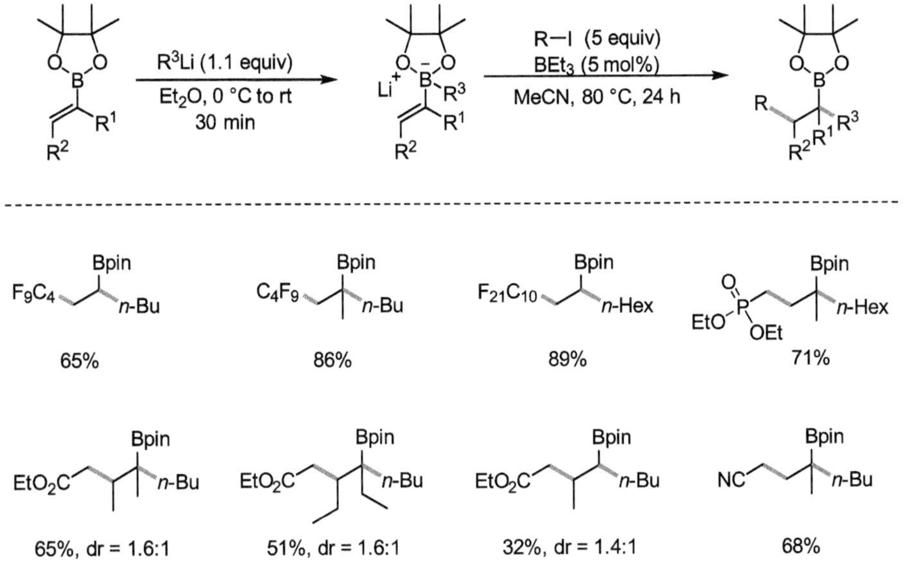

Scheme 2. Selected examples for radical-polar cross over reactions towards secondary and tertiary alkyl boronic esters[4]

To demonstrate the potential of the boronic ester moiety in versatile follow-up chemistry, we used the boron intermediate for the synthesis of alcohols and valuable γ-lactones. Furthermore, the alkyl boronic ester could be used for the construction of quaternary C-centers.[5]

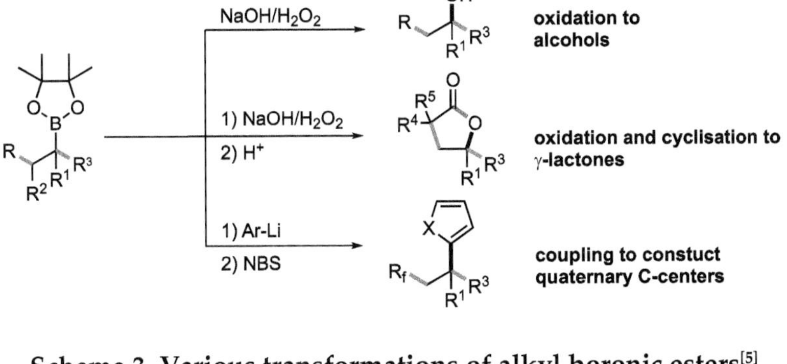

Scheme 3. Various transformations of alkyl boronic esters[5]

References

1. Organisch-Chemisches Institut, Westfälische Wilhelms-Universität, Corrensstraße 40, 48149, Münster (Germany). E-mail: studer@uni-muenster.de. We thank the European Research Council ERC (Advanced Grant agreement No. 692640) for supporting this work.
2. Miyaura, N.; Yamada, K.; Suzuki, A. *Tetrahedron Lett.* **1979**, *20*, 3437–3440.
3. Zweifel, G.; Arzoumanian, H.; Whitney, C. C. *J. Am. Chem. Soc.* **1967**, *89*, 3652–3653.
4. Zhang, L.; Lovinger, G. J.; Edelstein, E. K.; Szymaniak, A. A.; Chierchia, M. P.; Morken, J. P. *Science* **2016**, *351*, 70–74.
5. Kischkewitz, M.; Okamoto, K.; Mück-Lichtenfeld, C.; Studer, A. *Science* **2017**, *355*, 936–938.
6. Gerleve, C.; Kischkewitz, M.; Studer, A. *Angew. Chem. Int. Ed.* **2018**, *57*, 2441–2444.
7. Silvi, M.; Sanford, C.; V. K. Aggarwal, V. K.; *J. Am. Chem. Soc.* **2017**, *139*, 5736–5739.

Appendix
Chemical Abstracts Nomenclature (Registry Number)

Isopropenyl boronic acid pinacol ester; (126726-62-3)
n-Butyllithium; (109-72-8)
Nonafluoro-1-iodobutane; (423-39-2)
Sodium thiosulfate; (7772-98-7)
Sodium sulfate; (7757-82-6)

Marvin Kischkewitz obtained his B.Sc. degree in chemistry in 2013 from the Westfälische Wilhelms-University of Münster. During his following master studies in Münster, he did a research stay in the laboratory of Prof. Mark Lautens at the University of Toronto. In 2015 he completed his master thesis in the group of Prof. Armido Studer, where he is currently pursuing his doctoral studies in the field of organic chemistry. His research focuses on radical reactions and electron catalysis.

Armido Studer received his Diploma in 1991 and his Ph.D. in 1995 from ETH Zürich with Prof. Dieter Seebach. He then did postdoctoral studies at the University of Pittsburgh with Prof. Dennis P. Curran. In 1996 he started his independent career at the ETH Zürich. In 2000 he was appointed Associate Professor of Organic Chemistry at the Philipps-University in Marburg, and in 2004 Professor of Organic Chemistry at the Westfälische Wilhelms-University in Münster.

Helene Wolleb obtained her B.Sc. and M.Sc. degree in chemistry from ETH Zürich, conducting research with Prof. Erick M. Carreira and Prof. Antonio Togni at the same institution, and with Prof. Steven V. Ley at the University of Cambridge. After an internship at Bayer HealthCare AG in Wuppertal, she joined the group of Prof. Erick M. Carreira for her doctoral studies in 2015 to work on the synthesis of complex natural products.

Organic **S**yntheses

Preparation of (pin)B–B(dan)

Hiroto Yoshida,[1]* Yuya Murashige, and Itaru Osaka

Department of Applied Chemistry, Graduate School of Engineering, Hiroshima University, Higashi-Hiroshima 739-8527, Japan

Checked by Feng Peng and Kevin Campos

(pin)B–B(dan)
1

Procedure (Note 1)

A. *(pin)B–B(dan)* *(1)*. A flame-dried 50 mL two-necked round-bottomed flask equipped with a 1.5 cm Teflon-coated magnetic oval stir bar, a water condenser capped with rubber septum and a three-way stopcock, which is purged with argon, is charged with bis(pinacolato)diboron (6.35 g, 25.0 mmol) (Note 2) and toluene (15 mL) (Note 3). To the colorless solution (Figure 1), which is degassed by three freeze-pump-thaw cycles, is added 1,8-diaminonaphthalene (3.95 g, 25.0 mmol) (Note 4), and the resulting deep red solution is capped with a glass septum and heated in a silicon oil bath at 100 °C for 48 h under argon atmosphere (Figure 2) (Note 5). The mixture is allowed to cool to ambient temperature and an orange slurry forms. The solvent is removed by rotary evaporation (60 °C, 16 mmHg) to afford a red-brown solid (Figure 3). *n*-Hexane (100 mL) (Note 6) is added to the solid, and the resulting mixture is stirred at ambient temperature for 0.5 h. The suspension is filtered through a vacuum filter holder equipped with a PTFE type membrane filter (Figure 4) (Note 7), and the residue is washed with *n*-hexane (2 x 100 mL). The resulting pale brown solid (Figure 5) is dissolved in ethyl acetate (200 mL) (Note 8), treated with activated charcoal (1.5 g)

Org. Synth. **2018**, *95*, 218-230
DOI: 10.15227/orgsyn.095.0218

Published on the Web 7/30/2018
© 2018 Organic Syntheses, Inc.

(Note 9) at 20 °C for 2 h with stirring (Figure 6), and filtered through a Celite pad (5 g) with a Hirsch funnel (Figures 7 and 8). This activated charcoal treatment of the filtrate is repeated two more times (Figure 9). After removal of the solvent by rotary evaporation (40 °C, 16 mmHg) (Figure 10), the residue is stirred with 1:50 ethanol:n-hexane (100 mL) at 20 °C for 0.5 h (Figure 11) (Note 10). The first crop is collected by filtration on a vacuum filter holder equipped with a PTFE type membrane filter (Figure 4), and the filtrate is concentrated by rotary evaporation (40 °C, 16 mmHg). After the same treatment of the residue with 1:50 ethanol:n-hexane (100 mL), the second crop is collected by the filtration. Repetitive treatment of the filtrate with 1:50 ethanol:n-hexane (100 mL) gives a third crop, and the combined solid is dried overnight at 3.0 mmHg to afford the analytically pure (pin)B–B(dan) (5.20 g, 71%) as an off-white solid (Figure 12) (Notes 11, 12, 13, and 14).

Figure 1. Reaction setup

Figure 2. After heating

Figure 3. After evaporation

Figure 4. Filtration apparatus

Figure 5. After washing

Figure 6. Activated charcoal treatment

Figure 7. Filtration apparatus

Figure 8. After first treatment

Figure 9. After third treatment

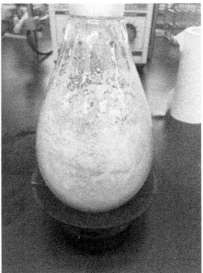

Figure 10. After evaporation

Figure 11. After ethanol-hexane treatment   **Figure 12. Product 1**

Notes

1. Prior to performing each reaction, a thorough hazard analysis and risk assessment should be carried out with regard to each chemical substance and experimental operation on the scale planned and in the context of the laboratory where the procedures will be carried out. Guidelines for carrying out risk assessments and for analyzing the hazards associated with chemicals can be found in references such as Chapter 4 of "Prudent Practices in the Laboratory" (The National Academies Press, Washington, D.C., 2011; the full text can be accessed free of charge at https://www.nap.edu/catalog/12654/prudent-practices-in-the-laboratory-handling-and-management-of-chemical. See also "Identifying and Evaluating Hazards in Research Laboratories" (American Chemical Society, 2015) which is available via the associated website "Hazard Assessment in Research Laboratories" at https://www.acs.org/content/acs/en/about/governance/committees/chemicalsafety/hazard-assessment.html. In the case of this procedure, the risk assessment should include (but not necessarily be limited to) an evaluation of the potential hazards associated with

bis(pinacolato)diboron, 1,8-diaminonaphthalene, toluene, *n*-hexane, ethyl acetate and ethanol.

2. Bis(pinacolato)diboron (>99%, ChemICHIBA) was used as received.

3. Toluene (99.5%, Wako Pure Chemical Industries, Ltd.) was dried over activated 4Å molecular sieves (1/16, Wako Pure Chemical Industries, Ltd.) before use. The submitters activated 4Å molecular sieves by using microwaves (500W, 3 x 90 s) and subsequent cycles of vacuum/argon purge.

4. 1,8-Diaminonaphthalene (>98%, Tokyo Chemical Industry Co., Ltd.) was used as received.

5. The reaction progress was monitored by GC. A reaction mixture aliquot was withdrawn and diluted with ethyl acetate. GC conditions: Column: TC-1 (GL Science), 30 m x 0.25 mm, film 0.25 μm; Flow rate: 1.89 mL/min; Injector temperature: 250 °C; Oven temperature: 100 °C to 250 °C at 20 °C/min, hold at 250 °C for 10 min; FID temperature: 250 °C; Retention times: (pin)B–B(pin) = 4.4 min, 1,8-diaminonaphthalene = 6.9 min, (pin)B–B(dan) = 12.5 min.

6. *n*-Hexane (95%, Kanto Chemical Co., Inc.) was used as received.

7. A vacuum filter holder (KGS-47) and a PTFE type membrane filter (T100A047A) were purchased from ADVANTEC. Pore size = 1 um.

8. Ethyl acetate (99%, Japan Alcohol Trading Co., Ltd.) was used as received.

9. Activated charcoal (Merck, 102186) was used as received.

10. Ethanol (99.5%, Wako Pure Chemical Industries, Ltd.) was used as received.

11. The product (1) exhibits the following analytical data: mp 191–193 °C; ^{1}H NMR (500 MHz, CDCl$_3$) δ: 1.31 (s, 12H), 6.21 (br s, 2H), 6.27 (d, *J* = 7.3 Hz, 2H), 7.00 (d, *J* = 8.2 Hz, 2H), 7.08 (dd, *J* = 7.7 Hz, 2H); ^{13}C NMR (125 MHz, CDCl$_3$) δ: 25.1, 83.3, 105.4, 117.6, 121.1, 127.5, 136.4, 140.5; [M + H] calcd for C$_{12}$H$_{20}$B$_2$N$_2$O$_2$: 293.1862 Found: 293.1866.

12. The purity was determined to be 99.8% wt. by quantitative ^{1}H NMR spectroscopy in CDCl$_3$ using 47.3 mg of the compound 1 and 76.5 mg of 1,3,5-trimethoxybenzene as an internal standard (D1 = 20 s).

13. A second reaction on identical scale provided 5.15 g (70%) of the identical product

14. Alternative isolation procedure: The end-of-reaction mixture is allowed to cool to ambient temperature (an orange slurry formed), 10 mL of toluene is distilled out via rotavap. The resulting orange slurry is filtered to afford a tan cake. Cake is then washed with hexane

(3 x 15 mL) and dried under vacuum (100 mmHg) via N_2 sweep for 3 h (tan solid, 6.1 g) (Figure 13). The resulting tan solid is dissolved in ethyl acetate (200 mL) treated with activated charcoal (1.5 g) at ambient temperature for 2 h with stirring, and filtered through a celite pad (5 g) with a Hirsch funnel. After removal of the solvent by rotary evaporation (40 °C, 16 mmHg), the residue dissolves in 50 mL of isopropyl acetate (IPAC) at 78 °C with a 1.5 cm Teflon-coated magnetic oval stir bar. The resulting solution is then cooled down to rt with stirring in 2 h and a slurry forms. The solvent (40 mL) is then distilled out via rotary evaporation and hexane (60 mL) is charged into the slurry in 1 h under stirring. This slurry is filtered and the cake is washed with hexane (2 x 10 mL). The cake is dried under vacuum with N_2 sweep for 3 h. Analytically pure (pin)B–B(dan) (5.1 g, 99.6 NMR wt%, 70%) as a white solid.

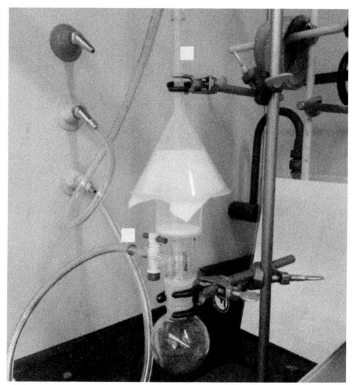

Figure 13. Drying under vacuum with nitrogen sweep

Working with Hazardous Chemicals

The procedures in *Organic Syntheses* are intended for use only by persons with proper training in experimental organic chemistry. All hazardous materials should be handled using the standard procedures for work with chemicals described in references such as "Prudent Practices in the Laboratory" (The National Academies Press, Washington, D.C., 2011; the full text can be accessed free of charge at http://www.nap.edu/catalog.php?record_id=12654). All chemical waste should be disposed of in accordance with local regulations. For general guidelines for the management of chemical waste, see Chapter 8 of Prudent Practices.

In some articles in *Organic Syntheses*, chemical-specific hazards are highlighted in red "Caution Notes" within a procedure. It is important to recognize that the absence of a caution note does not imply that no significant hazards are associated with the chemicals involved in that procedure. Prior to performing a reaction, a thorough risk assessment should be carried out that includes a review of the potential hazards associated with each chemical and experimental operation on the scale that is planned for the procedure. Guidelines for carrying out a risk assessment and for analyzing the hazards associated with chemicals can be found in Chapter 4 of Prudent Practices.

The procedures described in *Organic Syntheses* are provided as published and are conducted at one's own risk. *Organic Syntheses, Inc.*, its Editors, and its Board of Directors do not warrant or guarantee the safety of individuals using these procedures and hereby disclaim any liability for any injuries or damages claimed to have resulted from or related in any way to the procedures herein.

Discussion

The unsymmetrical diboron (**1**) has proven to be a powerful borylating reagent in organic synthesis, since the first synthesis of **1** and its synthetic application to the Pt- or Ir-catalyzed regioselective diboration of alkynes were reported in 2010 (Scheme 1).[2] The Pt-catalyzed diboration was also applicable to allenes,[3] in which a B(dan) moiety was regioselectively attached to a terminal carbon of allenes. One of the most characteristic

R—≡

Pt(dba)$_2$ (2 mol %)
P[3,5-(F$_3$C)$_2$C$_6$H$_3$]$_3$ (2.2 mol %)
or
[IrCl(cod)]$_2$ (1.5 mol %)

toluene, 80 °C

(pin)B   B(dan)

R

(pin)B–B(dan)

R
   C=•=
R

Pt(dba)$_2$ (4 mol %)
Sphos (6 mol %)

toluene, 80 °C

R   B(dan)

R   B(pin)

Scheme 1. Catalytic diboration using (pin)B–B(dan)

features of the diboration with **1** is that we can install a cross-coupling-active [B(pin)] and -inactive ("masked") [B(dan)] moieties into organic frameworks synchronously, leading to the chemoselective cross-coupling of the resulting diborylated compounds at the B(pin) moiety.[2,3] Furthermore, the high synthetic significance of **1** has also been demonstrated by its conversion into a masked borylcopper species [Cu–B(dan)] through selective σ-bond metathesis. The generated masked borylcopper species acted as a key intermediate in three-component B(dan)-installing reactions of unsaturated C–C bonds, including hydroboration,[4] borylstannylation,[5] aminoboration[6] and carboboration (Scheme 2).[7] In addition, various organic halides were directly convertible into R–B(dan) by copper-catalyzed substitution reaction.[8,9]

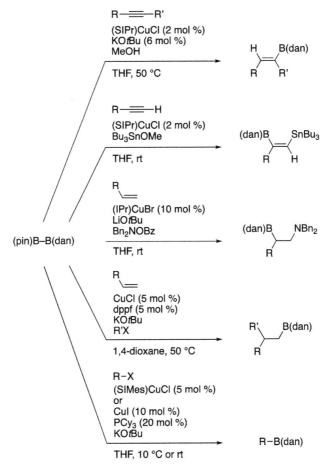

Scheme 2. Copper-catalyzed borylation using (pin)B–B(dan)

There have been two alternatives for synthesizing **1**, where equimolar amounts of pinacol, 1,8-diaminonaphthalene and tetrakis(dimethylamino)-diboron are employed in the presence or absence of 1N ethereal solution of hydrogen chloride (Scheme 3). The required tetrakis(dimethylamino)-diboron is prepared from tribromoborane[2] or bis(catecholato)diboron.[9c,d] These methods use relatively moisture-sensitive substrates, which have to be handled with care. In contrast, the present method enables **1** to be readily accessed by simply mixing bench-stable substrates of high availability in one step.

Scheme 3. Reported methods for synthesizing (pin)B–B(dan)

References

1. Department of Applied Chemistry, Graduate School of Engineering, Hiroshima University, Higashi-Hiroshima 739-8527, Japan; yhiroto@hiroshima-u.ac.jp.

2. Iwadate, N.; Suginome, M. *J. Am. Chem. Soc.* **2010**, *132*, 2548–2549.

3. Guo, X.; Nelson, A. K.; Slebodnick, C.; Santos, W. L. *ACS Catal.* **2015**, *5*, 2172–2176.

4. (a) Yoshida, H.; Takemoto, Y.; Takaki, K. *Chem. Commun.* **2014**, *50*, 8299–8302. (b) Yoshida, H.; Takemoto, Y.; Takaki, K. *Asian J. Org. Chem.* **2014**, *3*, 1204–1209.

5. Yoshida, H.; Takemoto, Y.; Takaki, K. *Chem. Commun.* **2015**, *51*, 6297–6300.

6. (a) Sakae, R.; Hirano, K.; Satoh, T.; Miura, M. *Angew. Chem. Int. Ed.* **2015**, *54*, 613–617. (b) Sakae, R.; Hirano, K.; Miura, M. *J. Am. Chem. Soc.* **2015**, *137*, 6460–6463. (c) Nishikawa, D.; Hirano, K.; Miura, M. *Org. Lett.* **2016**, *18*, 4856–4859.

7. Kageyuki, I.; Osaka, I.; Takaki, K.; Yoshida, H. *Org. Lett.* **2017**, *19*, 830–833.

8. Yoshida, H.; Takemoto, Y.; Kamio, S.; Osaka, I.; Takaki, K. *Org. Chem. Front.* **2017**, *4*, 1215–1219.

9. Other synthetic applications: (a) Cid, J.; Carbo, J. J.; Fernández, E. *Chem. Eur. J.* **2014**, *20*, 3616–3620. (b) Nagashima, Y.; Hirano, K.; Takita, R.; Uchiyama, M. *J. Am. Chem. Soc.* **2014**, *136*, 8532–8535. (c) Cuenca, A. B.; Cid, J.; García-López, D.; Carbó, J. J.; Fernández, E. *Org. Biomol. Chem.*

2015, *13*, 9659–9664. (d) Miralles, N.; Cid, J.; Cuenca, A. B.; Carbó, J. J.; Fernández, E. *Chem. Commun.* **2015**, *51*, 1693–1696. (e) Miralles, N.; Romero, R. M.; Fernández, E.; Muniz, K. *Chem. Commun.* **2015**, *51*, 14068–14071. (f) Xu, L.; Li, P. *Chem. Commun.* **2015**, *51*, 5656–5659. (g) Chen, K.; Zhang, S.; He, P.; Li, P. *Chem. Sci.* **2016**, *7*, 3676–3680. (h) Verma, A.; Snead, R. F.; Dai, Y.; Slebodnick, C.; Yang, Y.; Yu, H.; Yao, F.; Santos, W. L. *Angew. Chem. Int. Ed.* **2017**, *56*, 5111–5115. (i) Krautwald, S.; Bezdek, M. J.; Chirik, P. J. *J. Am. Chem. Soc.* **2017**, *139*, 3868–3875.

Appendix
Chemical Abstracts Nomenclature (Registry Number)

(pin)B–B(dan): 1*H*-Naphtho[1,8-*de*]-1,3,2-diazaborine, 2,3-dihydro-2-(4,4,5,5-tetramethyl-1,3,2-dioxaborolan-2-yl)-; (1214264-88-6)
Bis(pinacolato)diboron: 2,2'-Bi-1,3,2-dioxaborolane, 4,4,4',4',5,5,5',5'-octamethyl-; (73183-34-3)
1,8-Diaminonaphthalene: 1,8-Naphthalenediamine; (479-27-6)

Hiroto Yoshida was born in Fukuoka, Japan, in 1973. He graduated from Kyoto University in 1996 and received his Ph.D. from Kyoto University under the supervision of Professors Tamejiro Hiyama and Eiji Shirakawa in 2001. He then became an Assistant Professor at Hiroshima University in 2001 and was promoted to an Associate Professor in 2006. He received many awards including The Chemical Society of Japan Award for Young Chemists (2007) and The Young Scientists' Prize, The Commendation for Science and Technology by the Minister of Education, Culture, Sports, Science and Technology (2009). His research interests include (1) catalytic reactions for synthesis of main group organometallics containing boron, tin or silicon, and (2) aryne-based organic synthesis.

Yuya Murashige was born in Yamaguchi, Japan, in 1993. He graduated from Hiroshima University in 2017, and has been a master's course student in the same department, focusing on development of copper-catalyzed borylation reactions.

Itaru Osaka received his Ph.D. from University of Tsukuba in 2002 under the supervision of Professor Kazuo Akagi and Professor Hideki Shirakawa. After a 4-year research stint at Fujifilm, he joined Professor R. D. McCullough's group at Carnegie Mellon University as a postdoctoral fellow in 2006. He started his professional carrier at Hiroshima University as an assistant professor in 2009, and moved to RIKEN as a senior research scientist in 2013. He was then appointed as a professor at Hiroshima University in 2016. His research focuses the design and synthesis of π-conjugated polymers for organic electronics.

Feng Peng joined the Process Research Department of Merck & Co., Inc. in 2012. His research focuses on using state-of-art organic chemistry to address critical problems in drug development. He received his B. S. degree from Beijing Normal University. He obtained his M.S. under the supervision of Professor Dennis Hall at University of Alberta with a research focus on Boron Chemistry. Feng then moved to New York City, where he obtained Ph.D. in the area of total synthesis (maoecrystal V) with Professor Samuel Danishefsky at Columbia University.

Discussion Addendum for:
Preparation of a Carbazole-Based Macrocycle Via Precipitation-Driven Alkyne Metathesis

Christopher C. Pattillo, Morgan M. Cencer, and Jeffrey S. Moore*[1]

Department of Chemistry, University of Illinois at Urbana-Champaign, Urbana, IL, 61820

Original Article: Zhang, W.; Cho, H.M.; Moore, J.S. Org. Synth. **2007**, *84*, 177-191

Within the last two decades, the development of efficient alkyne metathesis catalytic systems has allowed for the application of this methodology in both organic synthesis and materials chemistry.[2–6] The utility of alkyne metathesis dynamic covalent chemistry (DC$_V$C) for the preparation of complex molecular architectures has resulted in the use of this method for the synthesis of discrete, shape persistent molecular structures.[3]

The inherent reversibility of alkyne metathesis means that undesired byproducts must be continually removed in order to drive the reaction forward to a desired product.[4,5] Classically, this may be achieved through the use of propynylated substrates that produce 2-butyne as a reaction byproduct (Figure 1A).[5,7] This volatile product may then be removed *via* continually subjecting the reaction mixture to high vacuum.[5,7] While this method can efficiently facilitate conversion to the desired product, the use of vacuum-driven metathesis conditions can lead to a number of drawbacks.[4] In order to minimize solvent loss, high-boiling solvents such as toluene or chlorobenzenes are often used, or solvent must be periodically added.[4] The efficiency of this method also decreases dramatically as the scale of the reaction is increased, greatly limiting the utility of vacuum-driven metathesis on large scale.[7]

DOI: 10.15227/orgsyn.095.0231

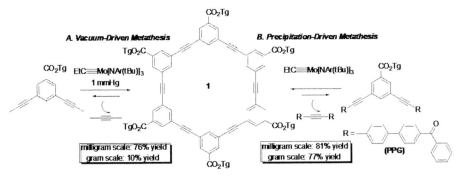

Figure 1. Comparison of vacuum and precipitation-driven alkyne metathesis

In order to address these limitations, we developed a *precipitation-driven* alkyne metathesis system (Figure 1B).[7,8] This strategy relies on the incorporation of a *precipitating group* (PPG) into the desired reaction precursor (Figure 1B).[7,8] Upon metathesis, the resulting diaryl acetylene byproduct precipitates from the reaction mixture thus driving the reaction equilibrium toward product formation.[7] The advantage of this method is that it is highly effective for the large-scale synthesis of a variety of macrocycles (i.e. **1**) *via* alkyne-metathesis that would afford low yields under classical vacuum-driven methods.[7,8]

In the years since we reported this method, the precipitating group strategy has been employed in both our own work and that of others.[3] In addition to our original report where we prepared a variety of phenylene ethynylene macrocycles, we have also reported the synthesis of cyclic trimer **2** *via* precipitation-driven alkyne metathesis (Figure 2).[7–9]

Upon mixing macrocycles **2** and **3** under metathesis conditions, tetrameric macrocycle **4** could be identified by mass spectrum analysis, demonstrating that arylene ethynylene macrocycles are dynamic under these conditions.[9] Our own group and others have further demonstrated the utility of precipitation-driven alkyne metathesis in the development of a variety of robust alkyne metathesis catalysts.[6,10–14] In order to simplify purification and catalyst lifetimes we have demonstrated that both solid silica-supported and polyhedral oligomeric silsesquioxane (POSS) ligands are highly active and selective catalysts for a variety of substrates, including those which rely on the use of the precipitation strategy.[13,14] Zhang and coworkers have also pioneered the use of multidentate triphenol ligands that also show high catalytic activity and functional group tolerance and can be extended to the use of our precipitation driven method in high yields and short reaction times.[10–12,15]

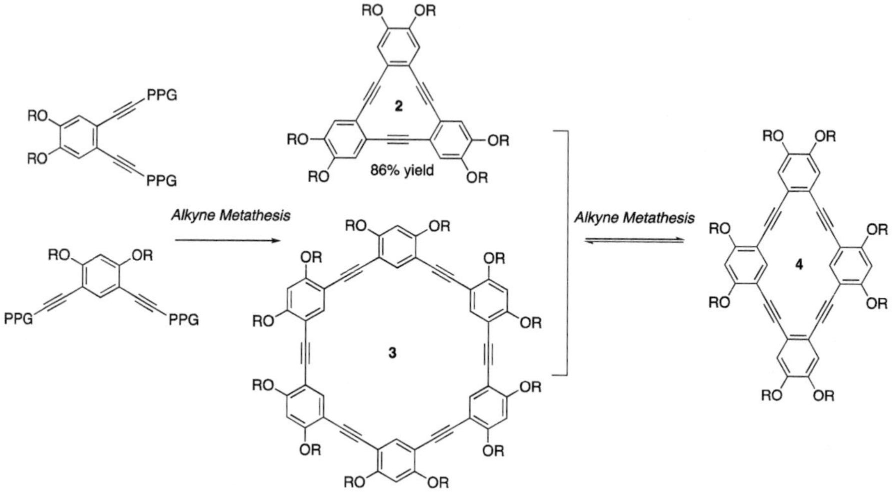

Figure 2. Dynamic trimeric and hexameric macrocycles
via **precipitation-driven metathesis**

The utility of precipitation-driven alkyne metathesis is particularly noteworthy for the synthesis of complex molecular cages.[16–20] Zhang and coworkers have made use of this methodology for the preparation of a number of molecular cages (Figure 3). Cage **5** was prepared in 56% yield *via* precipitation driven metathesis from a rigid porphyrin precursor.[17] Notably, this shape-persistent molecular prism shows high binding affinity for both C_{60} and C_{70} fullerenes, with the binding constant for encapsulation of C_{70} being three orders of magnitude greater than that for C_{60}.

Following this report, Zhang and coworkers prepared bisporphyrin macrocycle **6** in 60% yield through precipitation alkyne metathesis.[19] Fullerene binding studies demonstrated that this macrocycle shows the highest affinity for C_{84}. These results demonstrate the ability to prepare a number of structures with variable properties that can be accessed through efficient precipitation-driven metathesis.[19] In addition to facilitating efficient metathesis, the use of the precipitating group moiety can improve the purification of complex substrates.[16,20] Zhang and coworkers have demonstrated this in the synthesis of both carbazole[20] and porphyrin[16] -

based precursors. In both examples, the inclusion of the precipitating group was found to improve purification owing to its increased polarity over other alkyne substitutions. These results demonstrate the additional utility of this strategy as a handle for improved purification of metathesis precursors.

5
Selective for C_{70}

6
Selective for C_{84}

Figure 3. Molecular cages prepared by Zhang and coworkers using precipitation-driven metathesis

While precipitation driven alkyne metathesis is broadly applicable and has been widely used for many different substrates, this method requires additional synthetic overhead to include the precipitating moiety.[4,7] Additionally, the high molecular weight of the precipitating group results in poor atom economy for the overall reaction.[4] Methyl-capped alkynes produce volatile 2-butyne as a side product, which has classically been removed *via* vacuum to drive the reaction.[7] In 2010, Fürstner and coworkers reported that 5Å molecular sieves efficiently sequester 2-butyne with comparable yields as precipitation and vacuum-driven procedures.[21]

Molecular sieves have successfully been used to drive alkyne metathesis on a variety of different substrates, and can be used for large-scale synthesis in comparable yields to precipitation-based methods as demonstrated in the synthesis of carbazole macrocycle **7** (Figure 4).[8,21] We have also demonstrated that carbazole macrocycle **7** may be prepared *via* a depolymerization/macrocyclization route from polymeric carbazole precursors.[22]

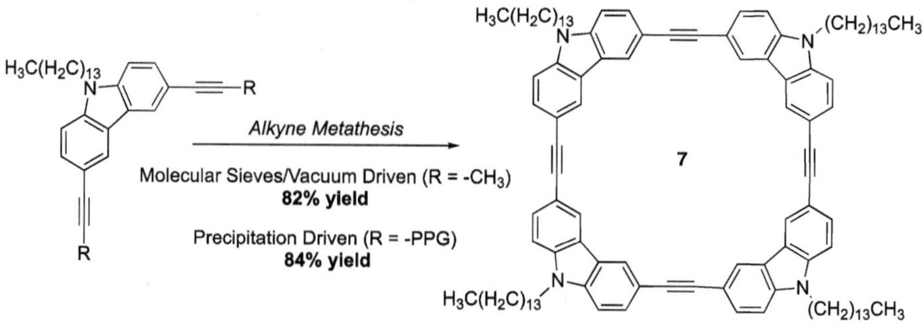

Figure 4. Comparison of molecular sieves and vacuum/precipitation driven methods

The use of molecular sieves in alkyne metathesis is an effective and general method that has been widely applied in both natural product and materials synthesis.[3,4] In recent years we have heavily relied on this method for the synthesis of a number of complex molecular architectures (Figure 5).[23-26] In 2016 we reported the preparation of kinetically trapped tetrahedral cages (**8**) *via* alkyne metathesis from methyl-capped tritopic precursors using molecular sieves to sequester 2-butyne.[25] This reaction proceeds through initial formation of oligomeric intermediates that 'self-correct' to the tetrahedral cage product. The use of propynylated substrates is particularly advantageous in terms of solubility for both precursors and reaction products, especially for reactions that proceed through initial formation of oligomeric intermediates along the pathway to a discrete product. We have also demonstrated the cyclooligomerization of bifunctional precursors to prepare cycloparaphenyleneacetylene precursors (**9**). Both of these processes proceed in near quantitative yields through the use of molecular sieves and can be performed on gram scales with similar efficiency.[26] Commercial grade 5Å powdered molecular sieves are used for these reactions, and must be thoroughly dried in a vacuum oven prior to use. We typically employ a ratio of 800 mg to 1 g of molecular sieves per millimole of propynyl groups, and recommend stirring the mixture vigorously to ensure distribution of powdered sieves.[25,26]

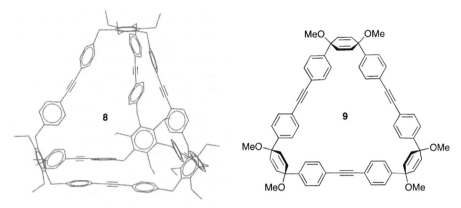

Figure 5. Tetrahedral cages and cyclooligomers prepared via alkyne metathesis using molecular sieves

The use of a precipitating group or a molecular sieves approach largely depends on both the synthetic overhead and structural characteristics of the system in question. The use of a precipitating-group strategy can be advantageous in terms of both scalability and purification of complex reaction precursors.[8,16] The use of this moiety also results in poor atom economy and can complicate solubility of reaction precursors. The use of methyl-capped precursors in conjunction with molecular sieves is an attractive and scalable method with high efficiency.[4,25] Installation of propyne groups is synthetically simple and significantly more atom economical. Overall, while both precipitation and molecular sieve strategies are viable methods that furnish very comparable yields and scalability across a wide range of precursors, the efficiency and synthetic ease of using a molecular-sieves driven approach make this method the most general for driving alkyne metathesis reactions.[4,7]

References

1. Department of Chemistry, University of Illinois at Urbana-Champaign, Urbana, IL, 61820. email: jsmoore@illinois.edu. This work was supported by the National Science Foundation, CHE Division under Grant Number 16-10328.
2. Fürstner, A. *Alkyne Metathesis*; 2014; Vol. 5.

3. Jin, Y.; Wang, Q.; Taynton, P.; Zhang, W. *Acc. Chem. Res.* **2014**, *47* (5), 1575–1586.
4. Fürstner, A. *Angew. Chem. Int. Ed.* **2013**, *52* (10), 2794–2819.
5. Fürstner, A. *Handb. Metathesis Second Ed.* **2015**, *2–3*, 445–501.
6. Gross, D. E.; Zang, L.; Moore, J. S. *Pure Appl. Chem.* **2012**, *84* (4), 869–878.
7. Zhang, W.; Moore, J. S. *J. Am. Chem. Soc.* **2004**, *126*, 12796.
8. Zhang, W.; Cho, H.M.; Moore, J.S. *Org. Synth.* **2007**, *84, 177–191*
9. Zhang, W.; Brombosz, S. M.; Mendoza, J. L.; Moore, J. S. *J. Org. Chem.* **2005**, *70* (24), 10198–10201.
10. Jyothish, K.; Zhang, W. *Angew. Chem. Int. Ed.* **2011**, *50* (15), 3435–3438.
11. Yang, H.; Liu, Z.; Zhang, W. *Adv. Synth. Catal.* **2013**, *355* (5), 885–890.
12. Du, Y.; Yang, H.; Zhu, C.; Ortiz, M.; Okochi, K. D.; Shoemaker, R.; Jin, Y.; Zhang, W. *Chem. Eur. J.* **2016**, *22* (23), 7959–7963.
13. Cho, H. M.; Weissman, H.; Moore, J. S. *J. Org. Chem.* **2008**, *73* (5), 4256–4258.
14. Hyeon, M. C.; Weissman, H.; Wilson, S. R.; Moore, J. S. *J. Am. Chem. Soc.* **2006**, *128* (46), 14742–14743.
15. Jyothish, K.; Wang, Q.; Zhang, W. *Adv. Synth. Catal.* **2012**, *354* (11–12), 2073–2078.
16. Yu, C.; Long, H.; Jin, Y.; Zhang, W. *Org. Lett.* **2016**, *18* (12), 2946–2949.
17. Zhang, C.; Wang, Q.; Long, H.; Zhang, W. *J. Am. Chem. Soc.* **2011**, *133* (51), 20995–21001.
18. Wang, Q.; Yu, C.; Zhang, C.; Long, H.; Azarnoush, S.; Jin, Y.; Zhang, W. *Chem. Sci.* **2016**, *7* (5), 3370–3376.
19. Zhang, C.; Long, H.; Zhang, W. *Chem. Commun.* **2012**, *48* (49), 6172.
20. Wang, Q.; Yu, C.; Long, H.; Du, Y.; Jin, Y.; Zhang, W. *Angew. Chem. Int. Ed.* **2015**, *54* (26), 7550–7554.
21. Heppekausen, J.; Stade, R.; Goddard, R.; Fürstner, A. *J. Am. Chem. Soc.* **2010**, *132* (32), 11045–11057.
22. Gross, D.; Moore, J. *Macromolecules* **2011**, 3685–3687.
23. Moneypenny, T. P.; Liu, H.; Yang, A.; Robertson, I. D.; Moore, J. S. *J. Polym. Sci. Part A Polym. Chem.* **2017**, 1–14.
24. Moneypenny, T. P.; Walter, N. P.; Cai, Z.; Miao, Y.-R.; Gray, D. L.; Hinman, J. J.; Lee, S.; Zhang, Y.; Moore, J. S. *J. Am. Chem. Soc.* **2017**, 3259–3264.
25. Lee, S.; Yang, A.; Moneypenny, T. P.; Moore, J. S. *J. Am. Chem. Soc.* **2016**, *138* (7), 2182–2185.
26. Lee, S.; Chénard, E.; Gray, D. L.; Moore, J. S. *J. Am. Chem. Soc.* **2016**, *138* (42), 13814–13817.

Christopher Pattillo received his B.S. in Chemistry from California Polytechnic State University, San Luis Obispo in 2013. He is currently a Ph.D. Candidate in Chemistry at the University of Illinois at Urbana-Champaign in the lab of Professor Jeffrey Moore. His research is focused on the synthesis of novel molecular structures *via* alkyne metathesis DCC and expanding the scope of orthogonal chemistries in dynamic synthesis.

Morgan Cencer received her B.S. in Chemistry from Michigan Technological University. She is currently pursuing her Ph.D. in Materials Chemistry at the University of Illinois at Urbana-Champaign under the supervision of Jeffery Moore. Her current research focuses on the use of discrete molecular cages as a substrate for solid-state lithium ion electrolytes.

Professor Jeffrey Moore received his B.S. in chemistry in 1984 from the University of Illinois, and his Ph.D. in Materials Science and Engineering with Samuel Stupp in 1989. After a NSF postdoctoral position at Caltech with Robert Grubbs, he began his independent career at the University of Michigan in Ann Arbor. He returned in 1993 to the University of Illinois, where he is currently the Director of the Beckman Institute for Advanced Science and Technology, Ikenberry Endowed Chair, and Professor in the Departments of Chemistry and Materials Science and Engineering.

Hydrodecyanation by a Sodium Hydride-Iodide Composite

Guo Hao Chan, Derek Yiren Ong, and Shunsuke Chiba*[1]

Division of Chemistry and Biological Chemistry, School of Physical and Mathematical Sciences, Nanyang Technological University, Singapore 637371, Singapore

Checked by Joyce Leung and Chris Senanayake

A.

$$\text{PhCH}_2\text{CN} + \text{Br}\diagdown\diagdown\text{Br} \xrightarrow[\substack{50\% \text{ aq. NaOH} \\ 65\,°C}]{\text{BnEt}_3\text{NCl (2.5 mol\%)}} \mathbf{1}$$

B.

$$\mathbf{1} \xrightarrow[\substack{\text{THF} \\ \text{reflux}}]{\text{NaH, LiI}} \mathbf{2}$$

Procedure (Note 1)

A. *1-Phenylcyclopentane-1-carbonitrile (1)* (Note 2). A 100 mL two-necked, round-bottomed flask is equipped with a 25 x 12 mm Teflon-coated, oval magnetic stir bar, a thermometer, rubber septum. Benzyltriethylammonium chloride (569 mg, 2.50 mmol, 0.025 equiv) (Note 3) and 50% (w/v) aqueous sodium hydroxide (25 mL) are added. The reaction vessel is placed in a water bath and benzyl cyanide (11.5 mL, 99.6 mmol, 1.0 equiv) (Note 4) is added in one portion with vigorous stirring (Note 5). 1,4-Dibromobutane (14.3 mL, 120 mmol, 1.2 equiv) (Note 6) is added portion wise (ca. 5 mL per portion) (Note 7) and the rubber septum is replaced with a Graham reflux condenser (Note 8) connected to a Schlenk

Org. Synth. **2018**, *95*, 240-255
DOI: 10.15227/orgsyn.095.0240

Published on the Web 9/7/2018
© 2018 Organic Syntheses, Inc.

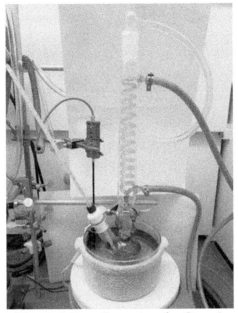

Figure 1. Reaction setup for Step A

line (Figure 1). The orange biphasic mixture is heated to an internal temperature of 65 °C (Note 9) for 24 h resulting in an orange biphasic suspension (Notes 10 and 11). The reaction is cooled to 23 °C (inner temperature). Water (50 mL) is added and the layers are partitioned in a 250-mL separatory funnel. The aqueous layer is extracted twice with diethyl ether (60 mL x 2) and the combined organic layer is washed with brine (50 mL), dried over anhydrous magnesium sulfate (5 g), filtered via gravity filtration and the filtrate is concentrated on a rotary evaporator under reduced pressure (40 °C, 14 mmHg) to afford an orange oil. The obtained residue is purified by column chromatography (Note 12) to afford (14.0 g, 81.9 mmol, 82%) (Note 13) of 1-phenylcyclopentane-1-carbonitrile as a pale yellow oil (Figure 2) (Notes 14 and 15).

Org. Synth. **2018**, *95*, 240-255   **241**   DOI: 10.15227/orgsyn.095.0240

Figure 2. Purified nitrile 1

B. *Cyclopentylbenzene (2).* An oven-dried, 250 mL, three-necked round-bottomed flask equipped with a 30 x 16 mm Teflon-coated, oval magnetic stir bar, a type-T thermocouple, rubber septum and a Graham reflux condenser is connected to a Schlenk line (Notes 16 and 17). Sodium hydride (NaH) (3.75 g, 93.7 mmol, 2.0 equiv) (Note 18) and lithium iodide (LiI) (6.27 g, 46.8 mmol, 1.0 equiv) (Note 19) is charged, after which the reaction vessel is evacuated and backfilled with nitrogen three times. Dry THF (80 mL) (Note 20) is added via syringe (Note 21) and 1-phenylcyclopentane-1-carbonitrile (*1*) (8.02 g, 46.8 mmol, 1.0 equiv) is added in one portion with a THF (14 mL) rinse (Note 22). Rubber septum is quickly replaced with a 24/40 glass stopper (Note 23). The grey suspension is then refluxed (Note 24) for 5 h resulting in tan suspension (Notes 25 and 26) (Figure 3). The reaction mixture is cooled to 4 °C (inner temperature) with an ice water bath. The glass stopper is quickly replaced with a rubber septum and cold water (12 mL) is added dropwise over 5 min (Note 27) before 88 mL of cold water is added in one portion. The layers are partitioned in a 500-mL separatory funnel. The aqueous layer (Note 28) is extracted twice with diethyl ether (100 mL x 2), and the combined organic layer is washed with brine (100 mL), dried over anhydrous magnesium sulfate (10 g), and filtered via Büchner filtration. The filtrate is concentrated on a rotary evaporator under reduced pressure (30 °C, 14 mmHg) before transferring to a 50-mL

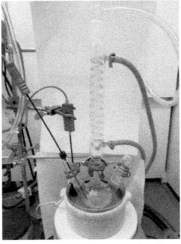

Figure 3. Reaction setup for Step B

round-bottomed flask to afford a dark yellow oil. The round-bottomed flask is equipped with a 15 x 6 mm Teflon-coated, oval magnetic stir bar and attached to the short-path distillation setup (Figure 4) (Notes 29 and 30). The product is distilled under vacuum in one fraction (0.92 mmHg, 56–57 °C) (Notes 31 and 32) affording cyclopentylbenzene (2) (Notes 33 and 34) as a colorless liquid (6.0 g, 41.1 mmol, 88%) (Figure 5) (Note 35).

Figure 4. Distillation setup for purification of cyclopentylbenzene (2)

Figure 5. Purified cyclopentylbenzene (2)

Notes

1. Prior to performing each reaction, a thorough hazard analysis and risk assessment should be carried out with regard to each chemical substance and experimental operation on the scale planned and in the context of the laboratory where the procedures will be carried out. Guidelines for carrying out risk assessments and for analyzing the hazards associated with chemicals can be found in references such as Chapter 4 of "Prudent Practices in the Laboratory" (The National Academies Press, Washington, D.C., 2011; the full text can be accessed free of charge at https://www.nap.edu/catalog/12654/prudent-practices-in-the-laboratory-handling-and-management-of-chemical. See also "Identifying and Evaluating Hazards in Research Laboratories" (American Chemical Society, 2015) which is available via the associated website "Hazard Assessment in Research Laboratories" at https://www.acs.org/content/acs/en/about/governance/committees/chemicalsafety/hazard-assessment.html. In the case of this procedure, the risk assessment should include (but not necessarily be limited to) an evaluation of the potential hazards associated with sodium hydride, lithium iodide, benzyl cyanide, 1,4-dibromobutane, sodium cyanide, hydrogen gas, magnesium sulfate, sodium hypochlorite solution, THF, hexane, ethyl acetate and diethyl ether.

2. The procedure is slightly modified from the report by M. Makosza and A. Jonczyk.[2]

3. Benzyltriethylammonium chloride was purchased from Oakwood and was further purified by reprecipitating it through addition of ether to its saturated hot solution in acetone. The submitter purchased the reagent from Alfa Aesar.

4. Benzyl cyanide 98% was purchased from Sigma-Aldrich and used as is.

5. Submitter conducted the reaction at the stirring rate of 1400 rpm.

6. 1,4-Dibromobutane 98+% was purchased from Sigma Aldrich and used as is. Submitter purchased the reagent from Alfa Aesar.

7. One portion was added over 10 seconds by syringe, with a 30 second interval to the next addition. There was 6 °C exotherm observed after the first portion, but subsequent portions do not increase the internal temperature further. Submitter observed an initial 22 °C exotherm after the first addition.

8. Submitter used a Liebig reflux condenser.

9. The temperature controller was set to 66 °C, while the oil bath temperature fluctuated between ±2 °C during the reaction. An initial increase of internal temperature to 72–73 °C was observed for the first 20 min.

10. The reaction was monitored by TLC (EMD Millipore™ TLC Silica Gel 60 F254, glass plates) on silica gel using hexanes:EtOAc (93:7) as eluent and visualization with UV light. Benzyl cyanide had R_f = 0.24 and the product (1) has R_f = 0.46.

11. The time for reaction completion varies with agitation speed. Submitter reported a reaction time of 7 h.

12. Column chromatography was performed using a Chemglass Life Science chromatography column with reservoir and 4.0 mm PTFE stopcock, 1000 mL, with coarse fritted disc, top joint: ST/NS 24/40, I.D. × L 2.5 in × 8.0 in. The column was packed with 250 g of silica gel, high-purity grade, pore size 60 Å, 230-400 mesh particle size purchased from Sigma-Aldrich and was conditioned with 1 L hexanes. Residual solvent was removed before crude compound (1) was loaded on the column neat and hexanes (10 mL) was used to wash the flask containing compound (1) before loading on the column. The column was eluted with 1 L of hexanes, followed by 1 L of hexanes/EtOAc (98:2). At this point all eluent was collected, the column was eluted with 1.4 L of hexanes/EtOAc (98:2). The compound (1) was then concentrated on a rotary evaporator under reduced pressure (40 °C, 84 mmHg) and further dried under vacuum (room temperature, 0.92 mmHg) overnight.

13. A second reaction on identical scale provided 14.3 g (84%) of the same product. The Submitter observed potential insufficient separation between the unreacted 1,4-dibromobutane and compound (**1**).

14. 1-Phenylcyclopentane-1-carbonitrile (**1**) is bench stable. It displays the following characterization data: R_f = 0.46 (hexanes/ethyl acetate 93:7); ^{1}H NMR (500 MHz, CDCl$_3$) δ: 1.90–1.98 (m, 2H), 2.01–2.12 (m, 4H), 2.46–2.51 (m, 2H), 7.29–7.30 (m, 1H), 7.36–7.39 (m, 2H), 7.45–7.47 (m, 2H); ^{13}C NMR (125 MHz, CDCl$_3$) δ: 24.3, 40.5, 47.8, 124.4, 126.0, 127.8, 128.9, 139.9. IR (film) 2964, 2876, 2232, 1600, 1465, 1448, 756, 696, 512 cm^{-1}. HRMS (DART) m/z calcd for C$_{12}$H$_{14}$N [M+H]$^+$: 172.11216. Found: 172.11208.

15. Purity was determined to be 99.4% on the first run and 97.3% on the second run using qNMR and 1,3,5-trimethoxybenzene as an internal standard.

16. Submitter further flame-dried all glassware under vacuum (5 mmHg) and backfilled with nitrogen once cooled to room temperature.

17. Submitter used a Liebig reflux condenser and a thermometer instead of the thermocouple.

18. Sodium hydride in 60% dispersion mineral oil was purchased from Sigma-Aldrich. Submitter stored the reagent in an argon-filled glovebox.

19. Lithium iodide bead, (anhydrous, −10 mesh, 99.99% trace metals basis) was purchased from Sigma-Aldrich and used as received. Submitter purchased lithium iodide beads (99%) from Sigma-Aldrich and the reagent was dried over P$_4$O$_{10}$ under reduced pressure (5 mmHg) at 120 °C before using.

20. Tetrahydrofuran, anhydrous, contains 250 ppm BHT as inhibitor, ≥99.9% was bought from Sigma-Aldrich and used as received. Submitter purchased tetrahydrofuran, HiPerSolv CHROMANORM® for HPLC from VWR Chemicals, which was further purified with Pure Solv MD-5 solvent purification system by Innovative Technology before use.

21. There was a 19-20 °C exotherm observed.

22. The reaction mixture was cooled to about 30 °C before compound (**1**) was added.

23. Submitter used rubber septum for the reaction. A build-up in pressure was observed during the course of the reaction; hence, a glass stopper was replaced prior to heating to reflux.

24. The temperature controller was set to 75 °C, while the oil bath temperature fluctuated between ±2 °C during the reaction.

25. The reaction was monitored by TLC (EMD Millipore™ TLC Silica Gel 60 F254, glass plates) on silica gel using hexanes:EtOAc (96:4) as eluent and visualization with UV light. Product (**1**) had R_f = 0.33 and the product (**2**) has R_f = 0.89.

26. A pinkish hue reaction mixture with white solids was observed by the submitter after 5 hours of heating.

27. There was a 14-15 °C exotherm observed and the reaction vessel should be connected to a bubbler to vent the hydrogen gas generated.

28. The aqueous layer containing NaCN as a co-product of the reaction was treated with sodium hypochlorite solution, available chlorine 14.5 % (100 mL) purchased from Alfa Aesar overnight before discarding.

29. Vacuum was applied slowly to prevent bumping of the compound in the distillation flask.

30. The distillation set-up was evacuated and purged with nitrogen gas three times before final vacuum was applied.

31. Residual solvent was removed before heating was applied, oil bath temperature was slowly raised from 70 °C to a final temperature of 85 °C.

32. The receiver flask was submerged in an ice-bath.

33. Cyclopentylbenzene (**2**) is bench stable. It displays the following characterization data: R_f = 0.89 (hexanes/ethyl acetate 96:4); ^{1}H NMR (500 MHz, CDCl$_3$) δ: 1.56–1.63 (m, 2H), 1.64–1.72 (m, 2H), 1.77–1.84 (m, 2H), 2.04–2.09 (m, 2H), 2.95–3.02 (m, 1H), 7.15–7.18 (m, 1H), 7.23–7.29 (m, 4H); ^{13}C NMR (125 MHz, CDCl$_3$) δ: 25.5, 34.6, 46.0, 125.7, 127.1, 128.2, 146.5. IR (film) 3060, 3027, 2950, 2867, 1603, 1492, 1451, 754, 696, 525 cm^{-1}. HRMS (DART) m/z calcd for C$_{11}$H$_{15}$ [M+H]$^+$: 147.11689. found: 147.11683.

34. Purity was determined to be ≥99.5% for both runs using qNMR and 1,3,5-trimethoxybenzene as an internal standard.

35. A yield of 90% was obtained on the second run at the same scale.

Working with Hazardous Chemicals

The procedures in *Organic Syntheses* are intended for use only by persons with proper training in experimental organic chemistry. All hazardous materials should be handled using the standard procedures for work with chemicals described in references such as "Prudent Practices in the Laboratory" (The National Academies Press, Washington, D.C., 2011; the full text can be accessed free of charge at http://www.nap.edu/catalog.php?record_id=12654). All chemical waste should be disposed of in accordance with local regulations. For general guidelines for the management of chemical waste, see Chapter 8 of Prudent Practices.

In some articles in *Organic Syntheses*, chemical-specific hazards are highlighted in red "Caution Notes" within a procedure. It is important to recognize that the absence of a caution note does not imply that no significant hazards are associated with the chemicals involved in that procedure. Prior to performing a reaction, a thorough risk assessment should be carried out that includes a review of the potential hazards associated with each chemical and experimental operation on the scale that is planned for the procedure. Guidelines for carrying out a risk assessment and for analyzing the hazards associated with chemicals can be found in Chapter 4 of Prudent Practices.

The procedures described in *Organic Syntheses* are provided as published and are conducted at one's own risk. *Organic Syntheses, Inc.*, its Editors, and its Board of Directors do not warrant or guarantee the safety of individuals using these procedures and hereby disclaim any liability for any injuries or damages claimed to have resulted from or related in any way to the procedures herein.

Discussion

Nitriles are versatile synthons for chemical synthesis. Hydride reduction of nitriles can be performed by various covalent hydride reagents to form aldehydes or amines, depending on the choice of covalent hydride reagents.[3] Several methods for hydrodecyanation of nitriles have also been developed.[4] We recently reported the use of ionic sodium hydride (NaH) in the presence of lithium iodide (LiI) for hydrodecyanation of α-quaternary benzyl cyanides,[5] that is contemporary to the existing procedures. This manuscript describes 2-step synthesis of cycloalkyl arenes from readily available benzyl cyanides.

The synthesis is started from α-dialkylation of benzyl cyanides with dihaloalkanes under phase-transfer conditions catalyzed by tetraalkylammonium salt in alkali aqueous media or under NaH-mediated reaction conditions in THF (Table 1). These processes worked smoothly to construct cyclohexane, -pentane, and -butane rings as well as a tetrahydropyrane ring α to the cyano group.

Table 1. Synthesis of α-quaternary nitriles

$$\text{benzyl cyanides} + \underset{\text{(1.2 equiv)}}{\text{dihalides}} \xrightarrow[\text{A or B}]{\text{conditions}} \text{products}$$

conditions **A**: Et₃BnCl (2.5 mol%), 50% aq. NaOH, 65 °C, 7 h
conditions **B**: NaH (2.5 equiv) in THF, 0 °C to reflux, 16 h

entry	benzyl cyanides	dihalides	products	conditions	yields[a]
1	MeO— ⟨phenyl⟩ —CH₂CN (20.3 mmol)	Br ⟨chain⟩ Br	MeO— cyclohexane–CN	B	73%
2	Cl— ⟨phenyl⟩ —CH₂CN (20.0 mmol)	Br ⟨chain⟩ Br	Cl— cyclohexane–CN	B	71%
3	⟨phenyl⟩ —CH₂CN (99.6 mmol)	Br ⟨chain⟩ Br	cyclopentane–CN	A	90%
4	⟨phenyl⟩ —CH₂CN (19.9 mmol)	Cl ⟨O-chain⟩ Cl	tetrahydropyran–CN	A	64%
5	Ph— ⟨phenyl⟩ —CH₂CN (8.01 mmol)	Br ⟨chain⟩ Br	Ph— cyclobutane–CN	B[b]	69%

[a] Isolated yields. [b] The reaction was conducted in DMSO-Et$_2$O (1:1) as the solvent at room temperature for 24 h.

We recently uncovered that unique hydride donor reactivity is installed onto sodium hydride (NaH) by its solvothermal treatment with NaI or LiI in THF and the resulting NaH-iodide composites could be used for a series of hydride reduction.[6,7] Treatment of α–quaternary benzyl cyanides with NaH and LiI in THF under reflux conditions[8] enables hydrodecyanation to give cycloalkylarenes in good yields (Scheme 1).

85%	92%	92%	91%
(4.56 mmol scale)	(5.34 mmol scale)	(46.8 mmol scale)	(4.30 mmol scale)

Scheme 1. Cycloalkyl arenes synthesized by hydrodecyanation

The process is initiated by hydride attack of the nitrile moiety of **1** from NaH to afford iminyl sodium intermediate **A**, which subsequently undergoes concerted C-C bond cleavage-1,2-proton shift to give hydrodecyanated product **2** and sodium cyanide (NaCN). In the transition state **B** for this C-C bond cleavage, the imine hydrogen atom has partial positive charge (δ+) and the benzylic carbon possesses partial negative charge (δ–), and thus the hydrogen is rearranged to the adjacent carbon (δ–) via 1,2-proton transfer.[9] This result demonstrates the unique *umpolung* nature of the decyanation, where the nucleophilic hydride originated from NaH is changed to the electrophilic proton in the later stage.

Scheme 2. Proposed reaction mechanism of hydrodecyanation

It is observed that an electron-rich 4-methoxyphenyl group renders the rate of the C-C bond cleavage slower (Scheme 2). The reaction under reflux conditions afforded not only 4-cyclohexylanisole in 54% yield but also provides the corresponding aldehyde in 31% yield even after running for 40 h. Higher reaction temperature (85 °C) in sealed conditions (Figure 6) allowed for completion of the process within 26 h, to give 4-cyclohexylanisole as a sole product in 89% yield. It should be noted that the present protocol is thus far proven unsuccessful for non-benzylic cyanides.

conditions		
reflux 40 h	54%	31%
85 °C (sealed tube) 26 h	89%	0%

Scheme 3. Reactions of 1-(4-methoxyphenyl)cyclohexane-1-carbonitrile

Figure 6. Reaction setup in sealed conditions (photo provided by submitter)

References

1. Division of Chemistry and Biological Chemistry, School of Physical and Mathematical Sciences, Nanyang Technological University, Singapore 637371, Singapore. shunsuke@ntu.edu.sg; this work was supported by Nanyang Technological University, the Singapore Ministry of Education (Academic Research Fund Tier 1: RG10/17), Singapore Economic Development Board (EDB), and Pfizer Asia Pacific Pte. Ltd. GHC thanks to EDB-Industrial Postgraduate Program (IPP) for the scholarship support.
2. Makosza, M.; Jonczyk, A. *Org. Synth.* **1976**, *55*, 91–95.
3. Barrett, A. G. M. In Comprehensive Organic Synthesis; Trost, B. M., Fleming, I., Eds.; Pergamon Press: Oxford, 1991; Vol. 8, pp 251–257.
4. For reviews on hydrodecyanation, see: (a) Mattalia, J. M.; Marchi-Delapierre, C.; Hazimeh, H.; Chanon, M. *Arkivoc* **2006**, *iv*, 90–118. (b) Fleming, F. F.; Zhang, Z. *Tetrahedron* **2005**, *61*, 747–789. (c) Sinz, C. J.; Rychnovsky, S. D. *Top. Curr. Chem.* **2001**, *216*, 51–92.

5. Too, P. C.; Chan, G. H.; Tnay, Y. L.; Hirao, H.; Chiba, S. *Angew. Chem. Int. Ed.* **2016**, *55*, 3719–3723.

6. (a) Ong, D. Y.; Tejo, C.; Xu, K.; Hirao, H.; Chiba, S. *Angew. Chem. Int. Ed.* **2017**, *56*, 1840–1844. (b) Hong, Z.; Ong, D. Y.; Muduli, S. K.; Too, P. C.; Chan, G. H.; Tnay, Y. L.; Chiba, S.; Nishiyama, Y.; Hirao, H.; Soo, H. S. *Chem. Eur. J.* **2016**, *22*, 7108–7114.

7. For use of NaH-iodide composites as the unprecedented Brønsted bases, see: (a) Kaga, A.; Hayashi, H.; Hakamata, H.; Oi, M.; Uchiyama, M.; Takita, R.; Chiba, S.; *Angew. Chem. Int. Ed.* **2017**, *56*, 11807–11811. (b) Huang, Y.; Chan, G. H.; Chiba, S. *Angew. Chem. Int. Ed.* **2017**, *56*, 6544–6547.

8. In our original communication (ref. 4), all the reactions for hydrodecyanation were examined under sealed conditions at 85 °C in THF, while in the present procedure, we conducted hydrodecyanation under reflux conditions in THF and found it reproducible and scalable (except for the substrate in Scheme 3).

9. A stepwise pathway including fragmentation of **A** to benzylic carbanion and hydrogen cyanide followed by deprotonation cannot be ruled out as a possibility of the reaction mechanism.

Appendix
Chemical Abstracts Nomenclature (Registry Number)

Sodium Hydroxide (1310-73-2)
Benzyltriethylammonium chloride (56-37-1)
Sodium hydride (7646-69-7)
Lithium iodide (10377-51-2)
Tetrahydrofuran (109-99-9)
Benzyl cyanide (140-29-4)
1,4-Dibromobutane (110-52-1)
1-Phenylcyclopentane-1-carbonitrile (77-57-6)
Cyclopentylbenzene (700-88-9)

Guo Hao Chan was born and raised in Singapore. He completed his undergraduate studies at Nanyang Technological University (NTU, Singapore) in 2014 before beginning his Ph.D. work in the lab of Shunsuke Chiba at the same university under EDB-Industrial Postgraduate Program (IPP) with Pfizer Asia Pacific Pte. Ltd. He is currently focusing on methodology development using sodium hydride-iodide composites.

Derek Yiren Ong was born and raised in Singapore. He completed his undergraduate studies at Nanyang Technological University (NTU, Singapore) in 2013. He started his Ph.D. work in the lab of Shunsuke Chiba at the same university in 2016. He is currently focusing on methodology development using sodium hydride-iodide composites.

Shunsuke Chiba earned his Ph.D. in March 2006 from the University of Tokyo under the supervision of Prof. Koichi Narasaka. He was appointed as a Research Associate at the University of Tokyo in May 2005. In April 2007, he joined Nanyang Technological University as an Assistant Professor. In March 2012, he was granted tenure and promoted to Associate Professor in the same university. In September 2016, he was promoted to full Professor. His research focuses on methodology development in the area of synthetic organic chemistry.

Joyce Leung received her B.S. in Chemical Biology from University of California, Berkeley in 2007. After working at Nanosyn, Inc. as a Research Associate for a year, she began her graduate studies at The University of Texas at Austin, where she received her Ph.D. in organic chemistry under the supervision of Professor Michael Krische in 2013. Then she conducted postdoctoral research in the lab of Professor John Wood at Baylor University. In 2017, she joined Boehringer Ingelheim Pharm. Inc. in Ridgefield, CT, where she is currently a Senior Scientist. Her research focuses on the development of efficient and practical synthetic methods for drug candidates.

Indole-Catalyzed Bromolactonization: Preparation of Bromolactone in Lipophilic Media

Zhihai Ke,* Tao Chen, and Ying-Yeung Yeung*[1]

Department of Chemistry, The Chinese University of Hong Kong, Shatin, NT, Hong Kong, China

Checked by Zhaobin Han and Kuiling Ding

Procedure (Note 1)

A. *4-Phenyl-4-pentenoic acid (2).* A cloudy white suspension of methyltriphenylphosphonium bromide (25.51 g, 70.0 mmol, 1.87 equiv) (Note 2) in anhydrous toluene (70 mL) (Note 3) is placed in an oven-dried, 250-mL reaction flask equipped with a magnetic stir bar (2.5 x 0.8 cm Teflon-coated) (Note 4) and a reflux condenser capped with a rubber septum. An inert gas inlet is inserted via a needle, and the solution is cooled to 0 °C (ice water bath temperature). Sodium bis(trimethylsilyl)amide (NaHMDS, 67.5 mL, 67.5 mmol, 1.8 equiv, 1 M solution in THF) (Note 5) is added dropwise by syringe to the suspension under a nitrogen atmosphere over 15 min. The resulting solution is allowed to stir for 45 min and is then cooled to –78 °C using a dry ice-acetone bath. Methyl 3-benzoylpropionate (7.21 g, 37.5 mmol, 1 equiv) (Note 6) is added dropwise by syringe over 5 min. The reaction mixture is warmed to 22 °C for 2 h and is the heated to reflux for 40 h under a nitrogen atmosphere (Figure 1). The reaction is monitored by ^{1}H NMR investigation of the crude sample (Note 7). Upon cooling to 22 °C, the reaction is quenched by the addition of saturated aqueous ammonium chloride (100 mL) and the resulting slurry is diluted with water (100 mL). The organic layer is separated and the aqueous layer is extracted with EtOAc (100 mL x 3). The combined organic extracts are washed with brine (100 mL), dried over anhydrous Na_2SO_4 (5.0 g), filtered, and concentrated by rotary evaporation. The residue is purified by flash column chromatography on silica gel (hexane/EtOAc : 10:1) to give 5.16 g of methyl 4-phenyl-4-pentenoate (**1**) as pale yellow oil (Note 7).

Figure 1. Reaction setup for synthesis of 1

Into a 500-mL reaction flask containing a magnetic teflon-coated stir bar (2.5 x 0.8 cm) is added a solution of methyl 4-phenyl-4-pentenoate (**1**) (5.49 g, 28.9 mmol, 1 equiv) in THF (100 mL) and H_2O (100 mL) containing LiOH•H_2O (12.1 g, 289.0 mmol, 10.0 equiv) (Note 8) at 22 °C. The mixture is stirred for 12 h (monitored by silica TLC using hexane/ethyl acetate : 5/1) (Note 7) and then diluted with water (150 mL). The aqueous fraction is washed with diethyl ether (3 x 100 mL) and acidified with 2 M HCl to pH 2. After extraction of the aqueous phase with EtOAc (3 x 150 mL), the combined organic extracts are washed with brine (200 mL), dried over Na_2SO_4 (5.0 g), filtered and concentrated under reduced pressure to give the desired alkenoic acid **2** as a white solid, which is purified by flash column chromatography (silica, pure EtOAc) to yield 4.77 g of 4-phenyl-4-pentenoic acid (**2**) (28.6 mmol, 72% from methyl 3-benzoylpropionate) as a white solid (Note 9).

B. *5-(Bromomethyl)-5-phenyldihydrofuran-2(3H)-one (3)*. An oven-dried, 1-L reaction flask equipped with a teflon-coated magnetic stir bar (5.0 x 2.0 cm, ovoid-shaped) and a glass stopper (Figure 2) is charged with a mixture of alkenoic acid **2** (3.97 g, 22.5 mmol, 1.0 equiv) and ethyl 2-methylindole-3-carboxylate (46.2 mg, 0.225 mmol, 0.01 equiv) (Note 10) in heptane (450 mL) (Note 11) at 22 °C, to which is added N-bromosuccinimide (4.81 g, 27 mmol, 1.2 equiv) (Note 12) in three portions with five min intervals in the absence of light (the flask is wrapped with aluminum foil tightly). The resulting mixture is vigorously stirred at 22 °C in the dark till completion. The reaction is monitored by ¹H NMR (Note 13). After 48 h the insoluble succinimide is filtered and washed with a mixture of diethyl ether/hexane (ratio 4:5) (3 x 45 mL). The combined filtrate is concentrated under reduced pressure. The remaining

Figure 2. Reaction setup for synthesis of 3

solvent is removed under a high vacuum (1.0 mm Hg), and without using column chromatography, 5.01 g (87% yield) of 5-(bromomethyl)-5-phenyldihydrofuran-2(3H)-one (**3**) is obtained as a pale-yellow oil with a purity of 99%, as determined by ^{1}H NMR spectroscopy (Note 13).

Notes

1. Prior to performing each reaction, a thorough hazard analysis and risk assessment should be carried out with regard to each chemical substance and experimental operation on the scale planned and in the context of the laboratory where the procedures will be carried out. Guidelines for carrying out risk assessments and for analyzing the hazards associated with chemicals can be found in references such as Chapter 4 of "Prudent Practices in the Laboratory" (The National Academies Press, Washington, D.C., 2011; the full text can be accessed free of charge at https://www.nap.edu/catalog/12654/prudent-practices-in-the-laboratory-handling-and-management-of-chemical). See also "Identifying and Evaluating Hazards in Research Laboratories" (American Chemical Society, 2015) which is available via the associated website "Hazard Assessment in Research Laboratories" at https://www.acs.org/content/acs/en/about/governance/committees/chemicalsafety/hazard-assessment.html. In the case of this procedure, the risk assessment should include (but not necessarily be limited to) an evaluation of the potential hazards associated with methyltriphenylphosphonium bromide, toluene, sodium bis(trimethylsilyl)amide, tetrahydrofuran, methyl 3-benzoylpropionate, ammonium chloride, ethyl acetate, silica gel, brine, sodium sulfate, hexane, lithium hydroxide, diethyl ether, hydrochloric acid, ethyl 2-methylindole-3-carboxylate, heptane, *N*-bromosuccinimide, and succinimide.
2. Methyltriphenylphosphonium bromide was purchased from Acros, (98% purity, white solid) and used as received.
3. Toluene was purchased from Merck, (≥99.9% purity, colorless liquid) and was dried over INERT Pure Solv Solvent Purification System before use.

4. All glassware was thoroughly washed and dried in an oven at 110 °C. Teflon-coated magnetic stirring bars were washed with acetone and dried.

5. Sodium bis(trimethylsilyl)amide was purchased from Sigma-Aldrich, (1.0 M solution in THF) and used as received.

6. Methyl 3-benzoylpropionate was purchased from Sigma-Aldrich, (CDS001561, colorless liquid) and used as received. The checkers purchased methyl 3-benzoylpropionate (>98%) from Tokyo Chemical Industry Co., Ltd. and used it as received.

7. The product was purified by flash chromatography on a column (5 x 40 cm) of 100 g of silica gel and eluted with hexane/EtOAc (10:1), R_f = 0.60 in hexane/EtOAc (5:1). A second reaction on the same scale provided 5.13 g (76%) of methyl 4-phenyl-4-pentenoate (**1**). The product exhibits the following characteristics: ^{1}H NMR (400 MHz, CDCl$_3$) δ: 2.48 (t, J = 8.0 Hz, 2H), 2.84 (t, J = 7.6 Hz, 2H), 3.66 (s, 3H), 5.09 (s, 1H), 5.31 (s, 1H), 7.25–7.43 (m, 5H) ; ^{13}C NMR (100 MHz, CDCl$_3$) δ: 30.4, 32.9, 51.5, 112.7, 126.0, 127.5, 128.3, 140.4, 146.7, 173.4 . IR (film): 2951, 1734, 1494, 1155, 897, 777, 701 cm^{-1}; HRMS (ESI, [M+H]$^+$) m/z calcd for C$_{12}$H$_{15}$O$_2$: 191.1067. Found: 191.1068. The purity of product **1** was determined using ^{1}H QNMR analysis. ^{1}H QNMR was performed using a mixture of methyl 4-phenyl-4-pentenoate (25.4 mg) and ethylene carbonate (7.1 mg) (Alfa Aesar, ≥99% purity, white solid) as an internal standard in CDCl$_3$. The purity was calculated according to standard method as 99 wt%.

8. LiOH•H$_2$O was purchased from Sigma-Aldrich, (≥98.0% purity, white solid) and used as received.

9. The product was purified by flash chromatography on a column (5 x 40 cm) of 100 g of silica gel and eluted with pure EtOAc, R_f = 0.65 in EtOAc. A second reaction on the same scale provided 4.71 g of **2** (71% from methyl 3-benzoylpropionate). The product exhibits the following characteristics: ^{1}H NMR (400 MHz, CDCl$_3$) δ: 2.54 (t, J = 7.6 Hz, 2H), 2.85 (t, J = 7.6 Hz, 2H), 5.11 (d, J = 1.2 Hz, 1H), 5.33 (s, 1H), 7.24–7.43 (m, 5H), 11.58 (br s, 1H) ; ^{13}C NMR (100 MHz, CDCl$_3$) δ: 30.0, 33.0, 112.9, 126.0, 127.6, 128.4, 140.3, 146.4, 179.9 IR (film): 3051, 2922, 1693, 920, 777, 701 cm^{-1}; HRMS (ESI, [M-H]$^-$) m/z calcd for C$_{11}$H$_{11}$O$_2$: 175.0765. Found: 175.0764; Anal. Calcd for C$_{11}$H$_{12}$O$_2$: C, 74.98; H, 6.86. Found: C, 75.02; H, 6.87. The purity of product **2** was determined using ^{1}H QNMR analysis.

¹H QNMR was performed using a mixture of **2** (20.1 mg) and 1,3,5-trimethoxybenzene (7.0 mg) (Alfa Aesar, ≥99% purity, white solid) as an internal standard in CDCl₃. The purity was calculated according to standard method as 99 wt%.

10. Ethyl 2-methylindole-3-carboxylate was purchased from Sigma-Aldrich, (99.0% purity, white solid) and used as received.

11. *n*-Heptane was purchased from Acros, (99.83% purity, colorless liquid) and used as received.

12. *N*-Bromosuccinimide was purchased from Alfa Aesar (99% purity, white solid) and recrystallized before use.

13. A second reaction on the same scale provided 5.00 g (88%) of **3**. The product **3** exhibits the following characteristics: R_f = 0.67 in CH_2Cl_2; ¹H NMR (400 MHz, CDCl₃) δ: 2.49–2.63 (m, 2H), 2.76–2.90 (m, 2H), 3.70 (d, *J* = 11.2 Hz, 1H), 3.75 (d, *J* = 11.6 Hz, 1H), 7.33–7.46 (m, 5H) ; ¹³C NMR (100 MHz, CDCl₃) δ: 28.8, 32.2, 40.9, 86.2, 124.7, 128.4, 128.6, 140.5, 175.4. IR (film): 1773, 1153, 928, 766, 700 cm⁻¹; HRMS (ESI, [M+H]⁺) *m/z* calcd for $C_{11}H_{12}BrO_2$: 255.0015. Found: 255.0015. The purity of product **3** was determined using ¹H QNMR analysis. ¹H QNMR was performed using a mixture of **3** (40.7 mg) and ethylene carbonate (5.5 mg) (Alfa Aesar, ≥99% purity, white solid) as an internal standard in CDCl₃. The purity was calculated according to standard method as 99.0 wt%.

Working with Hazardous Chemicals

The procedures in *Organic Syntheses* are intended for use only by persons with proper training in experimental organic chemistry. All hazardous materials should be handled using the standard procedures for work with chemicals described in references such as "Prudent Practices in the Laboratory" (The National Academies Press, Washington, D.C., 2011; the full text can be accessed free of charge at http://www.nap.edu/catalog.php?record_id=12654). All chemical waste should be disposed of in accordance with local regulations. For general guidelines for the management of chemical waste, see Chapter 8 of Prudent Practices.

In some articles in *Organic Syntheses*, chemical-specific hazards are highlighted in red "Caution Notes" within a procedure. It is important to recognize that the absence of a caution note does not imply that no significant hazards are associated with the chemicals involved in that procedure. Prior to performing a reaction, a thorough risk assessment should be carried out that includes a review of the potential hazards associated with each chemical and experimental operation on the scale that is planned for the procedure. Guidelines for carrying out a risk assessment and for analyzing the hazards associated with chemicals can be found in Chapter 4 of Prudent Practices.

The procedures described in *Organic Syntheses* are provided as published and are conducted at one's own risk. *Organic Syntheses, Inc.*, its Editors, and its Board of Directors do not warrant or guarantee the safety of individuals using these procedures and hereby disclaim any liability for any injuries or damages claimed to have resulted from or related in any way to the procedures herein.

Discussion

Lactone is a privileged heterocycle, as it is the essential unit in many natural products and drug molecules. As a result, synthesis of functionalized lactones has steadily attracted a great deal of interest among organic chemists. A significant amount of research has been well documented over the past few decades.

Halolactonization, which can be dated back to 1954, remains one of the superior ways to construct lactones with an easily manipulated halogen handle.[2] The resulting lactones are oftentimes pharmaceutically important drug cores. To avoid handling toxic liquid bromine, alternative electrophilic halolactonization reactions have become more popular in recent decades.[3] N-Bromosuccinimide (NBS) is one of the inexpensive electrophilic halogen sources that can be handled with ease.[4] However, the Br carrier (succinimide) is soluble in both polar and halogenated solvents, which complicates the purification process particularly for large-scale reactions. Halogenation in nonpolar solvents (e.g., heptane) is uncommon in literature due to the insolubility of polar halogenating sources, although some of the nonpolar solvents are attractive reaction media in industrial processes.[5]

Previously, we have exploited the use of a 1*H*-indole-3-carboxylate-based solid-liquid phase-transfer organocatalyst in the bromolactonization reaction of olefinic carboxylic acids in lipophilic solvent.[6] The 1*H*-indole-3-

carboxylate system can be readily constructed using a two-step sequence starting from aniline.[7]

This new type of halogen activation using a structurally simple indole organocatalyst can be applied to various electrophilic halogenation reactions. The major side product for halogenation is succinimide, which is insoluble in non-polar solvents. This methodology allows succinimide to be removed easily via simple filtration without the need to use column chromatography and thus saving time and money in the purification of the desired halogenation products.

In summary, 1*H*-indole-3-carboxylate has been identified to be an efficient organocatalyst for large-scale bromolactonization of alkenoic acids in green lipophilic media such as heptane. The reaction is operationally simple: the lactonization can be performed at room temperature and the workup process can be facilitated by filtration. Considering the practicality of this type of catalytic halogenation reaction, significant further applications are expected.

References

1. Department of Chemistry, The Chinese University of Hong Kong, Shatin, NT, Hong Kong, China. E-mail: chmkz@cuhk.edu.hk; yyyeung@cuhk.edu.hk. We thank the financial support from RGC General Research Fund of HKSAR (CUHK 14306916) and The Chinese University of Hong Kong Direct Grant (Project code: 4053193). Equipment was partially supported by the Faculty Strategic Fund for Research from the Faculty of Science of the Chinese University of Hong Kong, China.

2. (a) Tamelen, V.; Eugene, E.; Shamma, M. *J. Am. Chem. Soc.* **1954**, *76*, 2315–2317; (b) Klein, J. *J. Am. Chem. Soc.* **1959**, *81*, 3611–3614; c) House, H.; Carlson, R.; Babad, H. *J. Org. Chem.* **1963**, *28*, 3359–3361.

3. (a) Zhou, L.; Tan, C. K.; Jiang, X.; Chen, F.; Yeung, Y.-Y. *J. Am. Chem. Soc.* **2010**, *132*, 15474–15476; (b) Jiang, X.; Tan, C. K.; Zhou, L.; Yeung, Y.-Y. *Angew. Chem Int. Ed.* **2012**, *51*, 7771–7775; (c) Chen, J.; Zhou, L.; Tan, C. K.; Yeung, Y.-Y. *J. Org. Chem.* **2012**, *77*, 999–1009; (d) Whitehead, D. C.; Yousefi, R.; Jaganathan, A.; Borhan, B. *J. Am. Chem. Soc.* **2010**, *132*, 3298–3300; (e) Zhang, W.; Zheng, S.; Liu, N.; Werness, J. B.; Guzei, I.A.; Tang, W. *J. Am. Chem. Soc.* **2010**, *132*, 3664–3665.

4. (a) Chichester, E. H. *In Synthetic Reagents*; Pizey, J. S., Ed.; Wiley: New York, 1974; Vol 2, pp 1–311. (b) Carey, F. A.; Sundberg, R. J. *In Advanced Organic Chemistry Part A: Structure and Mechanisms*, 5th ed.; Carey, F. A., Sundberg, R. J., Eds.; Springer: New York, 2007; pp 473–577. (c) Tan, C. K.; Yeung, Y.-Y. *Chem. Commun.* **2013**, *49*, 7985–7996.

5. (a) Constable, D. J. C.; Dunn, P. J.; Hayler, J. D.; Humphrey, G. R.; Leazer, J. L., Jr.; Linderman, R. J.; Lorenz, K.; Manley, J.; Pearlman, B. A.; Wells, A.; Zaks, A.; Zhang, T. Y. *Green Chem.* **2007**, *9*, 411–420; (b) Constable, D. J. C.; Curzons, A. D.; Freitas dos Santos, L. M.; Geen, G. R.; Hannah, R. E.; Hayler, J. D.; Kitteringham, J.; McGuire, M. A.; Richardson, J. E.; Smith, P.; Webb, R. L.; Yu, M. *Green Chem.* **2001**, *3*, 7–9; (c) Curzons, A. D.; Constable, D. J. C.; Mortimer, D. N.; Cunningham, V. L. *Green Chem.* **2001**, *3*, 1–6; (d) Constable, D. J. C.; Curzons, A. D.; Cunningham, V. L. *Green Chem.* **2002**, *4*, 521–527.

6. Chen, T.; Yeung, Y.-Y. *ACS Catal.* **2015**, *5*, 4751–4756.

7. Würtz, S.; Rakshit, S.; Neumann, J. J.; Dröge, T.; Glorius, F. *Angew. Chem., Int. Ed.* **2008**, *47*, 7230–7233.

Appendix
Chemical Abstracts Nomenclature (Registry Number)

Methyltriphenylphosphonium bromide; (1779-49-3)
Toluene; (108-88-3)
NaHMDS: Sodium bis(trimethylsilyl)amide; (1070-89-9)
LiOH•H$_2$O: Lithium hydroxide monohydrate; (1310-66-3)
Ethyl 2-methylindole-3-carboxylate; (53855-47-3)
n-Heptane; (142-82-5)
N-Bromosuccinimide; (128-08-5)

Organic Syntheses

Zhihai Ke received his Ph.D. degree from The Chinese University of Hong Kong in 2012 under the direction of Prof. Hak-Fun Chow. He joined Prof. Ying-Yeung Yeung's research group at the National University of Singapore as a postdoctoral fellow in late 2012. In Aug 2015, he moved back to The Chinese University of Hong Kong as a Research Assistant Professor, dedicating his efforts to the development of novel organic synthetic methods.

Tao Chen was born in 1985 in Jiangsu, China. He obtained his B.S. degree from Soochow University in 2008. After receiving an M.S. degree from the same university in 2011, he started Ph.D. study under the supervision of Prof. Ying-Yeung Yeung in National University of Singapore in the same year. He received his Ph.D. degree in 2016, after which he joined the research group of Prof. Shunsuke Chiba as a postdoctoral fellow at the Nanyang Technological University.

Ying-Yeung Yeung received his B.Sc. (2001) at The Chinese University of Hong Kong. He continued his graduate research in the same university under the supervision of Prof. Tony K. M. Shing. After four years research dedicated to natural product synthesis, Dr. Yeung moved to the USA to conduct postdoctoral research with Prof. E. J. Corey at Harvard University. In 2008, he joined National University of Singapore, Department of Chemistry. In 2015, he moved to The Chinese University of Hong Kong as an Associate Professor. His research interests include asymmetric catalysis, green oxidation, and methodology development.

Dr. Zhaobin Han graduated from the Department of Chemistry, Nanjing University in 2003. He received his Ph.D. degree from Shanghai Institute of Organic Chemistry under the supervision of Prof. Kuiling Ding and Prof. Xumu Zhang in 2009, working on development of novel chiral ligands for asymmetric catalysis. He is currently an Associate Professor in the same institute and his research interests focus on the development of efficient catalytic methods for organic synthesis based on homogeneous catalysis.

Discussion Addendum for:

Stereoselective Synthesis of 3-Arylacrylates by Copper-Catalyzed Syn Hydroarylation [(E)-Methyl 3-phenyloct-2-enoate]

Yoshihiko Yamamoto[*1]

Department of Basic Medicinal Sciences, Graduate School of Pharmaceutical Sciences, Nagoya University, Chikusa, Nagoya 464-8601, Japan

Original Article: Kirai, N.; Yamamoto, Y. Org. Synth. 2010, 87, 53–58

Conjugate arylation of alkynes activated by an electron-withdrawing group is a practical method to prepare substituted alkenes. Conventionally, arylcopper reagents have been employed for this purpose,[2] and the copper-catalyzed conjugate arylation using stoichiometric amounts of arylmagenesium halides has also been developed.[3] The E/Z stereoselectivity of the arylation products strongly depends on the structures of the alkyne substrates and arylmetal reagents as well as the reaction conditions. Moreover, nucleophilic arylmetal reagents used in the conventional methods have limited functional group compatibility. In striking contrast, the transition metal-catalyzed hydroarylations of alkynes with a broad substrate scope have recently been developed using bench-top stable arylboron reagents.[4] In particular, hydroarylation of activated alkynes generally afford products in which the aryl group is introduced at the carbon β to the electron-withdrawing group.

Although transition metal-catalyzed hydroarylations of alkynes are very useful, they require expensive rhodium or palladium catalysts, as well as additional ligands and/or additives. A few examples of catalysts based on inexpensive first-row transition metals, such as nickel and cobalt, have been reported, but they have seen limited development as compared to rhodium and palladium catalysts.[5] We have independently developed a hydroarylation of internal alkynoates with arylboronic acids using inexpensive copper catalysts, such as CuOAc or Cu(OAc)$_2$, in methanol at ambient temperature, which selectively afforded *syn*-hydroarylation products.[6] Furthermore, neither ligands nor additives are required in this reaction. Thus, the Cu-catalyzed hydroarylation of alkynoates provides easy access to synthetically valuable β,β–disubstituted acrylates.

With the abovementioned features, the Cu-catalyzed hydroarylation of internal alkynoates has been applied to the synthesis of various heterocyclic motifs, which are found in natural products and pharmaceutically important compounds. For example, 4-arylbutenolides, 4-arylpentenolides, and 4-arylcoumarins have been synthesized from alkynoates having an alcohol or phenol moiety on the alkyne terminal *via* tandem Cu-catalyzed hydroarylation/lactonization processes.[7,8] Representative examples are shown in Scheme 1. Notably, hydroarylation/coumarin formation enabled the efficient synthesis of seven natural neoflavones.[7]

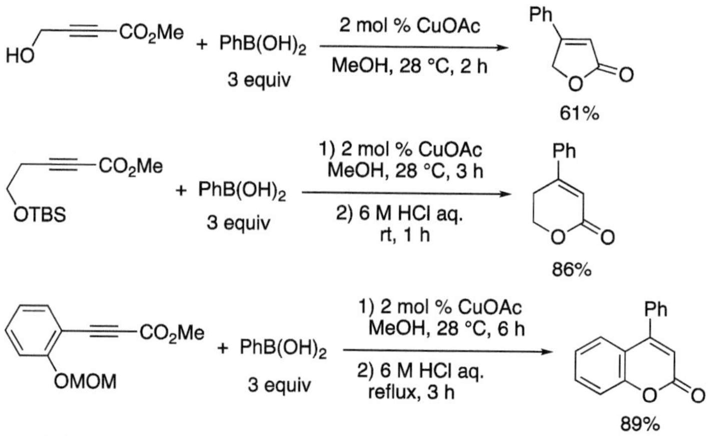

Scheme 1. Cu-catalyzed hydroarylation/lactonization

Nitrogen heterocycles have also been synthesized by Cu-catalyzed hydroarylation, as shown in Scheme 2.[9–11] The hydroarylation product derived from (*o*-nitrophenyl)alkynoate **1** was converted into 3-phenyl-indole-2-carboxylate **2** in high yield *via* Mo-catalyzed Cadogan cyclization.[9,12] Hydroarylation of the orthogonally protected (*o*-aminophenyl)alkynoate **3** could be performed under similar conditions to afford an *N*-benzyl-4-aryl-2-quinolone after acidic removal of the Boc group.[10] Because of the bulky protected *o*-aminophenyl moiety, protodeboration of the arylboronic acid proceeded faster than the desired hydroarylation. Thus, arylboronic acid 2,2-dimethyl-1,3-propanediol esters, such as **4**, should be used to suppress the undesired protodeboration. The use of a sterically more demanding pinacol ester retarded the reaction, thus diminishing the product yield. This tandem hydroarylation/lactamization process could be successfully applied to the synthesis of a natural alkaloid and relevant derivatives.[11]

Scheme 2. *N*-Heterocycles synthesis via Cu-catalyzed hydroarylation

Besides alkynoates, other electron-deficient alkynes have also been used as substrates for Cu-catalyzed hydroarylation. The hydroarylation of 3-aryl-2-propynenitriles stereoselectively affords 3,3-diarylacrylonitriles,[13] and this method has been successfully extended to the synthesis of antitumor agent CC-5079 (Scheme 3).[14] Since the biological activity of CC-5079 strongly depends on its olefin geometry,[15] Cu-catalyzed hydroarylation is quite useful as it affords both stereoisomers in a stereospecific manner. Trifluoromethyl groups also function as electron-withdrawing groups, as

Cu-catalyzed hydroarylation efficiently proceeds with (trifluoromethyl)-alkynes, affording valuable tri-substituted (trifluoromethyl)alkenes.[16] Thus, the Cu-catalyzed hydroarylation of (trifluoromethyl)alkyne **4** with (*o*-nitro)phenylboronic acid stereoselectively furnished trisubstituted alkene **5**, which was further transformed into 3-aryl-2-(trifluoromethyl)indole **6** in high yield *via* the modified Cadogan cyclization (Scheme 4).[17]

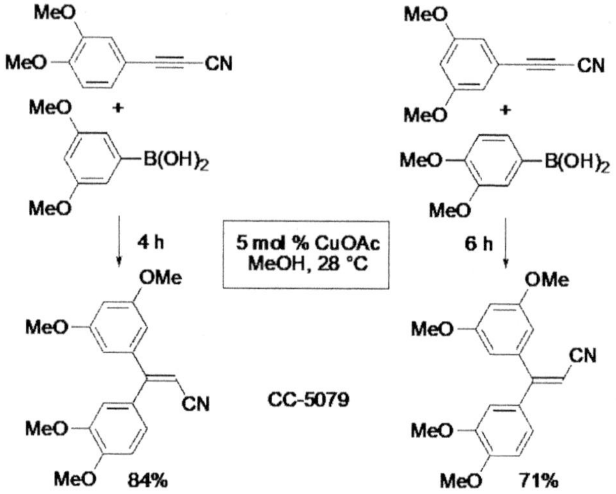

Scheme 3. Synthesis of CC-5079

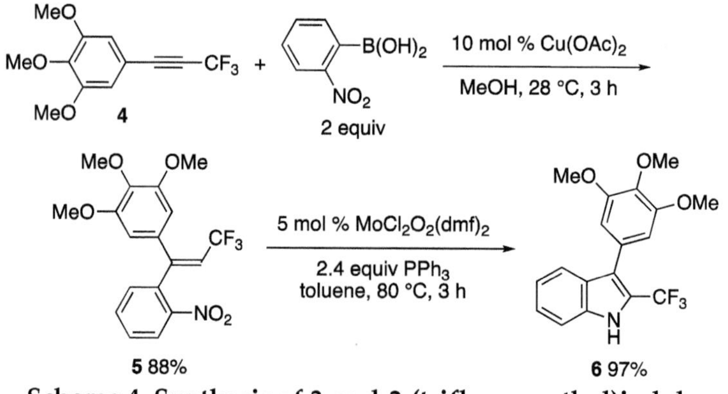

Scheme 4. Synthesis of 3-aryl-2-(trifluoromethyl)indole

Enantioselective conjugate reduction of β,β–diaryl-α,β–unsaturated carbonyl compounds provides efficient access to chiral 1,1-diarylalkyl motifs, which are frequently found in bioactive molecules. Therefore, stereoselective construction of the required unsaturated carbonyl precursors is crucial, and Cu-catalyzed hydroarylation has been used for this purpose.[18] As a demonstration of this strategy, Yun and co-workers reported the enantioselective synthesis of (R)-tolterodine,[19] a potent muscarinic antagonist, *via* Cu-catalyzed asymmetric conjugate reduction of a 3,3-diarylacrylonitrile (Scheme 5).[20] The required acrylonitrile precursor **8** was obtained in 71% yield *via* Cu-catalyzed hydroarylation of cyanoalkyne **7** bearing an unprotected phenol moiety with phenylboronic acid. Subsequent asymmetric conjugate reduction of **8** with polymethylhydrosiloxane (PMHS) as the reducing agent was performed in the presence of 2 mol % Cu(OAc)$_2$ and 2 mol % chiral ligand **9** ((R)-(S)-Josiphos) in toluene at room temperature, affording the desired saturated 3,3-diarylpropanenitrile **10** in 86% yield and with 96% ee. Finally, **10** was transformed to (R)-tolterodine in 63% yield over two steps.

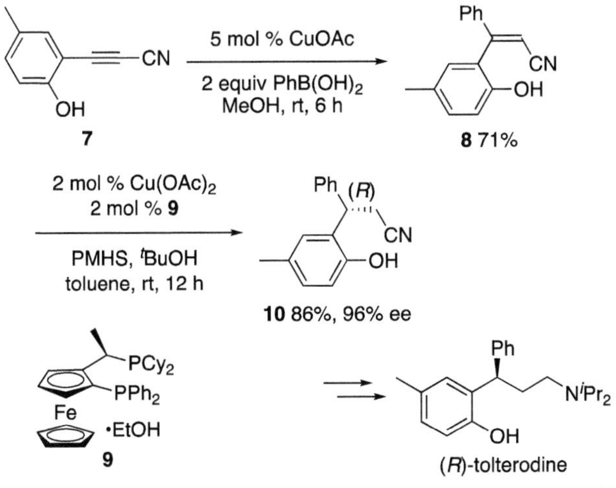

Scheme 5. Enantioselective synthesis of (R)-tolterodine

When allylboronate was used instead of arylboron reagents under the Cu-catalyzed hydroarylation conditions, hydroallylation products were obtained with high regio- and stereoselectivities. Notably, this mild Cu-catalyzed protocol selectively produced 1,4-dienes in good yields, although such a skipped diene tends to undergo isomerization to a more stable 1,3-diene (conjugate diene). Hydroallylation proceeded with alkynylamide **11b** and alkynylsulfone **11c**, in addition to alkynoates and cyanoalkynes

(Scheme 6).[21] The introduced allyl moiety could be employed as a synthetic handle for subsequent transformations, such as hydroboration/oxidation or hydroboration/Suzuki–Miyaura coupling. The interesting bicyclic butenolide **15** was also synthesized from 1,6-enynoate **13** via sequential hydroallylation/ring-closing metathesis.[22] Furthermore, Kong, Zhu, and co-workers reported the hydroallylation of thioalkynes, such as **16**.[23] In this case, modified conditions with a mixed solvent (MeOH/THF, 1:3) improved the product yield. The obtained product **17** could be used for nickel-catalyzed cross coupling with Grignard reagents via C–S bond cleavage.

Ph⎓Ewg + ⟍⟍Bpin →(n mol % Cu(OAc)₂, MeOH, 25 °C, t h) Ph⟍=⟍Ewg
11   1.5 equiv   **12**

11a/12a: Ewg = CO₂Et, n = 3, t = 3; 99%
11b/12b: Ewg = –N⟨O⟩, n = 8, t = 3: quant
11c/12c: Ewg = p-Ts, n = 3, t = 3: 84%

13 (HO) ⎓CO₂Me + ⟍⟍Bpin →
1) 5 mol % Cu(OAc)₂ MeOH, 25 °C, 3 h
2) 50 mol % p-TsOH 25 °C, 12 h
→ (90%)
→ 5 mol % **14** 10 mol % pBQ, CH₂Cl₂ reflux 15 min → **15** 80%

14 Mes: 2,3,5-Me₃C₆H₂

Ph⎓SEt + ⟍⟍Bpin →(10 mol % Cu(OAc)₂, MeOH/THF (1:3) 25 °C, 12 h) Ph⟍=⟍SEt
16   2 equiv   **17** 90%

→(10 mol % NiCl₂(dppe), RMgCl, THF reflux, 12 h) Ph⟍=⟍R
R = Me, 61%; R = Ph, 70%

Scheme 6. Cu-catalyzed hydroallylation of internal alkynes

A disadvantage of Cu-catalyzed hydroarylation and hydroallylation is that they are restricted to the synthesis of trisubstituted alkenes. Hence, an alternative method to synthesize tetrasubstituted alkenes was developed by Sawamura, Ohmiya, and co-workers: the three-component coupling of an alkynoate, alkyl-9-BBN **18**, and $^nBu_3SnOMe$ proceeded in the presence of catalytic amounts of CuOAc and ^tBuOK in dioxane at 60 °C, affording alkenylstannane **19** in 74% yield with a *syn/anti* ratio of 97:3 (Scheme 7).[24] The obtained product could be utilized as a substrate for various cross-coupling reactions. When adding ^tBuOH instead of $^nBu_3SnOMe$ as the proton source, hydroalkylation product **20** was also obtained quantitatively.[25] In this case, the use of $P(OPh)_3$ as the ligand was necessary to achieve a complete *syn* selectivity.

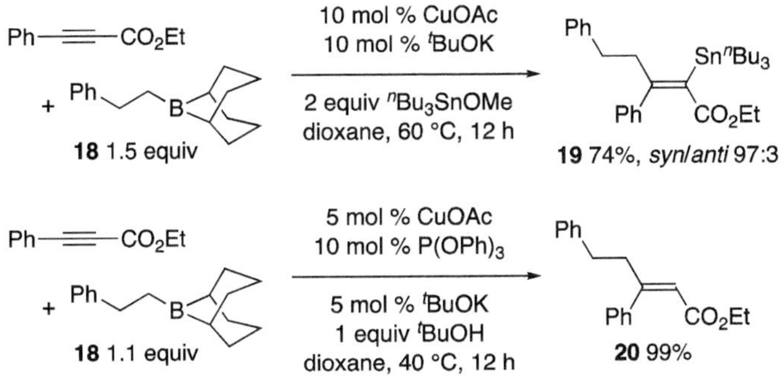

Scheme 7. Cu-catalyzed stannylalkylation and hydroalkylation

In summary, Cu-catalyzed hydroarylation of electron-deficient alkynes with arylboron reagents has been developed as an efficient protocol for the regio- and stereoselective synthesis of valuable trisubstituted alkenes with a functional group under mild conditions. This method is also intriguing because inexpensive copper acetates act as catalysts, and neither ligands nor additives are required. The broad substrate scope of this Cu-catalyzed hydroarylation has been exploited for the synthesis of important heterocyclic compounds, including butenolides, pentenolides, coumarins, indoles, and 2-quinolones. Moreover, the protocol was extended to hydroallylation and hydroalkylation/stannylalkylation using allylboronates or alkylborons, respectively, instead of arylboron reagents. In the future, these methods are expected to find broad application in the synthesis of natural products, functional materials, and bioactive molecules.

References

1. Department of Basic Medicinal Sciences, Graduate School of Pharmaceutical Sciences, Nagoya University, Chikusa, Nagoya 464-8601, Japan. Email: yamamoto-yoshi@ps.nagoya-u.ac.jp
2. Lipshutz, B. H.; Sengupta, S. *Org. React.* **1992**, *41*, 135–631.
3. (a) Xie, M.; Huang, X. *Synlett* **2003**, 477–480. (b) Jennings, M. P.; Swant, K. B. *Eur. J. Org. Chem.* **2004**, 3201–3204. (c) Mueller, A. J.; Jennings, M. P. *Org. Lett.* **2007**, *9*, 5327–5329.
4. Yamamoto, Y. *Catalytic Alkyne Hydroarylation Using Arylboron Reagents, Aryl Halides, and congeners*, in *Catalytic Hydroarylation of Carbon–Carbon Multiple Bonds* (Eds.: L. Ackermann, T. B. Gunnoe, L. G. Habgood), Wiley-VCH, Weinheim, **2017**, Chap. 1.7, pp. 305-359.
5. (a) Shirakawa, E.; Takahashi, G.; Tsuchimoto, T.; Kawakami, Y. *Chem. Commun.* **2001**, 2688–2689. (b) Robbins, D. W.; Hartwig, J. F. *Science* **2011**, *333*, 1423–1427. (c) Lin, P.-S.; Jeganmohan, M.; Cheng, C.-H. *Chem. Eur. J.* **2008**, *14*, 11296–11299.
6. Yamamoto, Y.; Kirai, N.; Harada, Y. *Chem. Commun.* **2008**, 2010–2012.
7. Yamamoto, Y.; Kirai, N. *Org. Lett.* **2008**, *10*, 5513–5516.
8. Yamamoto, Y.; Kirai, N. *Hetrocycles* **2010**, *80*, 269–279.
9. Yamamoto, Y.; Yamada, S.; Nishiyama, H. *Adv. Synth. Catal.* **2011**, *353*, 701–706.
10. Murayama, T.; Shibuya, M.; Yamamoto, Y. *Adv. Synth. Catal.* **2016**, *358*, 166–171.
11. Murayama, T.; Shibuya, M.; Yamamoto, Y. *J. Org. Chem.* **2016**, *81*, 11940–11949.
12. Sanz, R.; Escribano, J.; Pedrosa, M. R.; Aguado, R.; Arnáiz, F. J. *Adv. Synth. Catal.* **2007**, *349*, 713–718.
13. Yamamoto, Y.; Asatani, T.; Kirai, N. *Adv. Synth. Catal.* **2009**, *351*, 1243–1249.
14. Zhang, L.-H.; Wu, L.; Raymon, H. K.; Chen, R. S.; Corral, L.; Shirley, M. A.; Narla, R. K.; Gomez, J.; Muller, G. W.; Stirling, D. I.; Bartlett, J. B.; Schafer, P. H.; Payvandi, F. *Cancer Res.* **2006**, *66*, 951–959.
15. Fang, Z.; Song, Y.; Sarkar, T.; Hamel, E.; Fogler, W. E.; Agoston, G. E.; Fanwick, P. E.; Cushman, M. *J. Org. Chem.* **2008**, *73*, 4241–4244.
16. Yamamoto, Y.; Ohkubo, E.; Shibuya, M. *Green Chem.* **2016**, *18*, 4628–4632.

17. Yamamoto, Y.; Ohkubo, E.; Shibuya, M. *Adv. Synth. Catal.* **2017**, *359*, 1747–1751.

18. (a) Yoo, K.; Kim, H.; Yun, J. *Chem. Eur. J.* **2009**, *15*, 11134–11138; (b) Ebner, C.; Pfaltz, A. *Tetrahdron* **2011**, *67*, 10287–10290. (c) Itoh, K.; Tsuruta, A.; Ito, J.; Yamamoto, Y.; Nishiyama, H. *J. Org. Chem.* **2012**, *77*, 10914–10919. (d) Li, Y.; Dong, K.; Wang, Z.; Ding, K. *Angew. Chem. Int. Ed.* **2013**, *52*, 6748–6752.

19. Nilvebrant, L. *Rev. Contemp. Pharmacother.* **2000**, *11*, 13–27.

20. Yoo, K.; Kim, H.; Yun, J. *J. Org. Chem.* **2009**, *74*, 4232–4235.

21. Yamamoto, Y.; Yamamda, S.; Nishiyama, H. *Chem. Eur. J.* **2012**, *18*, 3153–3156.

22. Yamamoto, Y.; Shibano, S.; Kurohara, T.; Shibuya, M. *J. Org. Chem.* **2014**, *79*, 4503–4511. Also see, ref. 11a.

23. Kong, W.; Che, C.; Kong, L.; Zhu, G. *Tetrahedron Lett.* **2015**, *56*, 2780–2782.

24. Wakamatsu, T.; Nagao, K.; Ohmiya, H.; Sawamura, M. *Angew. Chem. Int. Ed.* **2013**, *52*, 11620–11623.

25. Wakamatsu, T.; Nagao, K.; Ohmiya, H.; Sawamura, M. *Beilstein J. Org. Chem.* **2015**, *11*, 2444–2450.

Professor Yoshihiko Yamamoto received his Ph.D. degree in 1996 from Nagoya University, where he was appointed Assistant Professor in 1996, promoted to Associate Professor in 2003. He moved to Tokyo Institute of Technology in 2006, and returned to Nagoya University in 2009. He began his independent career at the current position in 2012. His research interests are focused on organometallic catalysts and their application to the synthesis of biologically important molecules.

Modified McFadyen-Stevens Reaction for a Versatile Synthesis of Aromatic Aldehydes

Yuri Iwai[§] and Jun Shimokawa[‡*1]

[§]Graduate School of Pharmaceutical Sciences, University of Tokyo, 7-3-1 Hongo, Bunkyo-ku, Tokyo 113-0033, Japan; [‡]Graduate School of Science, Kyoto University, Sakyo-ku, Kyoto, 606-8502, Japan

Checked by Michael M. Yamano, Robert B. Susick, and Neil K. Garg

A.

B.

Procedure (Note 1)

A. *N-(1-Naphthoyl)-4-methylbenzenesulfonohydrazide (2)*. An oven-dried 1-L, three-necked round-bottomed flask is equipped with an internal thermometer, a rubber septum, argon gas inlet adaptor, a 50 mL dropping funnel, and a 3.0-cm rod-shape, Teflon-coated, magnetic stir bar. The flask is evacuated and refilled with argon twice, then charged with 4-dimethylaminopyridine (1.83 g, 15.0 mmol, 0.20 equiv) and *p*-

toluenesulfonyl hydrazide (14.0 g, 75.0 mmol, 1 equiv) in CH_2Cl_2 (350 mL)
(Figure 1) (Note 2). The reaction mixture is stirred in an ice bath at 1200 rpm
for 15 min (internal temperature is 1.0 °C), and triethylamine (15.8 mL,
113 mmol, 1.50 equiv) is added via syringe (Note 3). 1-Naphthoyl chloride
(11.8 mL, 78.8 mmol, 1.05 equiv) in 30 mL of CH_2Cl_2 is added via dropping
funnel over 10 min, such that the internal temperature does not exceed 5 °C
(Note 4). The pale yellow reaction mixture is stirred in an ice bath for 30 min
(Note 5). The reaction is quenched by the addition of saturated aqueous
NH_4Cl (100 mL) and the reaction mixture is poured into H_2O (300 mL) in a
2-L separatory funnel. The layers are separated, and the organic solution is
washed with 10% aqueous citric acid (200 mL), brine (100 mL) and dried
over sodium sulfate (30 g). After filtration through a cotton plug and rinsing
with 50 mL of CH_2Cl_2, the solution is concentrated on a rotary evaporator
(40 °C, 150 mmHg, water bath) to give 25.6 g of the crude product as a pale
yellow solid. The solid thus obtained is dissolved in boiling CH_2Cl_2
(350 mL), and n-hexane (250 mL) is slowly added. The mixture is then
cooled in an ice bath for 1 h and the white precipitate is collected by
Büchner funnel (diameter 90 mm) with a medium porosity fritted disk
using suction filtration. The precipitate is washed with ice-cooled
50% CH_2Cl_2 in n-hexane (50 mL) and dried under vacuum (50 °C,
6.0 x 10^{-2} mmHg, 3 h) to give N-(1-naphthoyl)-4-methylbenzene-
sulfonohydrazide (2) (15.8 g, 46.4 mmol, 62%) as white crystals (Figure 2)
(Notes 6, 7 and 8).

Figure 1. Reaction setup for the synthesis of 2

Figure 2. Sample of 2 after recrystallization

B. *1-Naphthaldehyde (3)*. An oven-dried 1-L, three-necked round-bottomed flask equipped with an internal thermometer, two rubber septa, argon gas inlet adaptor, and a 3.0-cm rod-shape, Teflon-coated, magnetic stir bar is evacuated and refilled with argon twice, then charged with *N*-(1-naphthoyl)-4-methylbenzenesulfonohydrazide (**2**) (20.0 g, 58.8 mmol, 1 equiv) in 260 mL of toluene (Note 9). At ambient temperature (23 °C), the stir rate is set to 600 rpm. The septum is removed temporarily and *N*-trimethylsilylimidazole (15.5 mL, 106 mmol, 1.8 equiv) is added via syringe (Note 10), and imidazole (7.21 g, 106 mmol, 1.8 equiv) is added in one portion. (Figure 3a) (Note 11). The resulting white suspension is stirred at 900 rpm and heated in an oil bath for 4.5 h, so that the internal temperature is maintained at 55 °C (Figure 3b) (Note 12). The pale-yellow, clear reaction mixture is cooled in an ambient temperature in water bath for 5 min and aqueous citric acid solution (2.0 M, 53.0 mL) is added and stirred vigorously at 900 rpm for 2.5 h. The reaction mixture is poured into H_2O (250 mL) in a 2-L separatory funnel. The layers are separated, and the aqueous layer is extracted with an additional 100 mL of toluene. The organic phase is combined, washed with H_2O (100 mL), brine (150 mL), and dried over sodium sulfate (20 g). Filtration through a cotton plug, rinsing with 20 mL of toluene and concentration on the rotary evaporator (45 °C, 50 mmHg, water bath) provides the crude oil (10.0 g). The crude oil is placed in 30 mL round-bottomed flask and attached to a vacuum distillation apparatus equipped with two tared receiving flasks (Figure 4). 1-Naphthaldehyde (**3**) is distilled in one fraction (0.15 mmHg, 105 °C) (Note 13). Upon completion of the distillation, the apparatus is refilled with argon. Compound **3** is obtained as a pale yellow liquid (8.40 g, 53.8 mmol, 92%) (Notes 14, 15, and 16).

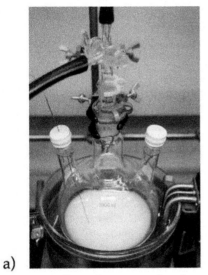

a)   b)

Figure 3. Reaction setup for the synthesis of 3. a) Immediately after the reaction is started; b) After the reaction is complete

Figure 4. Distillation of 3

Notes

1. Prior to performing each reaction, a thorough hazard analysis and risk assessment should be carried out with regard to each chemical substance and experimental operation on the scale planned and in the context of the laboratory where the procedures will be carried out. Guidelines for carrying out risk assessments and for analyzing the hazards associated with chemicals can be found in references such as Chapter 4 of "Prudent Practices in the Laboratory" (The National Academies Press, Washington, D.C., 2011; the full text can be accessed free of charge at https://www.nap.edu/catalog/12654/prudent-practices-in-the-laboratory-handling-and-management-of-chemical).
See also "Identifying and Evaluating Hazards in Research Laboratories" (American Chemical Society, 2015) which is available via the associated website "Hazard Assessment in Research Laboratories" at https://www.acs.org/content/acs/en/about/governance/committees/chemicalsafety/hazard-assessment.html. In the case of this procedure, the risk assessment should include (but not necessarily be limited to) an evaluation of the potential hazards associated with 4-dimethylaminopyridine, *p*-toluenesulfonyl hydrazide, dichloromethane, triethylamine, 1-naphthoyl chloride, ammonium chloride, citric acid, toluene, *N*-trimethylsilylimidazole, imidazole, and *n*-hexane.

2. The submitters purchased 4-dimethylaminopyridine (>99%) and *p*-toluenesulfonyl hydrazide (>98%) from Tokyo Chemical Industry Co., Ltd. and used as received. The submitters purchased anhydrous CH_2Cl_2 (>99.5%) from Wako Pure Chemical Industries, Ltd and used after passing through the prepacked alumina column.

3. The submitters purchased triethylamine (>99%) from Tokyo Chemical Industry Co., Ltd. and used as received.

4. The submitters purchased 1-naphthoyl chloride (>98%) from Tokyo Chemical Industry Co., Ltd. and used as received.

5. The consumption of *p*-toluenesulfonyl hydrazide was monitored by TLC analysis on Merck silica gel 60 F_{254} plates (0.25 mm, glass-backed, visualized with 254 nm UV lamp and stained with cerium phosphomolybdic acid) using 50% ethyl acetate in *n*-hexane as an eluant. *p*-Toluenesulfonyl hydrazide had R_f = 0.33 (UV active, black

after staining) and *N*-(1-naphthoyl)-4-methylbenzenesulfonohydrazide (**2**) had R_f = 0.66 (UV active, black after staining).

6. The product displayed the following physicochemical properties: mp 154.2–154.9 °C (decomp.); IR (film, cm^{-1}) 1685, 1594, 1356; ^{1}H NMR (CDCl$_2$, 400 MHz) δ: 2.47 (s, 3H), 4.48 (s, 2H), 7.35 (d, *J* = 8.2 Hz, 1H), 7.46–7.54 (m, 3H), 7.57 (dd, *J* = 7.1, 1.2, 1H), 7.72–7.78 (m, 1H), 7.85–7.91 (m, 3H), 7.96 (d, *J* = 8.2 Hz, 1H); ^{13}C NMR (CDCl$_3$, 100 MHz) δ: 21.9, 124.4, 124.6, 126.5, 126.6, 127.5, 128.8, 129.1, 129.8, 130.0, 131.2, 132.7, 133.3, 134.2, 145.6, 171.2; HRMS–APCI calcd for C$_{18}$H$_{17}$N$_2$O$_3$S ([M + H$^+$]) 341.09544, found 341.09448.

7. The purity was determined to be 98.1% wt. by quantitative ^{1}H NMR spectroscopy in CDCl$_3$ using 22.9 mg of compound **2** and 11.1 mg of 1,3,5-trimethoxybenzene as an internal standard.

8. A second run on full scale provided 15.8 g (62%) of **2**.

9. The submitters purchased anhydrous toluene (>99.5%) from Wako Pure Chemical Industries, Ltd and used after passing through the prepacked alumina column.

10. The submitters purchased *N*-trimethylsilylimidazole (>98%) from Tokyo Chemical Industry Co., Ltd. and used as received.

11. The submitters purchased imidazole (>98%) and *N*-trimethylsilylimidazole (>98%) from Tokyo Chemical Industry Co., Ltd. and used as received.

12. The consumption of **2** was monitored by TLC analysis on Merck silica gel 60 F$_{254}$ plates (0.25 mm, glass-backed, visualized with 254 nm UV lamp and stained with cerium phosphomolybdic acid) using 50% ethyl acetate in *n*-hexane as an eluant. **2** had R_f = 0.66 (UV active, black after staining) and 1-naphthaldehyde (**3**) had R_f = 0.80 (UV active, black after staining). The major byproduct of the reaction is TsSC$_6$H$_4$CH$_3$ generated by the disproportionation of sulfinate ion. This byproduct has almost the same R_f with **3**, making chromatographic purification difficult in this case.

13. The dimensions of the short path distillation apparatus used are 110 x 110 mm (height x width). The vapor temperature of the distillate was recorded as 105 °C, however, the heating bath for the distillation was heated from 135–190 °C. The distillation process was continued until no more 1-naphthaldehyde could be seen collecting into the tared flask.

14. The product displayed the following physicochemical properties: IR (film, cm-1) 1685, 1574, 1510; ^{1}H NMR (CDCl$_3$, 500 MHz) δ: 7.60 (dd,

J = 8.3, 7.0 Hz, 1H), 7.64 (dd, J = 8.3, 7.0 Hz, 1H), 7.70 (dd, J = 8.3, 7.0 Hz, 1H), 7.93 (d, J = 8.3 Hz, 1H), 8.00 (d, J = 7.0 Hz, 1H), 8.11 (d, J = 8.5 Hz, 1H), 9.26 (d, J = 8.5 Hz, 1H), 10.41 (s, 1H); ^{13}C NMR (CDCl$_3$, 100 MHz) δ: 125.02, 125.04, 127.1, 128.6, 129.2, 130.7, 131.6, 133.9, 135.4, 136.8, 193.7; HRMS–APCI calcd for C$_{11}$H$_9$O ([M + H$^+$]) 157.06479, found 157.06429.

15. The purity was determined to be 98.0% wt. by quantitative ^{1}H NMR spectroscopy in CDCl$_3$ using 21.8 mg of compound **3** and 21.6 mg of 1,3,5-trimethoxybenzene as an internal standard.

16. A second run on half scale provided 3.66 g (87%).

Working with Hazardous Chemicals

The procedures in *Organic Syntheses* are intended for use only by persons with proper training in experimental organic chemistry. All hazardous materials should be handled using the standard procedures for work with chemicals described in references such as "Prudent Practices in the Laboratory" (The National Academies Press, Washington, D.C., 2011; the full text can be accessed free of charge at http://www.nap.edu/catalog.php?record_id=12654). All chemical waste should be disposed of in accordance with local regulations. For general guidelines for the management of chemical waste, see Chapter 8 of Prudent Practices.

In some articles in *Organic Syntheses*, chemical-specific hazards are highlighted in red "Caution Notes" within a procedure. It is important to recognize that the absence of a caution note does not imply that no significant hazards are associated with the chemicals involved in that procedure. Prior to performing a reaction, a thorough risk assessment should be carried out that includes a review of the potential hazards associated with each chemical and experimental operation on the scale that is planned for the procedure. Guidelines for carrying out a risk assessment and for analyzing the hazards associated with chemicals can be found in Chapter 4 of Prudent Practices.

The procedures described in *Organic Syntheses* are provided as published and are conducted at one's own risk. *Organic Syntheses, Inc.*, its Editors, and its Board of Directors do not warrant or guarantee the safety of individuals using these procedures and hereby disclaim any liability for any

injuries or damages claimed to have resulted from or related in any way to
the procedures herein.

Discussion

Among the various known methods for the transformation of carboxylic
acid derivatives into aldehydes, the traditional McFadyen–Stevens reaction[2]
is a unique option. Under the original conditions reported in 1936, N,N'-
acylbenzenesulfonyl hydrazine **4** could be converted to the corresponding
aldehyde, by the treatment with potassium carbonate in ethylene glycol at
160 °C. The interesting point of this reaction is that the aldehyde **5** is
obtained from the corresponding carboxylic acid without using any
oxidants or reductants (Scheme 1a). Throughout the proposed mechanism
via **6**, **7**, and **8**, the high temperature required in the traditional McFadyen–
Stevens reaction is considered to be due to the slow N–H insertion process
from nitrene **7** to generate acyldiazene **8**.[3] Thus the more facile elimination
of the sulfinate ion from N,N-acylsulfonylhydrazine[4] **9** was expected to
afford the identical acyl diazene **8**, which would reduce the reaction
temperature. With the higher electron density of the neighboring nitrogen
atom facilitating the elimination of a sulfinate ion from **9**, an even weaker
base could be employed for the reaction, thereby suppressing the undesired
reactions. Among the bases examined, the reaction was most efficiently
performed when imidazole was used as the base. Since acceleration of the
reaction at higher temperature resulted in the formation of a small amount
of hydrazone **10**, in situ protection of the aldehyde was carried out with
TMS-imidazole. This treatment circumvents the formation of the hydrazone.
Facile formation of hemiaminal **11** masked the reactive aldehyde, which,
upon acidic workup, regenerated aldehyde **5** in high yield.[5]

Synthesis of the substrate N,N-acylsulfonylhydrazine could be
performed either by our originally reported two-step synthesis via the Boc-
protected intermediate, or by the direct condensation conditions between
acyl chloride with sulfonyl hydrazide in the presence of DMAP as an
indispensable catalyst. The latter conditions reported by Namba and Tanino
were employed in the Procedure A.[4e,4f] Substrate scope for the latter
transformation under the established reaction conditions is shown in Table
1. Benzoic acids with *para-* (entries 1–5) or *meta-* (entries 7–9) substitutions
could be efficiently converted to the corresponding aldehydes. The reaction

conditions were mild enough that even the pinacol borate moiety survived the transformation (entry 9). For the substrates with electron withdrawing substituents, the yields were either moderate (entries 3 and 8) or, in the case of the *p*-nitro group, zero in step B (entry 6) due to the formation of the acyl imidazole, resulting in the formation of the methyl ester after treatment with citric acid in methanol. The quinolone antibiotic nalidixic acid was successfully converted into the corresponding aldehyde **8** without affecting the pyridine and the carbonyl moieties (entry 10). 5-Bromo-2-thiophenecarboxylic acid was also a good substrate (entry 11).

a. Original McFadyen-Stevens Reaction

b. Our Modified McFadyen-Stevens Reaction

Scheme 1. Original and Modified McFadyen-Stevens Reactions

Table 1. Scope and Limitation of the Modified Conditions

Reaction conditions: imidazole (2.0 equiv), TMS-imidazole (2.0 equiv), toluene, 55 °C, then citric acid (10 equiv), CH$_3$OH, rt. Ar-C(=O)-N(Ts)-NH$_2$ → Ar-CHO

entry	sulfonyl hydrazide	aldehyde	yield (%)[a]
1	R$_1$ = I		93
2	R$_1$ = CO$_2$Me		82
3	R$_1$ = CF$_3$		80
4	R$_1$ = CN		60
5	R$_1$ = OMe		78
6	R$_1$ = NO$_2$		0[b]
7	R$_2$ = OMe		95
8	R$_2$ = NO$_2$		66
9	R$_2$ = Bpin		80
10[a]			80
11			93

[a]Yield of isolated product. [b]Methyl ester was obtained instead via the formation of acyl imidazole intermediate.

References

1. Graduate School of Science, Kyoto University, Sakyo-ku, Kyoto 606-8502, Japan; Email: shimokawa@kuchem.kyoto-u.ac.jp. Prof. Tohru Fukuyama (Nagoya Univ.) is deeply acknowledged for his helpful support. The authors thank Prof. Kosuke Namba (Tokushima Univ.) for the helpful discussions. Prof. Masanobu Uchiyama (Univ. of Tokyo, RIKEN) and Dr. Ryo Takita (RIKEN) are acknowledged for mechanistic

study of the original work. We are grateful for the Grant-in-Aid (21790009 and 20002004) from the Ministry of Education, Culture, Sports, Science, and Technology of Japan.

2. (a) McFadyen, J. S.; Stevens, T. S. *J. Chem. Soc.* **1936**, *0*, 584–587. (b) Mosettig, E. *Org. React.* **1954**, *8*, 218–257.
3. Matin, S.; Craig, J.; Chan, R. *J. Org. Chem.* **1974**, *39*, 2285–2289.
4. (a) Ruwet, A.; Renson, M. *Bull. Soc. Chim. Belg.* **1966**, *75*, 157–168. (b) Bihel, F.; Hellal, M.; Bourguignon, J.-J. *Synthesis* **2007**, 3791–3796. (c) Grehn, L.; Ragnarsson, U. *Tetrahedron* **1999**, *55*, 4843–4852. (d) Ragnarsson, U.; Grehn, L.; Koppel, J.; Loog, O.; Tšubrik, O.; Bredikhin, A.; Mäeorg, U.; Koppel, I. *J. Org. Chem.* **2005**, *70*, 5916–5921. (e) Namba, K.; Kaihara, Y.; Yamamoto, H.; Imagawa, H.; Tanino, K.; Williams, R. M.; Nishizawa, M. *Chem. Eur. J.* **2009**, *15*, 6560–6563. (f) Namba, K.; Shoji, I.; Nishizawa, M.; Tanino, K. *Org. Lett.* **2009**, *11*, 4970–4973.
5. Iwai, Y.; Ozaki, T.; Takita, R.; Uchiyama, M.; Shimokawa, J.; Fukuyama, T. *Chem. Sci.* **2013**, *4*, 1111–1119.

Appendix
Chemical Abstracts Nomenclature (Registry Number)

4-Dimethylaminopyridine; (1122-58-3)
p-Toluenesulfonyl hydrazide; (1576-35-8)
Triethylamine; (121-44-8)
1-Naphthoyl chloride; (879-18-5)
Imidazole; (288-32-4)
N-Trimethylsilylimidazole; (18156-74-6)

Organic
Syntheses

Yuri Iwai was born in Kyoto, Japan. She received her B.S. in 2009 from Kyoto Pharmaceutical University where she carried out undergraduate research under the supervision of Professor Jun'ichi Uenishi. She then moved to the graduate school of pharmaceutical sciences, the University of Tokyo and started her graduate study. She obtained her Ph.D. degree in 2014 under the direction of Professor Tohru Fukuyama.

Jun Shimokawa was born in 1980 in Tokyo, Japan. He received his B.S. (2003) and M.S. (2005) degrees at the University of Tokyo under the direction of Professor Yuichi Hashimoto. He performed his Ph.D. studies under the direction of Professor Tohru Fukuyama at the University of Tokyo where he conducted the research on total syntheses of complex natural products. In 2006, He started his academic carrier as an Assistant Professor in the same laboratory. In 2012, he moved to Nagoya University, where he was appointed as an Assistant Professor in Professor Masato Kitamura's laboratory. From 2018, he is an Associate Professor in Kyoto University in Professor Hideki Yorimitsu's laboratory. His research efforts focus on the development of novel synthetic methodology and applications to the multistep synthesis of complex molecules.

Robert Susick received his B.S. in chemistry from North Carolina State University in Raleigh, NC. After graduating, he spent two years working at Cirrus Pharmaceuticals, Inc. in the Research Triangle Park of NC developing analytical methods for API quantification and stability studies. He is currently a fourth-year graduate student in Professor Neil K. Garg's laboratory at the University of California, Los Angeles. His graduate studies are focused on the total synthesis of natural products.

Michael Yamano was born in La Habra, California. He received his B. S. in 2014 from the University of California, Irvine where he carried out research under the directions of Professors Kenneth C. Janda and Vy M. Dong. He then moved to the University of California, Los Angeles (UCLA) where he is currently a fourth-year graduate student in Professor Neil K. Garg's laboratory. His graduate studies are focused on developing methodologies to harness the high reactivity of strained cyclic allenes.

Discussion Addendum for:

Facile Syntheses of Aminocyclopropanes: *N,N*-Dibenzyl-*N*-(2-ethenylcyclopropyl)amine [(Benzenemethanamine, *N*-(2-ethenylcyclopropyl)-*N*-(phenylmethyl)]

Armin de Meijere[*1] and Sergei I. Kozhushkov[1]

Institut für Organische und Biomolekulare Chemie der Georg-August-Universität, Tammannstr. 2, D-37077 Göttingen, Germany

Original Article: de Meijere, A.; Winsel, H.; Stecker, B. Org. Synth. 2005, 81, 14-25

A. $Ti(OiPr)_4$ + $TiCl_4$ $\xrightarrow[\text{ether}]{\text{MeLi}}$ $MeTi(OiPr)_3$ **1**

B. (structure: H–C(=O)–NBn₂) + (allyl)MgBr $\xrightarrow[\text{THF, 25 °C, 1 h}]{MeTi(OiPr)_3}$ (vinylcyclopropyl)–NBn₂ **2**

Since the original report in 1996,[2] the transformations of *N,N*-dialkylcarboxamides with 1,2-dicarbanionic organometallics *in situ* generated from organomagnesium (Grignard) as well as organozinc reagents in the presence of stoichiometric or substoichiometric (semi-catalytic) quantities of a titanium alkoxide derivative of type $XTi(OR)_3$ with (X = OR, Cl, Me) has become a powerful tool in organic synthesis.[3] According to the generally accepted mechanism, the key intermediate is a titanacyclopropane-type complex **3a**, which can directly cycloadd to an amide carbonyl (Variant **A**) or undergo ligand exchange with alkenes to afford new titanacyclopropanes **3b** (pathway **B**). Both variants work well, both in their inter- as well as intramolecular versions. (Scheme 1).[3a,b,5]

The simplicity of the experimental handling and relatively low cost of the reagents favor these so-called Kulinkovich-de Meijere cyclopropanations for an increasing range of applications in organic synthesis.[4] The present Discussion Addendum is focused on the most remarkable new developments and synthetic employments of this reaction published since 2005.

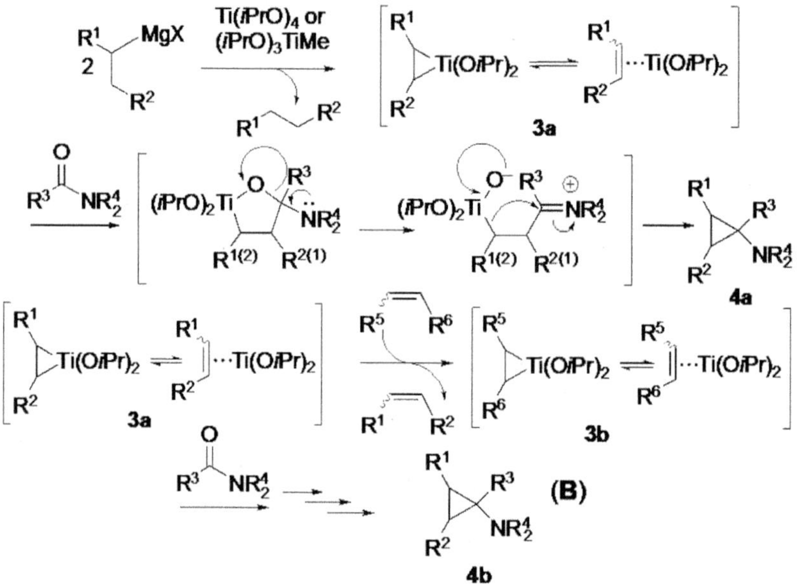

Scheme 1. Generally accepted mechanism of the Kulinkovich-de Meijere cyclopropanation

Attempted Further Extensions of the Reaction Scope

Several studies on the extension of the reaction scope appeared in this period. Thus, in an enantioselective version of the title reaction, compound **2** was obtained with *ee*s up to 80% in the presence of chirally modified Ti(TADDOL)$_2$.[5] A number of new spirocyclic **5** and ring-fused aminocyclopropanes **6** were prepared using an intermolecular cyclopropanation of lactams or applying cycloalkyl-Grignard reagents (Scheme 2).[5] Treatment of substituted 1*H*-benzo[*e*][1,4]diazepine-2,5-diones **7** with EtMgBr/MeTi(O*i*Pr)$_3$ resulted in selective cyclopropanation of only the anilide carbonyl group and afforded derivatives of spirobenzodiazepinone **8**.[6] Generally speaking, in most cases MeTi(O*i*Pr)$_3$ turned out to be more efficient. Indeed, with the employment of Ti(O*i*Pr)$_4$, **7** underwent decomposition.

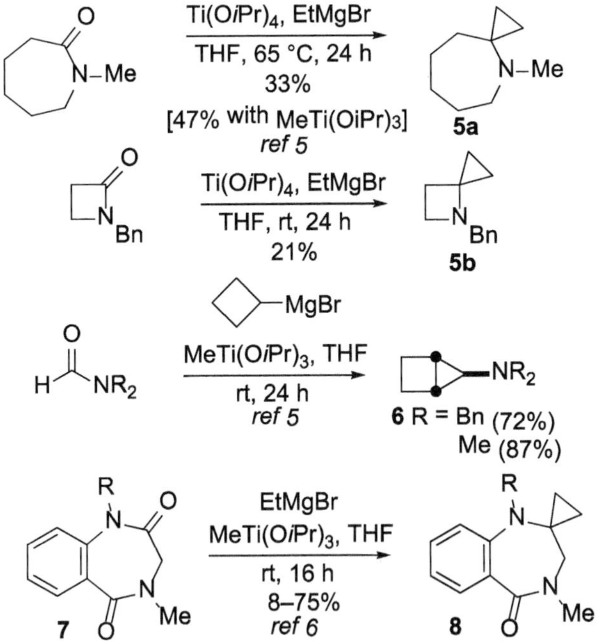

Organic Syntheses

Ti(O*i*Pr)₄, EtMgBr
THF, 65 °C, 24 h
33%
[47% with MeTi(OiPr)₃]
ref 5

5a

Ti(O*i*Pr)₄, EtMgBr
THF, rt, 24 h
21%

5b

MgBr

MeTi(O*i*Pr)₃, THF
rt, 24 h
ref 5

6 R = Bn (72%)
Me (87%)

EtMgBr
MeTi(O*i*Pr)₃, THF
rt, 16 h
8–75%
ref 6

7   **8**

R = Alk, Allyl, Ar, HetAr, cyclopropyl(cyclobutyl)methyl

Scheme 2. Preparation of spirocyclic and ring-fused aminocyclopropanes

Bertus and Szymoniak extended the Kulinkovich-de Meijere cyclopropanation towards imides **9** and developed a straightforward synthesis of α-spirocyclopropanated lactams **10** in 48–78% yields using MeTi(O*i*Pr)₃ as a titanium reagent and Et₂O•BF₃ as an activator for the second step of the transformation.[7] Notably, only one carbonyl group was converted under these conditions, yet the isolated product **10** could be cyclopropanated to give the bisspirocyclopropane derivative **11** with a larger excess of the EtMgBr/MeTi(O*i*Pr)₃ reagent and without addition of Et₂O•BF₃. Employing the former protocol, but with cyclohexylmagnesium instead of ethylmagnesium bromide, N-alkenylimides were converted to tricyclic lactams **12** with a cyclopropylamine moiety in reasonable yields (Scheme 3).

Scheme 3. Kulinkovich-de Meijere cyclopropanation of imides[7]

Formation of tricyclic cyclopropylamines of type **13** can be arrested when the nitrogen-assisted elimination of the titanium alkoxide moiety from the corresponding tricyclic intermediate would form an iminium ion with a bridgehead double bond that would violate Bredt's rule.[8] In these cases, hydrolysis of the intermediates with water without addition of Et₂O•BF₃ leads to carbocyclic amino ketones **14**, which are useful building blocks for the synthesis of certain alkaloids (Scheme 4).[9] Yet, with a large enough lactam ring in the starting material, i. e. an eight- ($n = 3$) or nine-membered ($n = 3$) allyllactame, quenching of the tricyclic N,O-acetal with water furnishes the corresponding tricyclic cyclopropylamines **13** ($n = 3$) and **13** ($n = 4$), respectively, as the sole products.[9] In further studies of intramolecular cyclopropanations with ligand-exchanged titanacyclo-propane intermediates, Six et al. have tested a range of amides **15** fitted with (E)- or (Z)-disubstituted alkene moieties, mostly containing a terminal oxygen functionality (Scheme 5).[10] Their intramolecular Kulinkovich-de Meijere reactions afforded predominantly *exo*-configured products *exo*-**16** from (Z)-**15** and *endo*-**16** from (E)-**15**, respectively.

Scheme 4. Intramolecular cyclopropanations of alkenyllactams towards tricyclic cyclopropylamines[9]

Scheme 5. Intramolecular Kulinkovich-de Meijere reactions of amides bearing disubstituted alkene moieties[10]

Under the typical conditions, simple thioamides **17** upon treatment with alkylmagnesium halides in the presence of Ti(O*i*Pr)$_4$ underwent a drastically different reaction, namely a reductive alkylation affording tertiary amines **18**, even in the presence of styrene as a favorable ligand-exchange candidate.

Scheme 6. Inter- and intramolecular transformations of thioamides upon treatment with alkylmagnesium halides and titanium tetraisopropoxide[11]

However, with an *N*-alkenyl group in the thioamide, such as in **19**, the compound undergoes the intramolecular cyclopropanation, albeit by a mechanism which is slightly different from that of the corresponding *N*-alkyl-*N*-alkenylamides. Thus, 2-azabicyclo[3.1.0]hexanes **20** were prepared from *N*-(but-3-enyl)thioamides **19** in good to very good yields (Scheme 6).[11] With a few exceptions, the thioamides are as efficiently converted to this framework as amides, but less productive for larger 2-azabicyclo[4.1.0]heptanes and 2-azabicyclo[5.1.0]octanes.

Several attempts to facilitate the generation of titanacyclopropane intermediates gave mixed results. Although active organometallic species formed from Ti(O*i*Pr)₄ and *n*BuLi possessed properties similar to those of a titanacyclopropane, it was surprisingly more stable as well as less reactive than the intermediate from Grignard reagents and Ti(O*i*Pr)₄. Thus, this protocol has only found rather limited synthetic applications.[12]

Selected New Examples of Kulinkovich-de Meijere Cyclopropanations towards Practically Useful Compounds

The majority of reductive cyclopropanations of amides was performed with the intention to obtain biologically active or other practically useful compounds. Although a number of competition experiments have disclosed that the reactivities towards reductive cyclopropanation decrease in the order nitriles > amides > esters, both the amide and the ester moiety in the suberic acid derivative **21** could be transformed with a large enough excess of reagents to yield **22** with cyclopropanol and cyclopropylamine fragments (Scheme 7).[13]

Scheme 7. Twofold cyclopropanation of a suberic acid amide ester[13]

Scheme 8. Titanium-mediated cyclopropanation of 3-benzyloxypropionic acid *N,N*-dibenzylamide (23)[14]

The titanium-mediated cyclopropanation of 3-benzyloxypropionic acid *N,N*-dibenzylamide (**23**) afforded *N,N*-dibenzyl-*N*-[1-(2-benzyloxyethyl)-2-ethenylcyclopropyl]amine (**24**) in 56% yield. The latter was further transformed into cyclopropyl analogues of β-homoornithine **26** and βhomoglutamic acid **25** in nine and six simple steps, respectively, as building blocks for potentially biologically relevant small peptide analogues (Scheme 8).[14] Cyclopropane-annelated amino-substituted pyrrolizidine derivatives **27**[15] as well as the key intermediate **28** for the synthesis of 3,4-(aminomethano)proline **29**[16] were obtained in good yields utilizing the titanium-mediated ligand exchange aminocyclopropanation methodology in its intra- and intermolecular version, respectively (Scheme 9).

Scheme 9. Preparation of aminosubstituted 3-azabicyclo[3.1.0]hexanes 27, 28

Scheme 10. Synthesis of a precursor to conformationally locked versions of L-deoxythreosyl phosphonate nucleosides 31[17]

Conformationally restricted versions of L-deoxythreosyl phosphonate nucleosides were synthesized by Marquez et al.[17] in order to investigate the conformational preference of the HIV reverse transcriptase. The key intermediates **30** en route to the enantiomeric diaxially disposed 4-(6-amino-9*H*-purin-9-yl)bicyclo[3.1.0]hexan-2-ol carbocyclic nucleoside **31** were assembled employing an intramolecular Kulinkovich-de Meijere reductive cyclopropanation of the appropriately substituted hexenoic acid *N,N*-dibenzylamide (Scheme 10).

Scheme 11. Reductive aminocyclopropanations in the preparation of building blocks for inhibitors of monoamine oxidase[18]

An intermolecular aminocyclopropanation employing ligand exchange with substituted styrenes or ethenylheteroarenes and ClTi(OR)$_3$ as a titanium source has been used by Joullié et al. to synthesize various 2-arylcyclopropylamines **32** as starting materials for conformationally constrained analogues of the neurotransmitters histamine and tryptamine as inhibitors of monoamine oxidase (Scheme 11).[18] Several other interesting applications of the Kulinkovich-de Meijere cyclopropanation in the synthesis of potentially biologically active compounds are summarized in Table 1.

Table 1. Selected Examples of the Kulinkovich-de Meijere Cyclopropanation Applied in Medicinal Chemistry*

Entry	Reaction/Biological target
1	1) EtMgBr, MeTi(O*i*Pr)$_3$ THF, 0–25 °C, 1.25 h 2) H$_2$O, rt 29% Compositions against stress granules (*ref 19*)
2	1) EtMgBr, MeTi(O*i*Pr)$_3$ THF, −78 °C, 0.5 h 2) −78 to 25 °C, 1 h (yields not reported) R = H, Me, OMe Inhibitors of the bromodomain BRD9 proteins (*ref 20*)
3	EtMgBr, Ti(O*i*Pr)$_4$ THF/Et$_2$O, rt 22% Potent inhibitors of M-tropic (R5) HIV-1 replication (*ref 21*)
4	1) EtMgBr, Ti(O*i*Pr)$_4$ THF/Et$_2$O, rt, 16 h 2) H$_2$O, rt, 0.5 h 33% Selective linear tachykinin NK2 receptor antagonists (*ref 22*)

5	1) EtMgBr, Ti(O*i*Pr)$_4$, THF, −78 to rt, 12 h 2) NH$_4$Cl/H$_2$O, rt 57%	Bacterial peptide deformylase inhibitors for the treatment of bacterial infections *(ref 23)*
6	1) EtMgBr, Ti(O*i*Pr)$_4$ THF/Et$_2$O, −78 °C, 0.5 h 2) −78 to 65 °C, 1 h 3) NH$_4$Cl, H$_2$O, rt R = Me: 38%; R = Bn: 61% Pharmaceutical compositions against drug-resistant microbes *(ref 24)*	
7	1) EtMgBr, MeTi(O*i*Pr)$_3$ THF/Et$_2$O, −78 °C, 7 min 2) −78 to 25 °C, 0.5 h 3) Rochelle salt, H$_2$O, rt 40% *Hsp70* ATPase modulators as potential therapeutics for Alzheimer's and other neurodegenerative diseases *(ref 25)*	
8	1) EtMgBr, MeTi(O*i*Pr)$_3$ THF/Et$_2$O, −78 °C, 10 min 2) −78 to 25 °C, 0.5–3 h 3) Rochelle salt, H$_2$O, rt 70% each Hydroxysteroid dehydrogenase inhibitors *(ref 25b)*	
9	1) EtMgBr, Ti(O*i*Pr)$_4$ THF, −78 °C, 5 h 2) NH$_4$Cl, H$_2$O, rt (yield not reported)	Antitumor agents *(ref 26)*

10	Ph [structure] N–Bn, O	1) EtMgBr, MeTi(OiPr)₃ THF, –78 to rt, 0.75 h 2) rt, 22 h	Ph [structure] N–Bn (+ by-products)	Inhibitors of HIV-1 attachment (*ref 27*)
11	O [structure] N–Bn Bn–N R TBDMSO	1) EtMgBr, Ti(OiPr)₄, THF rt, then 65 °C, 30 min 2) H₂O, 0 °C, 30 min 56%	Bn [structure] N–Bn Bn–N R TBDMSO	HIV protease inhibitors (ref 28)
12	O [structure] N–Bn S TBDMSO	1) EtMgBr, MeTi(OiPr)₃ THF, rt, 12 h 2) H₂O, rt 45%	[structure] N–Bn S TBDMSO	Transient receptor potential ankyrin 1 (TRPA1) antagonists useful in treatment of TRPA1-mediated diseases (*ref 29*)
13	O [structure] N–Me S THP–O	1) EtMgBr, Ti(OiPr)₄ THF, rt, 12 h 2) H₂O, rt, 20 min 36%	[structure] N–Me S THP–O	Tyrosine kinase inhibitors for treatment of hyperproliferative disorders (*ref 30*)
14	O [structure] N–Bn S O R	1) EtMgBr, MeTi(OiPrO)₃ THF/Et₂O, rt, 24h 2) H₂O, 0 °C R = Bn: 29% R = TBDMS: 26%	[structure] N–Bn S O R	Compounds having PGD2 receptor antagonist activity (*ref 31*)
15	[structure] N–Bn Boc–N O	1) EtMgBr, Ti(OiPr)₄, THF –78 °C, 1 h, then to 65 °C 2) rt, 16 h, then NH₄Cl/H₂O, rt, 15 min $n = m = 1$: 58%; $n = 1$, $m = 2$: 64%; $n = 2$, $m = 1$: 37%	[structure] N–Bn Boc–N	Protein tyrosine kinase inhibitors (*ref 32*)

16	![structure] Ph, N Boc, N-Me, O 1) EtMgBr, Ti(O*i*Pr)₄, THF/Et₂O, −78 to rt, 3 h 2) NH₄Cl/H₂O, rt 11%	![structure] Ph, N Boc, N-Me	Glycine transporter 1 (GLYT-1) inhibitors (*ref 33*)
17	TBDMSO ![structure] N, O, PMB + cyclopentyl-MgBr 1) MeTi(O*i*Pr)₄ Et₂O/THF, rt, 3 h 2) Rochelle salt, H₂O 52%	![structure] N-PMB, TBDMSO	Interleukin-1 receptor associated kinase (IRAK4) modulators (*ref 34*)

*Not in all cases were the reaction and work-up conditions optimized; therefore, moderate or low yields in several cases can be attributed to the non-optimized procedures; for details see ref [5].

Synthetically Useful and Theoretically Interesting Transformations of Dialkylaminocyclopropanes

The 2-azabicyclo[3.1.0]hexanes and 2-azabicyclo[4.1.0]heptanes prepared by the Kulinkovich-de Meijere reaction as described above, are susceptible to undergo cleavage of a vicinal (with respect to the nitrogen atom) C–C bond and thus enter [3+2] cycloaddition or intramolecular aromatic electrophilic substitution reactions to afford polycyclic systems with potential synthetic utility towards biologically active compounds (Scheme 12). For example, 2,3,3a,4-tetrahydro-6(5H)-indolones **33** were obtained upon heating of bicycles **16** with acetic anhydride.[35] Aerobic electrochemical oxidation of this family of bicyclic aminocyclopropanes afforded bicyclic α-amino endoperoxides **34**, which exhibited moderate antimalarial activity against the parasite *Plasmodium falciparum*.[10a,36] The ruthenium-catalyzed [3+2] cycloadditions of aminocyclopropanes **20** onto styrenes are not only of theoretical interest, but constitute a new highly diastereoselective synthetic approach to octahydrocyclopenta[b]pyrrole derivatives **35**.[37] At last, an intramolecular aromatic electrophilic substitution furnished oligocyclic 1-methyl-1,2,3,5,6,10b-hexahydro-pyrrolo[2,1-a]isoquinoline derivatives **36**.[38]

Scheme 12. Transformations of 2-azabicyclo[3.1.0]alkanes with scission of a cyclopropyl C–C bond adjacent to nitrogen

On the other hand, the aminocyclopropyl moiety can play the role of a directing group in the formation of C(sp³)–C(sp²) bonds in palladium(0)-catalyzed C–H functionalizations of cyclopropanes with retention of the three-membered ring. This remarkable transformation was developed by Cramer et al. and applied towards the synthesis of the cyclopropane-annelated γ-lactams.[39] Employment of the bulky TADDOL phosphonate ligand **L** in combination with adamantane-1-carboxylic acid as a cocatalyst afforded γ-lactams **38** in excellent yields and with high enantioselectivities from achiral precursors. Applying this protocol, azaspirooctane and -

nonane **5**, after debenzylation and acylation with chloroacetyl chloride were converted into enantiomerically enriched tricyclic γ-lactams **12** (Scheme 13). This sequence complements the alternative approach to such tricyclic aminocyclopropanes involving direct cyclopropanation of imides (cf. Scheme 3).[7]

Scheme 13. Enantioselective palladium(0)-catalyzed cyclopropane C–H activations in aminocyclopropanes[39]

Additional Syntheses of Various Cyclopropylamines

Some alternative synthetic approaches to aminocyclopropanes have been reported in the last 15 years; however, the utilizations of most of them (such as cyclopropanations of enamines[40] or aminations of cyclopropenes/methylenecyclopropanes[41]) appear to be rather limited. Yet, the high-yielding zinc-mediated diastereo- and enantioselective direct transformation of cyclopropanols **39** into N,N-dialkylated cyclopropylamines **4** (Scheme 14), as recently reported by Rousseaux et al.,[42] has a great potential, albeit its synthetic applications have not yet been fully developed.

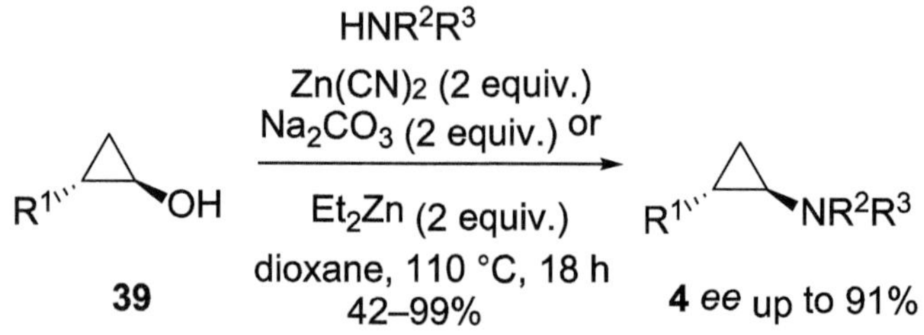

HNR²R³

Zn(CN)₂ (2 equiv.)
Na₂CO₃ (2 equiv.) or
———————————————
Et₂Zn (2 equiv.)
dioxane, 110 °C, 18 h
42–99%

R¹ˈˈ △ OH   **39**   →   R¹ˈˈ △ NR²R³   **4** *ee* up to 91%

Scheme 14. Zinc-mediated direct transformation of cyclopropanols into N,N-dialkylcyclopropylamines[42]

As was discovered in 2002 by Szymoniak et al., carbonitriles can be directly converted into primary cyclopropylamines **40** by treatment with appropriate Grignard reagents in the presence of Ti(O*i*Pr)₄ and subsequent addition of Et₂O·BF₃ (Scheme 15).[3,43] For 1-arylcyclopropylamines **41**, an alternative protocol developed by de Meijere et al. turned out to be more favorable (Scheme 15).[44] For primary cyclopropylamines, these latter methods may be superior to the original Kulinkovich-de Meijere protocol, as the debenzylation step is saved. By the number of publications, the conversion of nitriles has surpassed that of N,N-dialkylcarboxamides (347 *vs* 282 according to SciFinder).

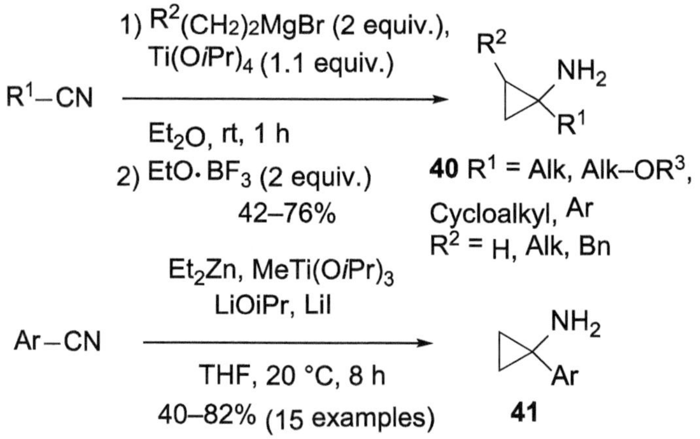

R¹–CN

1) R²(CH₂)₂MgBr (2 equiv.),
   Ti(O*i*Pr)₄ (1.1 equiv.)
   ———————————————————
   Et₂O, rt, 1 h
2) EtO·BF₃ (2 equiv.)
   42–76%

R² ⟍ NH₂ / R¹ (triangle)

40 R¹ = Alk, Alk–OR³,
Cycloalkyl, Ar
R² = H, Alk, Bn

Ar–CN

Et₂Zn, MeTi(O*i*Pr)₃
LiOiPr, LiI
———————————————
THF, 20 °C, 8 h
40–82% (15 examples)

NH₂ / Ar (triangle)   **41**

Scheme 15. Conversion of carbonitriles into primary cyclopropyl-amines[43,44]

However, since the *N*-alkylated aminocyclopropane derivatives have found growing employment in medicinal chemistry research (see Table 1), the Kulinkovich-de Meijere cyclopropanation remains as one of the most important methods for the preparation of aminocyclopropanes, while the Kulinkovich-Szymoniak reaction has found its own important applications.

References

1. Institut für Organische und Biomolekulare Chemie der Georg-August-Universität, Tammannstr. 2, D-37077 Göttingen, Germany. Email: Armin.deMeijere@chemie.uni-goettingen.de Email: skozhus@gwdg.de
2. Chaplinski, V.; de Meijere, A. *Angew. Chem. Int. Ed. Engl.* **1996**, *35*, 413–414.
3. Reviews: (a) de Meijere, A.; Kozhushkov, S. I.; Savchenko, A. I. *J. Organomet. Chem.* **2004**, *689*, 2033–2055; (b) de Meijere, A.; Kozhushkov, S. I.; Savchenko, A. I. in *Titanium and Zirconium in Organic Synthesis*, (Ed.: I. Marek), Wiley-VGH, Weinheim, **2002**, 390–434; (c) Cha, J. K.; Kulinkovich, O. G. *Org. React.* **2012**, *77*, 1–159.
4. Review: Miyamura, S.; Itami, K.; Yamaguchi, J. *Synthesis* **2017**, *49*, 1131–1149.
5. de Meijere, A.; Chaplinski, V.; Winsel, H.; Kordes, M.; Stecker, B.; Gazizova, V.; Savchenko, A. I.; Boese, R.; Schill (née Brackmann), F. *Chem. Eur. J.* **2010**, *16*, 13862–13875.
6. Lack, O.; Martin, R. E. *Tetrahedron Lett.* **2005**, *46*, 8207–8211.
7. Bertus, P.; Szymoniak, J. *Org. Lett.* **2007**, *9*, 659–662.
8. (a) Bredt, J. *Justus Liebigs Ann. Chem.* **1924**, *437*, 1–13; (b) review: Fawcett, F. S. *Chem. Rev.* **1950**, *47*, 219–274. For an updated version of Bredt's rule applicable in these cases, see: (c) Krenske, E. H.; Williams. C. M. *Angew. Chem. Int. Ed.* **2015**, *127*, 10754–10758.
9. Finn, P. B.; Derstine, B. P.; Sieburth, S. McN. *Angew. Chem. Int. Ed.* **2016**, *55*, 2536–2539.
10. (a) Ouhamou, N.; Six, Y. *Org. Biomol. Chem.* **2003**, *1*, 3007–3009; (b) Madelaine, C.; Ouhamou, N.; Chiaroni, A.; Vedrenne, E.; Grimaud, L.; Six, Y. *Tetrahedron* **2008**, *64*, 8878–8898.
11. (a) Hermant, F.; Nicolas, E.; Six, Y. *Tetrahedron* **2014**, *70*, 3924–3930; (b) Augustowska, E.; Boiron, A.; Deffit, J.; Six, Y. *Chem. Commun.* **2012**, *48*, 5031–5033.

12. Rassadin, V. A.; Six, Y. *Tetrahedron* **2014**, *70*, 787–794.

13. Gensini, M.; Quartara, L.; Altamura, M. *Lett. Org. Chem.* **2008**, *5*, 328–331.

14. Brackmann, F.; de Meijere, A. *Synthesis* **2005**, 2008–2014.

15. Gensini, M.; de Meijere, A. *Chem. Eur. J.* **2004**, *10*, 785–790.

16. Brackmann, F.; Colombo, N.; Cabrele, C.; de Meijere, A. *Eur. J. Org. Chem.* **2006**, 4440–4450.

17. Saneyoshi, H.; Deschamps, J. R.; Marquez, V. E. *J. Org. Chem.* **2010**, *75*, 7659–7669.

18. Faler, C. A.; Joullié, M. M. *Org. Lett.* **2007**, *9*, 1987–1990.

19. Larsen, G. R.; Weigelle, M.; Vacca, J. P.; Burnett, D. A.; Ripka, A. PCT Int. Appl. WO 002017066705 A1, April 20, 2017.

20. Martin, L.; Steurer, S.; Cockcroft, X.-L. PCT Int. Appl. WO 002016139361 A1, September 09, 2016.

21. Skerlj, R.; Bridger, G.; Zhou, Y.; Bourque, E.; Langille, J.; DiFluri, M.; Bogucki, D.; Yang, W.; Li, T.; Wang, L.; Nan, S.; Baird, I.; Metz, M.; Darkes, M.; Labrecque, J.; Lau, G.; Fricker, S.; Huskens, D.; Schols, D. *Bioorg. Med. Chem. Lett.* **2011**, *21*, 2450–2455.

22. Gensini. M.; Altamura, M.; Dimoulas, T.; Fedi, V.; Giannotti, D.; Giuliani, S.; Guidi, A.; Harmat, N. J. S.; Meini, S.; Nannicini, R.; Pasqui, F.; Tramontana, M.; Triolo, A.; Maggi, C. A. *ChemMedChem* **2010**, *5*, 65–78.

23. Qin, D.; Norton, B.; Liao, X.; Knox, A. N.; Fang, Y.; Lee, J.; Dreabit, J. C.; Christensen, S. B.; Benowitz, A. B.; Aubart, K. M. PCT Int. Appl. WO 002009061879 A1, May 14, 2009.

24. (a) Ma, Z.; Jin, Y.; Li, J.; Ding, C. Z.; Minor, K. P.; Longgood, J. C.; Kim, I. H.; Harran, S.; Combrink, K.; Morris, T. W. PCT Int. Appl. WO 002005070940 A2, August 04, 2005; (b) Ding, C. Z.; Jin, Y.; Longgood, J. C.; Ma, Z.; Li, J.; Kim, I. H.; Minor, K. P.; Harran, S. PCT Int. Appl. WO 002005070941 A1, August 04, 2005.

25. (a) Chiosis, G.; Kang, Y.; Patel, H. J.; Patel, M.; Ochiana, S.; Rodina, A.; Done, T.; Shrestha, L. PCT Int. Appl. WO 002015175707 A1, November 19, 2015. (b) Aertgeerts, K.; Brennan, N. K.; Cao, S. X.; Chang, E.; Kiryanov, A.; Liu, Y. PCT Int. Appl. 002006105127 A2, October 05, 2006.

26. He, W. PCT Int. Appl. WO 002017101791 A1, Juni 22, 2017.

27. Swidorski, J. J.; Liu, Z.; Yin, Z.; Wang, T.; Carini, D. J.; Rahematpura, S.; Zheng, M.; Johnson, K.; Zhang, S.; Lin, P.-F.; Parker, D. D.; Li, W.; Meanwell, N. A.; Hamann, L. G.; Regueiro-Ren, A. *Bioorg. Med. Chem. Lett.* **2016**, *26*, 160–167.

Organic
Syntheses

28. Williams, P. D.; McCauley, J. A.; Bennett, D. A.; Bungard, C. J.; Chang, L.; Chu, X.-J.; Dwyer, M. P.; Holloway, M. K.; Keertikar, K. M.; Loughran, H. M.; Manikowski, J. J.; Morriello, G. J.; Shen, D.-M.; Sherer, E. C.; Schulz, J.; Wandell, S. T.; Wiscount, C. M.; Zorn, N.; Satyanarayana, T.; Vijayasaradhi, S.; Hu, B.; Ji, T.; Zhong, B. PCT Int. Appl. WO 002015017393 A2, February 05, 2015.

29. Estrada, A.; Volgraf, M.; Chen, H.; Kolesnikov, A.; Villemure, E.; Verma, V.; Wang, L.; Shore, D.; Do, S.; Yuen, P.-w.; Hu, B.; Wu, G.; Lin, X.; Lu, A. PCT Int. Appl. WO 002016128529 A1, August 18, 2016.

30. (a) Xi, N. PCT Int. Appl. WO 002013148537 A1, October 03, 2013; (b) Xi, N. Pat. Appl. US 020100093727 A1, April 15, 2010.

31. Hata, K.; Masuda, M.; Taniyama, D.; Dobinaga, H.; Hato, Y.; Fujiu, M. PCT Int. Appl. WO 002013061977 A1, May 02, 2013.

32. (a) Schou, S. C.; Greve, D. R.; Nielsen, S. F.; Jensen, J. B.; Dack, K. N. PCT Int. Appl. WO 002012093169 A1, July 12, 2012; (b) Nielsen, S. F.; Greve, D. R.; Groe-Sørensen, G.; Ryttersgaard, C.; Schou, S. C.; Sams, A. G. PCT Int. Appl. WO 002012003829 A1, January 12, 2012; (c) Nielsen, S. F.; Greve, D. R.; Ryttersgaard, C.; Groe-Sørensen, G.; Ottosen, E. R.; Poulsen, T. D.; Schou, S. C.; Murray, A. PCT Int. Appl. WO 002011003418 A1, January 13, 2011.

33. Kolczewski, S.; Pinard, E.; Stadler, H. PCT Int. Appl. WO 002010086251 A1, August 05, 2010.

34. Lee, K. L.; Allais, C. P.; Dehnhardt, C. N.; Garvin, L. K.; Han, S.; Hepworth, D.; Lee, A.; Lovering, F. E.; Mathias, J. P.; Owen, D. R.; Papaioannou, N.; Saiah, A.; Strohbach, J. W.; Trzupek, J. D.; Wright, S. W.; Zaph, C. W. PCT Int. Appl. WO 002017033093 A1, March 02, 2017.

35. Larquetoux, L.; Kowalska A. J.; Six, Y. *Eur. J. Org. Chem.* **2004**, 3517–3525.

36. Madelaine, C.; Buriez, O.; Crousse, B.; Florent, I.; Grellier, P.; Retailleau, P.; Six, Y. *Org. Biomol. Chem.* **2010**, *8*, 5591-5601. Noteworthy, the families of 2-azabicyclo[3.1.0]hexanes and 2-azabicyclo[4.1.0]heptanes displayed significant antimalarial activities themselves as well.

37. S. Maity, M. Zhu, R. S. Shinabery, N. Zheng, *Angew. Chem. Int. Ed.* **2012**, *51*, 222–226.

38. Larquetoux, L.; Ouhamou, N.; Chiaroni, A.; Six, Y. *Eur. J. Org. Chem.* **2005**, 4654-4662.

39. (a) Pedroni, J.; Cramer, N. *Angew. Chem. Int. Ed.* **2015**, *54*, 11826–11829; (b) Review: Pedroni, J.; Cramer, N. *Chem. Commun.* **2015**, *51*, 17647–17657.

40. (a) Racine, S.; Hegedüs, B.; Scopelliti, R.; Waser, J. *Chem. Eur. J.* **2016**, *22*, 11997–12001, and references cited therein; (b) Tsai, C.-C.; Hsieh, I.-L.; Cheng, T.-T.; Tsai, P.-K.; Lin, K.-W.; Yan, T.-H. *Org. Lett.* **2006**, *8*, 2261–2263.

41. (a) Simaan, M.; I. Marek, I. *Angew. Chem. Int. Ed.* **2018**, *57*, 1543–1546; (b) Teng, H.-L.; Luo, Y.; Nishiura, M.; Hou, Z. *J. Am. Chem. Soc.* **2017**, *139*, 16506–16509; (c) Teng, H.-L.; Luo, Y.; Wang, B.; Zhang, L.; Nishiura, M.; Hou, Z. *Angew. Chem. Int. Ed.* **2016**, *55*, 15406–15410; (d) Sakae, R.; N. Matsuda, N.; Hirano, K.; Satoh, T.; Miura, M. *Org. Lett.* **2014**, *16*, 1228–1231. (e) Review: Müller, D. S.; Marek, I. *Chem. Soc. Rev.* **2016**, *5*, 4552-4566.

42. (a) Mills, L. R.; Arbelaez, L. M. B.; Rousseaux, S. A. L. *J. Am. Chem. Soc.* **2017**, *139*, 11357–11360; (b) Mills, L. R.; Rousseaux, S. A. L. *Synlett* **2018**, *29*, 683–686.

43. Review: (a) Bertus, P.; Szymoniak, J. *Synlett* **2007**, 1346–1356. Recent communications: (b) Forcher, G.; Clousier, N.; Beauseigneur, A.; Setzer, P.; Boeda, F.; Pearson-Long, M. S. M.; Karoyan, P.; Szymoniak, J.; Bertus, P. *Synthesis* **2015**, *47*, 992–1006; (c) Declerck, D.; Josse, S.; Van Nhien, A. N.; Szymoniak, J.; Bertus, P. Postel, D. *Tetrahedron* **2012**, *68*, 1145–1152; (d) Declerck, D.; Van Nhien, A. N.; Josse, S.; Szymoniak, J.; Bertus, P.; Bello, C.; Vogel, P.; Postel, D. *Tetrahedron* **2012**, *68*, 1802–1809.

44. Wiedemann, S.; Frank, D.; Winsel, H.; de Meijere, A. *Org. Lett.* **2003**, *5*, 753–755.

Armin de Meijere, born 1939, studied chemistry in Freiburg and Göttingen, receiving a doctoral degree (Dr. rer. nat.) in 1966 in Göttingen, completing postdoctoral training at Yale University in 1967-1969, and receiving Habilitation in 1971 in Göttingen. He was appointed full Professor of Organic Chemistry in Hamburg, 1977-1989 and ever since has been in Göttingen. He was visiting professor at 16 research institutions around the world including the University of Wisconsin, the Technion in Haifa, Israel, and Princeton University. His awards and honors include member of the Norwegian Academy of Sciences, Alexander von Humboldt/Gay Lussac prize, Honorary Professor of St. Petersburg State University in Russia, Adolf von Baeyer Medal of the German Chemical Society, *Dr. honoris causa* of the Russian Academy of Sciences. He has been or still is Editor or member of the editorial board of a number of scientific publications including Houben-Weyl, Chemical Reviews, Science of Synthesis, and Chemistry - A European Journal. His scientific achievements have been published in over 720 original publications, review articles, and book chapters as well as 25 patents.

Sergei I. Kozhushkov was born in 1956 in Kharkov, USSR. He studied chemistry at Lomonosov Moscow State University, where he obtained his doctoral degree in 1983 under the supervision of Professor N. S. Zefirov and performed his Habilitation in 1998. From 1983 to 1991, he worked at Moscow State University and then at Zelinsky Institute of Organic Chemistry. In 1991, he joined the research group of Professor A. de Meijere (Georg-August-Universität Göttingen, Germany) as an Alexander von Humboldt Research Fellow; since 1993 he has worked as a Research Associate, since 1996 he has held a position of a Scientific Assistant and since 2001 a permanent position as a Senior Scientist at the University of Göttingen. His current research interests focus on the chemistry of highly strained small ring compounds. The results of his scientific activity have been published in over 200 original publications, review articles, book chapters and patents.

(R)-2,2,2-Trichloro-1-phenylethyl (methylsulfonyl)-oxycarbamate

Hélène Lebel*[1], Henri Piras, and Johan Bartholoméüs

Chemistry Department, Université de Montréal, PO 6128, Station downtown, Montréal, Qc, Canada H3C 3J7

Checked by Mario Leypold and Mohammad Movassaghi

A.

B.

C.

D.

Procedure (Note 1)

A. *2,2,2-Trichloro-1-phenylethan-1-one* (**1**). A 2000-mL Erlenmeyer-flask equipped with a 7 cm flat magnetic stirrer is charged with (±)-2,2,2-trichloro-1-phenylethan-1-ol (38.3 g, 170 mmol, 1.00 equiv) (Note 2), dichloromethane (350 mL) (Note 3) and saturated NaHCO₃ (700 mL) (Note 4). The solution is stirred vigorously (900 rpm). To this solution is added

TEMPO (2.66 g, 17.0 mmol, 0.10 equiv) (Note 5) at 23 °C. The resulting orange/red solution is cooled to 0 °C (Note 6) and pyridinium tribromide (81.6 g, 255 mmol, 1.50 equiv) (Note 7) is carefully added portion wise (Note 8) (Figure 1A). After the addition is complete, the solution is allowed to warm to 23 °C and stirred vigorously for 2 h (Note 9). After completion, the red solution is cooled to 0 °C and excess oxidant quenched by addition of 5% Na$_2$S$_2$O$_3$ solution (250 mL) (Notes 10 and 11) (Figure 1B).

Figure 1A. Reaction set-up; 1B. After quench
(Photos provided by submitters)

The magnetic stir bar is removed and the solution is poured in a 2000-mL separatory funnel. The layers are separated and the aqueous layer is extracted with dichloromethane (2 x 200 mL) and EtOAc (2 x 200 mL). The combined organic layers are dried over MgSO$_4$ (30 g), filtered through an 8 cm diameter fritted glass funnel into a 2000-mL round bottomed flask, washed with EtOAc (75 mL) (Note 12) and concentrated using a rotary evaporator (38 °C, 375 mmHg to 35 mmHg) (Notes 13 and 14). The crude material is purified by flash chromatography using hexanes as eluent (200 g silica gel, eluent: hexanes, dimension: 14.0×6.5 cm, fraction size: 50 mL) to afford the pure product as a slightly yellow liquid (28.5 g, 75% yield) (Notes 15, 16, 17, and 18).

B. *(R)-2,2,2-Trichloro-1-phenylethan-1-ol* (**2**). A 1000-mL three-necked round-bottomed flask equipped with a 4 cm egg-shaped magnetic stirred is flame-dried under argon, then charged, after cooling back to 23 °C, with **1**

(27.9 g, 125 mmol, 1.00 equiv) dissolved in dry toluene (320 mL) (Note 19). (S)-(−)-2-Butyl-CBS-oxazaborolidine ((S)-CBS-Bu catalyst) in toluene (1.00 M, 6.25 mL, 6.25 mmol, 5.0 mol %) (Notes 20, 21, and 22) is added via a cannula to the reaction mixture and the resulting mixture is cooled to −78 °C (Note 23). A 1.05 M solution of catecholborane in THF (190 mL, 190 mmol, 1.60 equiv) (Note 24) is then carefully added dropwise over 7–9 h using a 500 mL dried pressure-equalized addition funnel to the stirred reaction mixture (500–600 rpm) (Note 25) (Figure 2).

Figure 2. Reaction set-up (Photo provided by submitters)

After 12 h of reaction (including the addition time), the solution is allowed to slowly warm to 23 °C (Note 26) and stirred at room temperature for 16 h. The solution is cooled to 0 °C (Note 6) and quenched by slow addition of water (230 mL) (Note 27) followed by EtOAc (230 mL). The biphasic mixture is poured in a 2000-mL separatory funnel and the layers are separated. The aqueous layer is extracted with EtOAc (140 mL) and the combined organic layers are washed with 2 M NaOH solution (5 x 180 mL) (Note 28) (Figure 3), 1 M HCl solution (3 x 140 mL) and saturated NaCl solution (140 mL). The organic layer is then dried over MgSO$_4$ (~20 g), filtered through a 4 cm diameter fritted glass funnel, washed with EtOAc (50 mL) and concentrated under reduced pressure (35–40 °C, 150 to 15 mmHg). The crude (yellow oily liquid) is then purified by filtration through a short pad of silica gel (135 g, eluent: Et$_2$O:hexanes (1:9), dimension: 13.0×6.0 cm, fraction size: 50 mL for the first eight fractions, then 150 mL for the next eight fractions) (Note 29). The fractions containing the product are collected in a 2000-mL round-bottomed flask and evaporated (225 to 15 mmHg, 40 °C). The resulting liquid is then dried *in vacuo* to afford the pure chiral alcohol as a colorless liquid (28.0 g, 99% yield, 94% ee) (Notes 30 and 31).

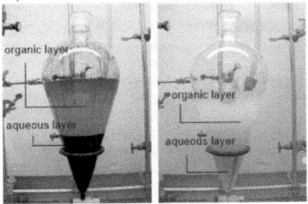

Figure 3. After the first and the last basic wash
(Photos provided by submitters)

C. *(R)-2,2,2-Trichloro-1-phenylethyl hydroxycarbamate* (**3**). A 1000-mL two-necked round-bottomed flask equipped with a 4 cm egg-shaped magnetic stirrer is flame-dried under argon atmosphere, then charged, after cooling back to room temperature, with **2** (27.3 g, 121 mmol, 1.00 equiv) and dry acetonitrile (300 mL) (Note 32). 1,1'-Carbonyldiimidazole (CDI) (21.6 g, 133 mmol, 1.10 equiv) (Note 33) is then added portion wise (Note 34) to the stirred solution (400–500 rpm) (Note 35) (Figure 4).

Figure 4. Reaction set-up (Photo provided by submitters)

The resulting mixture is stirred at 23 °C for 2 h or until the reaction reached completion (Note 36). The solution is then cooled to 0–5 °C (Note 5) before hydroxylamine hydrochloride (33.6 g, 484 mmol, 4.00 equiv) (Note 37 and 38) and imidazole (24.7 g, 363 mmol, 3.00 equiv) (Note 39) are successively added to the heterogeneous solution (Note 40). The resulting mixture is stirred at 0–5 °C for 45 min (Note 41). After the reaction is complete, acetonitrile is removed by evaporation under reduced pressure (180 to 15 mmHg, 40 °C). The residue is dissolved in a (5:1) mixture of 10% w/w HCl solution and EtOAc (450 mL total volume: 375 mL 10% w/w HCl and 75 mL EtOAc). The biphasic mixture is poured in a 1000-mL separatory funnel and the layers are separated. The aqueous layer is extracted with EtOAc (2 x 100 mL). The combined organic layers are washed with a saturated NaCl solution (200 mL), dried over Na$_2$SO$_4$ (20 g), filtered, washed with EtOAc (50 mL) and concentrated (35–40 °C, 150 to 15 mmHg) (Note 42). The resulting pale yellow oil is purified by precipitation in hexanes: the crude N-hydroxycarbamate is dissolved in chloroform (10 mL) (Note 43), and the resulting solution is added dropwise to hexanes (1000 mL) (Note 15) that is vigorously stirred (700 rpm) in a 2000-mL Erlenmeyer-flask with an 8 cm magnetic stir bar. A white solid is formed during addition. The solid is then collected by filtration through a 4 cm diameter fritted glass funnel, washed with hexanes (75 mL) and dried *in vacuo* to afford the desired product as a colorless solid (29.5 g, 86% yield) (Notes 44, 45, 46, and 47).

D. *(R)-2,2,2-Trichloro-1-phenylethyl (methylsulfonyl)oxycarbamate (4).* A 250-mL round-bottomed flask equipped with a 3 cm egg-shaped magnetic stirrer is flame-dried under argon atmosphere, then charged, after cooling back to 23 °C, with **3** (4.20 g, 13.8 mmol, 1.15 equiv) and dry dichloromethane (30 mL) (Note 48). The solution is stirred at 0 °C (500 rpm) and triethylamine (1.67 mL, 12.0 mmol, 1.00 equiv) is added to the solution with a syringe (Note 49). Methanesulfonyl chloride (933 µL, 12.0 mmol, 1.00 equiv) (Note 50) is then added dropwise (over 5 min) to the mixture at 0 °C. The resulting mixture is then stirred at 23 °C for 1 h (Note 51). After the reaction is complete, the reaction is quenched by addition of water (20 mL). The biphasic mixture is poured in a 250-mL separatory funnel and the layers are separated. The organic layer is washed with water (25 mL) and re-extracted with dichloromethane (5 mL). The combined organic layers are washed with brine (25 mL) (Note 52), dried over Na$_2$SO$_4$ (~8 g), filtered through cotton, washed with dichloromethane (10 mL), and concentrated under reduced pressure (325 to 15 mmHg at 40 °C) to afford a pale yellow sticky oil. The crude material is purified by flash chromatography (470 g silica gel, eluent: Et₂O:pentane (1:4), dimension: 28.0 × 6.5 cm, fraction size: 1600 mL for the first three fractions, then 80 mL for the next eighty fractions) (Notes 53 and 54). A total of 2.87–3.27 g (66–75% yield) of the desired product is obtained as a pure colorless solid (Notes 55, 56, and 57).

Notes

1. Prior to performing each reaction, a thorough hazard analysis and risk assessment should be carried out with regard to each chemical substance and experimental operation on the scale planned and in the context of the laboratory where the procedures will be carried out. Guidelines for carrying out risk assessments and for analyzing the hazards associated with chemicals can be found in references such as Chapter 4 of "Prudent Practices in the Laboratory" (The National Academies Press, Washington, D.C., 2011; the full text can be accessed free of charge at https://www.nap.edu/catalog/12654/prudent-practices-in-the-laboratory-handling-and-management-of-chemical. See also "Identifying and Evaluating Hazards in Research Laboratories" (American Chemical Society, 2015) which is available via the associated website "Hazard Assessment in Research Laboratories" at https://www.acs.org/content/acs/en/about/governance/committees/chemicalsafety/hazard-assessment.html. In the case of this procedure, the risk assessment should include (but not necessarily be limited to) an evaluation of the potential hazards associated with (±)-2,2,2-trichloro-1-phenylethan-1-ol, sodium bicarbonate, dichloromethane, sodium thiosulfate, TEMPO, pyridinium tribromide, ethyl acetate, magnesium sulfate, hexanes, silica gel, toluene, (S)-CBS-Bu, catecholborane, tetrahydrofuran, sodium hydroxide, hydrochloric acid, sodium chloride, diethyl ether, acetonitrile, 1,1'-carbonyldiimidazole, hydroxylamine hydrochloride, imidazole, sodium sulfate, chloroform, triethylamine, and methanesulfonyl chloride.

2. (±)-2,2,2-Trichloro-1-phenylethan-1-ol is prepared using the procedure described in *Org. Synth.* **1968**, *48*, 27–29. The submitters prepared and provided the sample of (±)-2,2,2-trichloro-1-phenylethan-1-ol (~98% purity) used for checking. (±)-2,2,2-Trichloro-1-phenylethan-1-ol is also commercially available.

3. Dichloromethane (certified ACS) is purchased from Caledon Company and used as received.

4. Sodium bicarbonate is purchased from Caledon Company and used as received. The saturated solution is prepared by adding 90 g of sodium bicarbonate to 1000 mL of distilled water.

5. TEMPO is purchased from Sigma-Aldrich Fine Chemicals Company Inc.

6. An ice/water bath is used.

7. Pyridinium tribromide (90% grade) is purchased from Sigma-Aldrich Fine Chemicals Company Inc and is used as received.

8. Portions of 5-10 g are added every 2 minutes, for a total addition of 15–20 minutes (a foam can form if the addition is too fast). The solution must be vigorously stirred (900 rpm or more).

9. The reaction is monitored by TLC analysis on silica gel using a mixture of Et_2O:hexanes (1:9) and visualization with UV light and $KMnO_4$ (R_f of starting material = 0.16, R_f of product = 0.68).

10. Sodium thiosulfate is purchased from Merck KGaA. The 5% w/w solution is prepared by dissolving 78 g of sodium thiosulfate pentahydrate ($Na_2S_2O_3\cdot5H_2O$) in 1000 mL of distilled water.

11. The red solution becomes yellow upon the addition of sodium thiosulfate.

12. A TLC analysis is performed on the last drop of filtration to be sure that all the product has been collected.

13. The checkers recommend the concentration of the crude on a rotary evaporator in a fume hood due to the presence of pyridine.

14. Submitters suggested purification by distillation using the following conditions: The crude material is suspended in hexanes (200 mL) (Note 15) and silica gel (30 g) (Note 16) is added to trap the colored impurities. The mixture is filtered through a pad of silica gel (15 g), washed with hexanes (600 mL) and the filtrate is evaporated *in vacuo* (38 °C, 375 to 35 mmHg). The resulting pale yellow liquid is distilled under reduced pressure (oil-bath: 80 °C to 120 °C, 0.2 mmHg, main fraction: bp 55–64 °C at 0.2 mmHg). The distillation short path is equipped with a 15 cm Vigreux column and grease is used for all joints. Checkers found that purification by distillation as described did not reliably provide the highest purity product (Note 17).

15. Hexanes (certified ACS) is purchased from Fisher Scientific Company and used as received.

16. Silica gel F60 type, 40-63 μm (230-400 mesh) is purchased from Zeochem AG Inc.

17. The checkers found that an additional purification by flash chromatography was necessary after distillation to obtain analytically pure product.

18. A second reaction performed on half-scale provided 15.0 g (79%) of the same product, after purification by distillation and chromatography. Analytical data for 2,2,2-trichloro-1-phenylethan-1-one (**1**). R_f = 0.66–0.69 (Et$_2$O:hexanes (1:9)); ^{1}H NMR (400 MHz, CDCl$_3$) δ: 7.48–7.53 (m, 2H), 7.62–7.66 (m, 1H), 8.25–8.28 (m, 2H); ^{13}C NMR (100 MHz, CDCl$_3$) δ: 95.6, 128.5, 129.2, 131.6, 134.4, 181.4; IR (film, ATR-FTIR) 1709, 1596, 1448, 1221, 1005, 820, 650 cm^{-1}; HRMS (ESI+) calc. for C$_8$H$_6$Cl$_3$O [M+H]$^+$: 222.9479; found: no exact mass found by ESI+; Anal. calc. for C$_8$H$_5$Cl$_3$O: C, 43.00, H, 2.26; found: C, 43.21, H, 2.19.

19. Dry toluene from a column purification solvent system (using activated alumina and CuO (treated with H$_2$) columns) under argon atmosphere is employed.

20. (S)-CBS-Bu catalyst (1.00 M in toluene) is purchased from Sigma-Aldrich Fine Chemicals Company Inc.

21. The checkers used commercially available solution of catalyst (Note 20). The submitters prepared a solution of (S)-CBS-Bu according to the following procedure: A 100 mL round-bottomed flask, equipped with a magnetic stirrer, is charged with n–butylboronic acid (780 mg, 7.70 mmol, 1.10 equiv), (S)–diphenyl-prolinol (1.77 g, 7.00 mmol, 1.00 equiv) and toluene (35 mL). The flask is then equipped with a Dean-Stark apparatus filled with toluene, put under argon, stirred and heated to reflux overnight. The solution is then used directly in the reaction without any purification.

22. Checkers found that the reaction with commercially available (S)-CBS-Bu catalyst (1.00 M in toluene) proceeded with the same efficiency, with respect to yield and level of enantioselection, as described by the submitters.

23. A dry ice/acetone bath is used. A period of 20–30 min is typically needed to reach –78 °C inside the reaction mixture.

24. A 1 M solution of catecholborane in THF (1.05 M according to the specification sheet) is purchased from Sigma-Aldrich Fine Chemicals Company Inc.

25. An internal thermostat probe is used to monitor the temperature of the solution. Addition rate (one drop every 5 seconds approximately) is carefully controlled to maintain the temperature below –65 °C.

26. The dry ice/acetone bath should not be removed during the warming to allow a slow increase of the temperature of the solution mixture. The checkers note that the warming of the solution to room temperature as described in this note by the submitters took 24–36 h.

27. Water is added dropwise (two drops per second) as rapid evolution of H_2 is observed.

28. The aqueous layer becomes green/black during the washing. The aqueous washings are performed until the aqueous layer becomes light brown. The checkers observed that the basic wash slightly warmed and slow phase separation was noted in the 4th and 5th wash.

29. Approx. 2000 mL of eluent is required.

30. The checkers noted a slightly higher level of enantioselection on half-scale experiment (14.5 g, 65.0 mmol of 1), which afforded 14.6 g of alcohol 2 as colorless liquid (>99% yield, 96% ee).

31. Analytical data for (R)-2,2,2-trichloro-1-phenylethan-1-ol (2). R_f = 0.24–0.27 (Et₂O:hexanes (2:8)); ¹H NMR (400 MHz, CDCl₃) δ: 3.41 (d, J = 4.0 Hz, 1H), 5.21 (d, J = 3.9 Hz, 1H), 7.38–7.45 (m, 3H), 7.62–7.65 (m, 2H); ¹³C NMR (100 MHz, CDCl₃) δ: 84.6, 103.2, 127.9, 129.3, 129.6, 134.9; IR (film, ATR-FTIR) 3450, 1454, 1059, 817, 743 cm⁻¹; $[\alpha]_D^{24}$ –39.2 (c 1.00, CHCl₃); Enantiomeric ratio is determined to be 97.1:2.9 by analytical HPLC, using the following conditions: Chiracel-OD chiralpak column (4.6 mm x 250 mm, particle size 10 µm, part #14025); 10% isopropanol in hexanes for 30 min, flow 1.0 mL/min, 210 nm detection; retention time: t_{major} = 8.5 min and t_{minor} = 11.4 min; HRMS (ESI+) calc. for $C_8H_7Cl_3NaO$ [M+Na]⁺: 246.9455; found: no exact mass found by ESI+; Anal. Calcd for $C_8H_7Cl_3O$: C, 42.61, H, 3.13, found: C, 42.99, H, 3.05.

32. Dry acetonitrile from a solvent purification system (using two neutral activated alumina columns) under argon atmosphere is employed.

33. 1,1'-Carbonyldiimidazole is purchased from Alfa Aesar and stored in glovebox prior to use.

34. Portions of 5 g are added every minute.

35. The solution becomes heterogeneous over time. A white precipitate appears a few minutes after the addition of CDI.

36. The reaction is monitored by TLC analysis on silica gel using EtOAc:hexanes (2:8) and visualization with UV light and KMnO₄ (R_f of starting material = 0.5, R_f of intermediate = 0.2).

37. The hydroxylamine hydrochloride salt is purchased from Sigma-Aldrich Fine Chemicals Company Inc.

38. Hydroxylamine hydrochloride is hygroscopic and is dried in an oven (110 °C) overnight before use.

39. Imidazole (99% grade) is purchased from Alfa Aesar and used as received.

40. The precipitate disappears for a while before another beige precipitate appears after the addition.

41. The reaction is monitored by TLC analysis on silica gel using EtOAc in hexanes (3:7) and visualization with UV light and KMnO$_4$ (R_f of product = 0.34).

42. Checkers observed some solid formation in the crude material.

43. Chloroform is purchased from Fisher Chemicals and used as received.

44. Checkers found that upon concentration of the mother liquor to approx. 100 mL, additional product precipitates. This second crop of product is an equal mixture of product **3** and starting material **2** and can give another ~1 g of the product **3** upon purification by flash chromatography (Note 46).

45. Checkers noted formation of product **3** with similar efficiency and with the similar purity for first and second crops on half-scale experiment (13.5 g, 60.0 mmol of **2**), affording 14.8 g of product **3** as colorless solid (87%) in first crop.

46. Analytical data for (*R*)-2,2,2-trichloro-1-phenylethyl hydroxycarbamate (**3**). R_f = 0.34-0.37 (EtOAc:hexanes (3:7)); ^{1}H NMR (400 MHz, CDCl$_3$) δ: 6.30 (s, 1H), 6.88 (br, 1H), 7.36-7.45 (m, 3H), 7.56-7.59 (m, 3H) (OH under this band); ^{13}C NMR (100 MHz, CDCl$_3$) δ: 84.2, 99.1, 128.1, 129.7, 130.1, 132.6, 156.8; IR (solid, ATR-FTIR) 3326, 1743, 1453, 1245, 1118, 751 cm^{-1}; mp 93–96 °C (checkers note the mp for purified product is 97-100 °C); [α]$_D^{25}$ –35.8 (c 1.00, CHCl$_3$); HRMS (ESI+) calc. for C$_9$H$_8$Cl$_3$NNaO$_3$ [M+Na]$^+$: 305.9462; found: 305.9464; Anal. calc. for C$_9$H$_8$Cl$_3$NO$_3$: N, 4.92, C, 37.99, H, 2.83; found: N, 4.81, C, 37.99, H, 2.71. (A sample is purified by flash chromatography eluting with MeOH in DCM (3:97) before performing the elemental analysis. NMR spectra of both, the product before and after purification are provided. The crude product (93.5% w/w) is directly used for subsequent reaction.).

47. Consistent with the submitters observations, the checkers observed no difference in use of product **3** directly from first crop or after chromatographic purification (Note 46). Checkers found that the purity of the isolated product in the first crop is 92.0% n/n, 93.5% w/w,

exactly the same as provided by submitters. The product **3** is isolated along with remaining starting alcohol **2**.

48. Dry DCM from a column purification solvent system (using two neutral activated alumina columns) under argon atmosphere is employed.

49. Anhydrous triethylamine from a solvent purification system (using two neutral activated alumina columns) under argon atmosphere is employed.

50. Methanesulfonyl chloride is purchased from Sigma-Aldrich Fine Chemicals Company Inc. and is freshly distilled under vacuum over P_2O_5 prior to use.

51. The reaction is monitored by TLC analysis on silica gel using EtOAc:hexanes (1:2) and visualization with UV light and $KMnO_4$ (R_f of product = 0.47).

52. Shaking too violently will produce an emulsion.

53. The checkers isolated analytically pure product **4** reliably by flash column chromatography, and careful analysis of the fractions prior to collection. Due to close elution of a byproduct (^{1}H NMR resonance at 6.28 ppm) with product **4**, fractions containing product were subject to ^{1}H NMR analysis in addition to TLC analysis. Only fractions with >99% purity of product are combined, concentrated, and dried *in vacuo*. Any mixed fractions are combined, concentrated, and re-purified by a second flash column chromatography using the same eluent with the 100-fold mass of silica gel compared to the mass of the mixed fraction. Checkers recommend a height-to-diameter ratio of the column between 6-8 for the second flash chromatography.

54. While the submitters described purification of **4** by crystallization, after significant experimentation the checkers were not able to confirm this method as a reliable means for purification of **4**. Thus, the checkers used flash chromatography conditions based on those described in Lebel, H.; Piras, H; Bartholoméüs, J *Angew. Chem. Int. Ed.* **2014**, *53*, 7300–7304.

55. Checkers found that concentration of the pure product from dichloromethane is necessary to remove trace amounts of Et_2O. The sample is dried *in vacuo* in a 500-mL round-bottomed flask for 48 h to afford solvent-free product **4**.

56. Checkers noted that the purification of compound **3** had no impact on success of reaction step D, consistent with the submitters' observations. Checkers performed reaction step D on ~1 g scale with both crude (first crop, 92% purity) and purified compound **3** (Note 46) and observed

similar yields, and overall outcome consistent with results described here for the multi-gram scale.

57. Analytical data for (R)-2,2,2-trichloro-1-phenylethyl (methylsulfonyl)-oxycarbamate (**4**): R_f = 0.47–0.50 (EtOAc:hexanes (3:7)); ^{1}H NMR (400 MHz, CDCl$_3$) δ: 3.20 (s, 3H), 6.35 (s, 1H), 7.40-7.49 (m, 3H), 7.60-7.63 (m, 2H), 8.46 (s (br), 1H); ^{13}C NMR (100 MHz, CDCl$_3$) δ: 36.7, 85.2, 98.6, 128.4, 129.7, 130.5, 131.7, 153.8; IR (solid, ATR-FTIR) 3276, 2940, 1753, 1455, 1374, 1234, 1182, 1092, 969, 787, 699 (cm^{-1}); mp = 75–81 °C; [α]$_D^{24}$ –15.9 (c 1.47, CHCl$_3$); HMRS (ESI+) calc. for C$_{10}$H$_{10}$Cl$_3$NNaO$_5$S [M+Na]+: 383.9259; found: 383.9240; Anal. calc. for C$_{10}$H$_{10}$Cl$_3$NO$_5$S: N, 3.86, C, 33.12, H, 2.78, S, 8.84; found: N, 3.91, C, 33.38, H, 2.74, S, 8.87.

Working with Hazardous Chemicals

The procedures in *Organic Syntheses* are intended for use only by persons with proper training in experimental organic chemistry. All hazardous materials should be handled using the standard procedures for work with chemicals described in references such as "Prudent Practices in the Laboratory" (The National Academies Press, Washington, D.C., 2011; the full text can be accessed free of charge at http://www.nap.edu/catalog.php?record_id=12654). All chemical waste should be disposed of in accordance with local regulations. For general guidelines for the management of chemical waste, see Chapter 8 of Prudent Practices.

In some articles in *Organic Syntheses*, chemical-specific hazards are highlighted in red "Caution Notes" within a procedure. It is important to recognize that the absence of a caution note does not imply that no significant hazards are associated with the chemicals involved in that procedure. Prior to performing a reaction, a thorough risk assessment should be carried out that includes a review of the potential hazards associated with each chemical and experimental operation on the scale that is planned for the procedure. Guidelines for carrying out a risk assessment and for analyzing the hazards associated with chemicals can be found in Chapter 4 of Prudent Practices.

The procedures described in *Organic Syntheses* are provided as published and are conducted at one's own risk. *Organic Syntheses, Inc.*, its Editors, and its Board of Directors do not warrant or guarantee the safety of individuals using these procedures and hereby disclaim any liability for any injuries or damages claimed to have resulted from or related in any way to the procedures herein.

Discussion

N-Sulfonyloxycarbamates are electrophilic nitrogen reagents that are used in numerous reactions involving the formation of C-N bonds.[2,3,4,5,6,7,8,9,10,11] These reagents have emerged as alternative metal nitrene precursors, namely to perform aziridination and C-H amination reactions.[12,13,14,15,16,17] Among these reagents, (*R*)-2,2,2-trichloro-1-phenylethyl (methylsulfonyl)oxycarbamate is reported to undergo stereoselective C-H amination,[18] thioether amination[19,20] and aziridination reactions,[21] in the presence of Rh₂{(*S*)-4-Br-nttl}₄ or Rh₂{(*S*)-nttl}₄.

The chiral alcohol ((*R*)-2,2,2-trichloro-1-phenylethan-1-ol) is prepared via the Corey CBS reduction of the corresponding ketone.[22,23] It should be noted that the facial selectivity is inverted in comparison with the result obtained for the asymmetric CBS reduction of acetophenone. Other enantioselective reduction methods failed to afford the desired alcohol with high enantiomeric excess.[24,25,26,27]

2,2,2-Trichloroacetophenone is not commercially available, but is accessible via the oxidation of racemic 2,2,2-trichloro-1-phenylethan-1-ol. At the outset, a mixture of sodium dichromate, sulfuric acid in glacial acetic was used, as reported in the literature.[28] Recently a novel procedure to oxidize highly electrophilic halogenated ketones was reported, using a mixture of pyridinium tribromide and catalytic amount of a modified TEMPO.[29] This procedure has been adapted for the standard TEMPO reagent and is now used to prepare 2,2,2-trichloroacetophenone in high yields and under more environmentally friendly reaction conditions (compared to chromium (VI) oxidation reaction conditions). The procedure could also be used to oxidize other aryl trichloroethanol derivatives (Table 1, entries 1–3). Aliphatic trichloroethanol derivatives were less reactive and only low yields were obtained, even under refluxing reaction conditions for a longer period of time (entry 4).

The *N*-mesyloxycarbamate reagent is prepared in two steps from (*R*)-2,2,2-trichloro-1-phenylethan-1-ol via the corresponding *N*-hydroxy-carbamate. Both products are solids and can be purified by flash column chromatography. (*R*)-2,2,2-trichloro-1-phenylethyl (methylsulfonyl)oxy-carbamate is highly stable (DSC/TGA analysis shows decomposition starting at 180 °C) and can be kept on the bench for an unlimited amount of time. The *N*-mesyloxycarbamate reagent is highly hygroscopic and up to 1 equiv of water may be associated with the reagent, as shown in the reported X-ray crystal structure.[18] The described procedure can be used to prepared a number of *N*-mesyloxycarbamates derived from primary and secondary alcohols.[30]

Table 1. Synthesis of various trichloroacetophenones

entry	conditions	product	yield (%)
1	25 °C, 12 h		75
2	25 °C, 4 h		91
3	25 °C, 4 h		87
4	45 °C, 14 h		28

References

1. Département de Chimie, Center for Green Chemistry and Catalysis, Université de Montréal, C.P. 6128, Succursale Centre-ville, Montréal, Québec, Canada H3C 3J7. Email: helene.lebel@umontreal.ca. This research was supported by NSERC (Canada), the Canadian Foundation for Innovation, the Canada Research Chair Program, the Université de Montréal and the Centre in Green Chemistry and Catalysis (CGCC).

2. Lwowski, W.; Maricich, T. J. *J. Am. Chem. Soc.* **1965**, *87*, 3630–3637.
3. Greck, C.; Genet, J. P. *Synlett* **1997**, 741–748.
4. Fioravanti, S.; Morreale, A.; Pellacani, L.; Tardella, P. A. *Eur. J. Org. Chem.* **2003**, 4549–4552.
5. Fioravanti, S.; Morreale, A.; Pellacani, L.; Tardella, P. A. *Tetrahedron Lett.* **2003**, *44*, 3031–3034.
6. Colantoni, D.; Fioravanti, S.; Pellacani, L.; Tardella, P. A. *Org. Lett.* **2004**, *6*, 197–200.
7. Burini, E.; Fioravanti, S.; Morreale, A.; Pellacani, L.; Tardella, P. A. *Synlett* **2005**, 2673–2675.
8. Colantoni, D.; Fioravanti, S.; Pellacani, L.; Tardella, P. A. *J. Org. Chem.* **2005**, *70*, 9648–9650.
9. Pellacani, L.; Fioravanti, S.; Tardella, P. A. *Curr. Org. Chem.* **2011**, *15*, 1465–1481.
10. Donohoe, T. J.; Chughtai, M. J.; Klauber, D. J.; Griffin, D.; Campbell, A. D. *J. Am. Chem. Soc.* **2006**, *128*, 2514–2515.
11. Liu, R.; Herron, S. R.; Fleming, S. A. *J. Org. Chem.* **2007**, *72*, 5587–5591.
12. Lebel, H.; Huard, K.; Lectard, S. *J. Am. Chem. Soc.* **2005**, *127*, 14198–14199.
13. Lebel, H.; Huard, K. *Org. Lett.* **2007**, *9*, 639–642.
14. Lebel, H.; Lectard, S.; Parmentier, M. *Org. Lett.* **2007**, *9*, 4797–4800.
15. Huard, K.; Lebel, H. *Chem. Eur. J.* **2008**, *14*, 6222–6230.
16. Huard, K.; Lebel, H. *Org. Synth.* **2009**, *86*, 59–69.
17. Lebel, H.; Parmentier, M.; Leogane, O.; Ross, K.; Spitz, C. *Tetrahedron* **2012**, *68*, 3396–3409.
18. Lebel, H.; Trudel, C.; Spitz, C. *Chem. Commun.* **2012**, *48*, 7799–7801.
19. Lebel, H.; Piras, H.; Bartholoméüs, J. *Angew. Chem. Int. Ed.* **2014**, *53*, 7300–7304.
20. Lebel, H.; Piras, H. *J. Org. Chem.* **2015**, *80*, 3572–3585.
21. Lebel, H.; Spitz, C.; Leogane, O.; Trudel, C.; Parmentier, M. *Org. Lett.* **2011**, *13*, 5460–5463.
22. Mellin-Morliere, C.; Aitken, D. J.; Bull, S. D.; Davies, S. G.; Husson, H. P. *Tetrahedron: Asymmetry* **2001**, *12*, 149–155.
23. Corey, E. J.; Helal, C. J. *Tetrahedron Lett.* **1993**, *34*, 5227–5230.
24. Gamble, M. P.; Smith, A. R. C.; Wills, M. *J. Org. Chem.* **1998**, *63*, 6068–6071.
25. Jiang, B.; Feng, Y.; Hang, J.-F. *Tetrahedron: Asymmetry* **2001**, *12*, 2323–2329.

26. Ramachandran, P. V.; Gong, B.; Teodorovic, A. V. *J. Fluorine Chem.* **2007**, *128*, 844–850.

27. Perryman, M. S.; Harris, M. E.; Foster, J. L.; Joshi, A.; Clarkson, G. J.; Fox, D. J. *Chem. Commun.* **2013**, *49*, 10022–10024.

28. Gallina, C.; Giordano, C. *Synthesis* **1989**, 466–468.

29. Mei, Z.-W.; Omote, T.; Mansour, M.; Kawafuchi, H.; Takaguchi, Y.; Jutand, A.; Tsuboi, S.; Inokuchi, T. *Tetrahedron* **2008**, *64*, 10761–10766.

30. Lebel, H.; Mamani Laparra, L.; Khalifa, M.; Trudel, C.; Audubert, C.; Szponarski, M.; Dicaire Leduc, C.; Azek, E.; Ernzerhof, M. *Org. Biomol. Chem.* **2017**, *15*, 4144–4158.

Appendix
Chemical Abstracts Nomenclature (Registry Number)

2,2,2-Trichloro-1-phenylethan-1-one; 2,2,2-Trichloroacetophenone:
Ethanone, 2,2,2-trichloro-1-phenyl-; (2902-69-4)
(±)-2,2,2-Trichloro-1-phenylethan-1-ol: Benzenemethanol,
α–(trichloromethyl)-; (2000-43-3)
(R)-2,2,2-Trichloro-1-phenylethan-1-ol: Benzenemethanol,
α–(trichloromethyl)-, (αR)-; (53432-39-6)
TEMPO: 1-Piperidinyloxy, 2,2,6,6-tetramethyl-; (2564-83-2)
Pyridinium tribromide: Hydrogen tribromide, compd. with pyridine (1:1);
(39416-48-3)
n-Butylboronic acid: Boronic acid, *B*-butyl-; (4426-47-5)
(*S*)-Diphenylprolinol: 2-Pyrrolidinemethanol, α,α–diphenyl-, (2*S*)-;
(112068-01-6)
Catecholborane: 1,3,2-Benzodioxaborole; (274-07-7)
(*R*)-2,2,2-Trichloro-1-phenylethyl hydroxycarbamate: Carbamic acid, *N*-hydroxy-, (1*R*)-2,2,2-trichloro-1-phenylethyl ester; (1391854-32-2)
1,1′-Carbonyldiimidazole (CDI): Methanone, di-1*H*-imidazol-1-yl-;
(530-62-1)
Hydroxylamine hydrochloride: Hydroxylamine, hydrochloride (1:1);
(5470-11-1)
(*R*)-2,2,2-Trichloro-1-phenylethyl (methylsulfonyl)oxycarbamate:
Methanesulfonic acid, [[(1*R*)-2,2,2-trichloro-1-phenylethoxy]carbonyl]azanyl
ester; (1391853-96-5)
Triethylamine: Ethanamine, *N*,*N*-diethyl-; (121-44-8)
Mesyl chloride: Methanesulfonyl chloride; (124-63-0)

Prof. Hélène Lebel obtained a BSc degree from Université Laval (1993) and a Ph.D. from Université de Montréal (1998). She then joined the group of Eric N. Jacobsen at Harvard University as a NSERC Postdoctoral Fellow. She began her academic career at the Université de Montréal in 1999, under a NSERC University Faculty Award. She was promoted to the rank of Full Professor in 2010. Her research interests focus on the development of new synthetic methodologies in organic chemistry based on transition metal-catalyzed processes.

Henri Piras was born in Paris and raised in l'île de la Réunion, in France. He obtained in 2011, an Engineer degree in synthetic and industrial organic chemistry from the École National Supérieure de Chimie de Clermont-Ferrand, and an MSc degree from Université Blaise-Pascal under the supervision of Prof. Yves Troin. Since January 2012, he has been a Ph.D. student with Prof. Hélène Lebel at Université de Montréal, working on the stereoselective synthesis of chiral sulfilimines and sulfoximines.

Johan Bartholoméüs was born and raised in Dunkerque, France. He received a Licence de chimie from Université du Littoral Côte d'Opale in Dunkerque in 2007 and a Master 1 in sciences from Université des Sciences et Technologies in Lille in 2008. He then completed a Master 2 in organic synthesis at Université de Bordeaux 1 under the supervision of Prof. Stéphane Quideaux. In September 2011, he joined the group of Prof. Hélène Lebel as a Ph.D. student and is currently writing his thesis on stereoselective amination of C-H bonds to synthesize propargylic amines.

Mario Leypold was born in 1987 in Austria. He studied chemistry at the Graz University of Technology and University of Graz (Austria), and performed his Ph.D. project under the guidance of Professor Rolf Breinbauer. He is currently working as a postdoctoral researcher with an Erwin-Schrödinger-Fellowship in the laboratory of Professor Mohammad Movassaghi (Massachusetts Institute of Technology, USA).

Large Scale Synthesis of Enantiomerically Pure (S)-3-(4-Bromophenyl)butanoic Acid

J. Craig Ruble,[#] H. George Vandeveer,[¶] and Antonio Navarro*[#1]

[#]Eli Lilly and Company, Lilly Research Laboratories, Lilly Corporate Center, Indianapolis, Indiana 46285, USA
[¶]Albany Molecular Research, Inc., 26 Corporate Circle, Albany, NY 12203, USA

Checked by Vignesh Palani and Richmond Sarpong

Procedure (Note 1)

A. *(S)-Ethyl 3-(4-bromophenyl)butanoate (1).* A 1-L, three-necked, round-bottomed flask equipped with a heating mantle, a 5 x 2 cm, Teflon coated egg-shaped magnetic stir bar, a thermometer through a rubber adapter, septum with an attached nitrogen line, and a 50 mL addition funnel capped with a septum (Figure 1) is charged sequentially with (4-bromophenyl)boronic acid (25.1 g, 125 mmol, 1.00 equiv), bis(norbornadiene)rhodium(I) tetrafluoroborate (467 mg, 1.25 mmol, 0.01 equiv), (R)-(+)-2,2′-bis(diphenylphosphino)-1,1′-binaphthyl (780 mg, 1.25 mmol, 0.01 equiv) and 1,4-dioxane (250 mL) (Note 2).

328

Published on the Web 10/8/2018
© 2018 Organic Syntheses, Inc.

Figure 1. Glassware set up

The mixture is stirred at 23 °C for 30 min under a positive pressure of nitrogen resulting in a light pink slurry (Figure 2A). Then, water (38 mL) is added in one portion and the reaction mixture becomes a clear red solution (Figure 2B). Upon subsequent addition of triethylamine (17.5 mL, 125 mmol, 1.00 equiv), a slight increase in internal temperature is observed (Δ~6 °C) and the mixture becomes darker (Figure 2C).

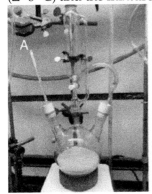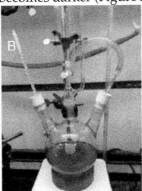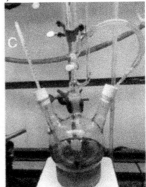

Figure 2. (A) Catalyst formation; (B) After water addition; (C) After TEA addition

The reaction is actively heated to 30 °C, then neat ethyl (*E*)-but-2-enoate (18.6 mL, 150 mmol, 1.20 equiv) is added through an addition funnel over 5 min (Note 3), and the mixture is stirred at this temperature for 21 h (Note 4), after which the reaction is complete (Note 5). The dark red mixture is allowed to cool to ambient temperature, transferred to a 1-L round-bottomed flask and concentrated in a rotary evaporator (Notes 6 and 7). The crude material is diluted with diethyl ether (200 mL), transferred to a 1-L separatory funnel (Figure 3a and 3b) and washed with water twice (200 mL and 100 mL, respectively) (Note 8). The combined aqueous layer is back extracted with diethyl ether (3 × 100 mL) until no product was observed in the organic layer (as judged by TLC) (Note 4) and the combined organic layers are dried over anhydrous Na_2SO_4 (7 g), filtered, and concentrated to give a dark red/brown oil (Figure 3c) (35.3 g) (Note 9).

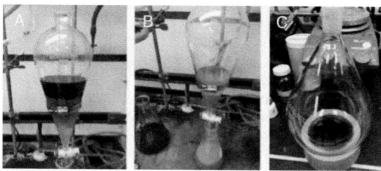

Figure 3. (A and B) Reaction workup; (C) crude product (Submitter's photo)

The crude product is filtered through a plug of SiO_2 (100 g in a glass fritted funnel, height = 8.5 cm, diameter = 9 cm), eluting with hexanes (400 mL) followed by 10% EtOAc in hexanes (1.5 L). Fractions containing product (Note 10) are combined and concentrated to dryness to give a clear light yellow oil (Figure 4) (33.75 g, >99%) (Notes 11 and 12).

Figure 4. (a) Silica gel plug; (b) Purified product (Submitter's photos)

B.   *(3S)-3-(4-Bromophenyl)butanoic acid (2)*. (a) A 1-L, three-necked, round-bottomed flask equipped with a heating mantel, a 5 × 2 cm Teflon coated egg-shaped magnetic stir bar, a thermometer through a septum (Note 1), a nitrogen line connected through a septum, and reflux condenser is charged consecutively with ethyl (3S)-3-(4-bromophenyl)butanoate (**1**, 33.75 g, 125 mmol), methanol (250 mL) (Note 13) and an aqueous solution of 5M NaOH (50 mL) while stirring at 18 °C internal temperature (Note 14) (Figure 5).

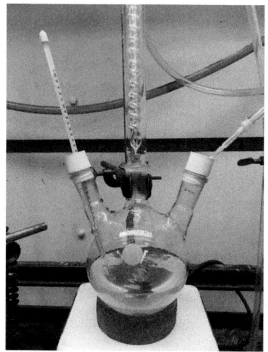

Figure 5. Reaction assembly for Step B

The system is then set to heat to 50 °C (Note 15). The reaction is stirred at the same temperature for an additional 1 h to reach completion to the desired acid. The heating system is removed, and the reaction mixture is allowed to cool to 30 °C. The contents are transferred to a 1-L round-bottomed flask, diluted with water (100 mL), and concentrated to remove methanol. The aqueous residue is transferred to a 1-L separatory funnel and washed with dichloromethane (100 mL) (Note 16), cooled down to 10 °C (ice-water bath), acidified with 12 M HCl (25 mL) solution to reach pH 1-2 (Note 17). The solution is then extracted with dichloromethane (3 × 100 mL) until no product was observed in the organic layer (as judged by TLC) (Note 16). The organics are dried over anhydrous Na₂SO₄, filtered, and concentrated in the rotavapor to give an off-white solid (29.95 g. 99%) (Note 18) (Figure 6).

(b) *Crystallization.* To the crude (3*S*)-3-(4-bromophenyl)butanoic acid (29.95 g, 125 mmol) is added heptane (250 mL), and this mixture is heated in an oil bath until complete dissolution of all the solids (60–65 °C) (Note 19). The hot solution is transferred to a 1-L round-bottomed flask. Additional warm heptane (200 mL) is used to rinse the residual product into the 1-L flask. The mixture is allowed to cool slowly to 20 °C over the course of 4–5 h. As the temperature reaches 25–30 °C, as determined by an inserted thermometer, crystals begin to form on the side of the flask (Note 20) (Figure 7). The liquid phase is decanted, concentrated to dryness, and placed under reduced pressure (0.25 mmHg) for 2 h to give (3S)-3-(4-bromophenyl)butanoic acid (**2**) as an off-white solid (21.93 g, 90.6 mmol) in a 73% yield (Notes 21 and 22).

Figure 6. Product prior to recrystallization

Figure 7. Crystallization after 16 h (A) Run 1 (B) Run 2

Notes

1. Prior to performing each reaction, a thorough hazard analysis and risk assessment should be carried out with regard to each chemical substance and experimental operation on the scale planned and in the context of the laboratory where the procedures will be carried out. Guidelines for carrying out risk assessments and for analyzing the hazards associated with chemicals can be found in references such as Chapter 4 of "Prudent Practices in the Laboratory" (The National Academies Press, Washington, D.C., 2011; the full text can be accessed free of charge at https://www.nap.edu/catalog/12654/prudent-practices-in-the-laboratory-handling-and-management-of-chemical. See also "Identifying and Evaluating Hazards in Research Laboratories" (American Chemical Society, 2015) which is available via the associated website "Hazard Assessment in Research Laboratories" at https://www.acs.org/content/acs/en/about/governance/committees/chemicalsafety/hazard-assessment.html. In the case of this procedure, the risk assessment should include (but not necessarily be limited to) an evaluation of the potential hazards associated with, as well as the proper procedures for (4-bromophenyl)boronic acid, bis(norbornadiene)rhodium(I) tetrafluoroborate, (R)-(+)-2,2'-bis(di-phenylphosphino)-1,1'-binaphthyl, 1,4-dioxane, triethylamine, ethyl (E)-

but-2-enoate, diethyl ether, anhydrous sodium sulfate, silica, hexanes, ethyl acetate, methanol, sodium hydroxide, dichloromethane, hydrochloric acid, and heptane.

2. (4-Bromophenyl)boronic acid (>99%, Chem Impex; *a known impurity is the (4'-bromobiphenyl)boronic acid*), ethyl crotonate (>98%, TCI America), bis(norbornadiene)rhodium(I) tetrafluoroborate (>98%, Aldrich) , (*R*)-BINAP (>99%, Acros Organics), triethylamine (≥99%, Sigma-Aldrich), 1,4-dioxane (>99%, EMD), deionized water, methanol (Fisher), 5.0 M sodium hydroxide (Fisher), and heptane (Fisher) were used as received.

3. The addition of ethyl (*E*)-but-2-enoate creates an increase in temperature that can reach ~40 °C; so, the heating mantel is turned off during the addition. Then it is turned back on to keep the temperature at ~35 °C during the whole reaction time.

4. During this time the process can be monitored by TLC analysis of an aliquot of the reaction mixture. TLC of 4-bromophenylboronic acid (50% ethyl acetate in hexanes): $R_f = 0.52$ (KMnO$_4$, UV), TLC of (*S*)-ethyl 3-(4-bromophenyl)butanoate (10% ethyl acetate in hexanes): $R_f = 0.57$ (KMnO$_4$, UV).

5. The submitters report that LCMS analysis showed ~95% (300 nm) conversion based on the ratio of (4-bromophenyl)boronic acid remaining and (*S*)-ethyl 3-(4-bromophenyl)butanoate formed.

6. The reaction flask is rinsed with EtOAc (50 mL), and this rinse is added to the reaction mixture.

7. The pressure of the rotavapor is set to 65 mmHg and the water bath at 40 °C.

8. A pink insoluble material is formed and can be separated (Figure 3B).

9. The isolated crude oil presents 92%wt/wt potency determined by quantitative ^{1}H NMR with the presence of 1,3-dimethoxybenzene as an internal standard.

10. Four fractions are collected (~1 L each) and analyzed by TLC (10% EtOAc in hexanes). Fractions 1 and 2 show product; fractions 3 and 4 show no product.

11. Quantitative ^{1}H NMR using 1,3-dimethoxybenzene as an internal standard revealed the purity to be 92.7% by weight. GCMS shows only one peak (R_t = 8.32 min, area = 100%, M+ *m/z*=270/272); ^{1}H NMR (400 MHz, CDCl$_3$) δ: 1.18 (t, *J* = 7.1 Hz, 3H), 1.27 (d, *J* = 7.0 Hz, 3H), 2.47–2.61 (m, 2H), 3.24 (tq, *J* = 7.2, 7.2 Hz, 1H), 4.02–4.11 (m, 2H), 7.05–7.13 (m, 2H), 7.38–7.44 (m, 2H); ^{13}C NMR (100 MHz, CDCl$_3$) δ: 14.3, 21.9, 36.1, 42.9, 60.5, 120.2, 128.7, 131.7, 144.8, 172.2, HRMS (ESI): calcd

for ([M+H], $C_{12}H_{16}BrO_2)^+$: 271.0328, found: 271.0326. The submitters estimated ee of **1** to be 88%, but this was not determined in checking.

12. A same scale reaction (125 mmol) was conducted again to provide 33.74 g (>99%) of **1**.

13. After the addition of MeOH, a fine white precipitate forms.

14. Upon addition of the NaOH solution, the reaction exotherms from 18 °C to 28 °C. The reaction mixture remained slightly cloudy after addition of NaOH. It turned homogeneous only after heating at 50 °C.

15. Approximately 20–30 min are required to reach the desired temperature, at which point the reaction conversion is ~80% (300 nm) (LCMS analysis).

16. The organic layer is analyzed (TLC or LCMS) to make sure there is no product. Analysis by TLC of (3S)-3-(4-bromophenyl)butanoic acid (50% ethyl acetate in hexanes): R_f = 0.65 (KMnO$_4$, UV).

17. The acid is added in one portion and the mixture quickly reaches a temperature of 25 °C. A white cloudy precipitate forms.

18. The solids obtained are analyzed by GCMS (R_t = 8.57 min, area = 35%, M+ m/z=242/244), LCMS (Rt = 0.63 min, M+NH$_4$ m/z= 260/262), ^{1}H NMR (400 MHz, CDCl$_3$): 10.84 (br s, 1H), 7.46 – 7.39 (m, 2H), 7.14 – 7.06 (m, 2H), 3.24 (tq, J = 7.2, 7.2 Hz, 1H), 2.67 – 2.52 (m, 2H), and 1.30 (d, J = 7.0 Hz, 3H), and Chiral HPLC (95/5 Heptane/IPA, 1 mL/min, Chiralcel OJ-H 4.6 x 150 mm, 73.3% e.e.).

19. The solution is analyzed by Chiral LC (95/5 Heptane/IPA, 1 mL/min, Chiralcel OJ-H 4.6 x 150 mm) to get the starting point, which is 73.3% ee.

20. Approximately 30 minutes after first crystals appear (solution shows 89.9% ee). After 4 hours, the amount of crystals visually increase (solution shows 97.9% ee). After standing overnight (~16h), chiral LC analysis (note 18) of the mother liquor shows 99.1% ee.

21. The solids are analyzed by GCMS (Rt = 8.57 min, area = 100%, M+ m/z=242/244), mp 50–53 °C, ^{1}H NMR (400 MHz, CDCl$_3$) δ: 1.30 (d, J = 7.0 Hz, 3H), 2.52 – 2.67 (m, 2H), 3.24 (tq, J = 7.2, 7.2 Hz, 1H), 7.06 – 7.14 (m, 2H), 7.39 – 7.46 (m, 2H), 10.84 (br s, 1H); ^{13}C NMR (100 MHz, CDCl$_3$) δ: 22.0, 35.8, 42.5, 120.4, 128.7, 131.8, 144.5, 178.6; HRMS (ESI): calcd for ([M–H], $C_{10}H_{10}BrO_2)^-$: 240.9870, found: 240.9870, which are all consistent with desired product. Purity was determined by quantitative ^{1}H NMR as 99.2% in the presence of an internal standard. The analytical solution is prepared by dissolving 24.6 mg of the product with 14.6 mg of 1,3-dimethoxybenzene in 0.7 mL CDCl$_3$.

22. The hydrolysis was repeated on a same scale (125 mmol) to provide 20.39 g (67%, 98.9% ee after 16 h crystallization) of **2**.

Working with Hazardous Chemicals

The procedures in *Organic Syntheses* are intended for use only by persons with proper training in experimental organic chemistry. All hazardous materials should be handled using the standard procedures for work with chemicals described in references such as "Prudent Practices in the Laboratory" (The National Academies Press, Washington, D.C., 2011; the full text can be accessed free of charge at http://www.nap.edu/catalog.php?record_id=12654). All chemical waste should be disposed of in accordance with local regulations. For general guidelines for the management of chemical waste, see Chapter 8 of Prudent Practices.

In some articles in *Organic Syntheses*, chemical-specific hazards are highlighted in red "Caution Notes" within a procedure. It is important to recognize that the absence of a caution note does not imply that no significant hazards are associated with the chemicals involved in that procedure. Prior to performing a reaction, a thorough risk assessment should be carried out that includes a review of the potential hazards associated with each chemical and experimental operation on the scale that is planned for the procedure. Guidelines for carrying out a risk assessment and for analyzing the hazards associated with chemicals can be found in Chapter 4 of Prudent Practices.

The procedures described in *Organic Syntheses* are provided as published and are conducted at one's own risk. *Organic Syntheses, Inc.*, its Editors, and its Board of Directors do not warrant or guarantee the safety of individuals using these procedures and hereby disclaim any liability for any injuries or damages claimed to have resulted from or related in any way to the procedures herein.

Discussion

Progress in the discovery of new therapeutic targets in the pharmaceutical industry demands access to an increasingly higher degree of molecular complexity that allows filling the space in three dimensions. Often times, these molecules contain chiral motifs that require either an asymmetric synthesis or a separation of enantiomers. The latter technique becomes expensive when synthesizing compounds at kilo-scale. Therefore, simple and cost effective access to enantiomerically pure substrates as intermediates in the synthesis of pharmaceutical active compounds is of high importance. In this sense, one strategy is to use readily available chiral materials from the chiral pool or chiral auxiliaries. Another strategy is the use of asymmetric catalysis to generate a stereocenter with high enantioselectivity.

The conjugate addition of organometallic reagents to α,β-unsaturated esters is one of the most useful processes for carbon–carbon bond formation giving β-substituted esters, which are versatile synthons to further organic transformations. One of the most successful and widely used is the *"rhodium-catalyzed asymmetric addition"* (RCAA), first described in 1997 by Miyaura et al.[2]

In this work, we have succeeded in applying the RCAA in multi-gram scale by carrying out the reaction in the presence of commercially available cost effective bis(norbornadiene)rhodium(I) tetrafluoroborate (S)-BINAP or (R)-BINAP as a catalyst in dioxane:H_2O (10:1) at "near ambient" temperature with very high yield (90%) and enantiomeric excess (88%). Furthermore, we have been able to increase the already high enantiomeric excess provided by the RCAA by crystallization of the racemate out of the nonracemic mixture of the β–substituted carboxylic acid after hydrolysis of the ester (>98%). This process allows for live monitoring of the enantiopurity enrichment of the product in the mother liquor by chiral HPLC.

The first step consists of the enantioselective Michael addition of *p*-bromophenyl boronic acid to ethyl crotonate under the RCAA conditions to make (R) or (S)-3-(4-bromophenyl)butanoic esters[3] with almost quantitative yields. The second step consists of the hydrolysis of the ester followed by *reverse crystallization* to further enhance chiral purity.[4] This process is simple, reproducible, reliable, scalable, safe, and cost effective to produce very valuable building blocks of high enantiomeric purity for the pharmaceutical industry.

We also determined the absolute stereochemistry of the chiral center generated by an independent method.[5] To do this, we chose to correlate the stereochemistry of our substrate with that of commercially available (S)-3-phenylbutyric acid. This material can be purchased from Aldrich (product number 78240, CAS[772-15-6], lot BCBF9385V). The Aldrich certificate of analysis for this lot indicated a specific rotation of +57 at a concentration of 1 in benzene. The reported positive specific rotation for the (S)-isomer in benzene is consistent with several literature preparations of optically active 3-phenylbutyric acid.[6]

To allow unambiguous assignment of configuration, we first sought to identify a simple derivative for which diastereomers could be resolved by NMR. We chose to examine amides derived from phenylglycinol because of the possible internal hydrogen bonding between –OH and carbonyl could act to rigidify the structures; and we found that the diastereomeric products derived from condensing racemic 3-phenylbutyric acid with (S)-phenylglycinol are resolved by ^{1}H NMR, ^{13}C NMR and LCMS, Scheme 1.[7]

racemic 3-phenylbutyric acid (3)
[4593-90-2]
Alfa Aesar

TBTU
(S)-phenylglycinol
NEt₃

DMF

Amide diastereomers (4) signals appear resolved in ^{1}H NMR, ^{13}C NMR and LCMS.

Scheme 1. Preparation of diastereomeric phenylglycinol amides

Utilizing the same bottle of (S)-phenylglycinol, we prepared the authentic (S)-N-((S)-2-hydroxy-1-phenylethyl)-3-phenylbutanamide (4) (Scheme 2, *Route A*) from the commercial (S)-3-phenylbutyric acid.[7] Then, we derivatized the single enantiomer of *3-(4-bromophenyl)butanoic acid* (2) to the corresponding single diastereoisomer amide by hydrogenation of the aryl bromide after amide formation (61% yield, Scheme 2, *Route B*).[9]

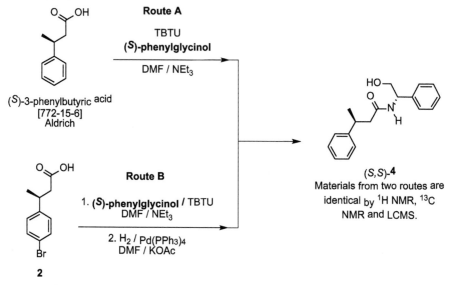

Scheme 2. Correlation of authentic (*S,S*)-amide with material from conjugate addition

The analytical data (¹H NMR, ¹³C NMR and LCMS) of *N*-((*S*)-2-hydroxy-1-phenylethyl)-3-phenylbutanamide was identical to the (*S,S*) standard (Figures 8 and 9). Our sample also co-eluted with the (*S,S*)-standard on both TLC and LCMS under conditions where the diastereomers were shown to separate.

From these data, we can unambiguously assign the *S* absolute configuration to the title compound 3-(4-bromophenyl)butanoic. Therefore, we can conclude that using (*R*)-BINAP as ligand in the RCAA produces the *S* isomer enantioselectively.[10]

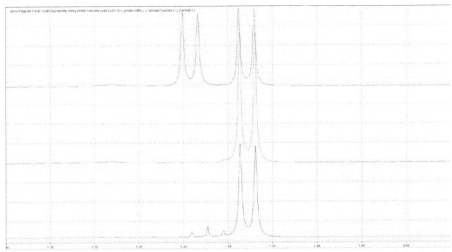

Figure 8. ¹H NMR overlay of mixture of diastereomeric amides(top), authentic (*S,S*)-amide (middle) and amide ultimately derived from using (*R*)-BINAP in the conjugate addition (bottom)

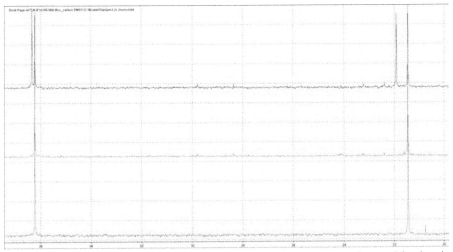

Figure 9. ¹³C NMR overlay of mixture of diastereomeric amides(top), authentic (*S,S*)-amide (middle) and amide ultimately derived from using (*R*)-BINAP in the conjugate addition (bottom)

References

1. Eli Lilly and Company, Lilly Research Laboratories, Lilly Corporate Center, Indianapolis, Indiana 46285. Email: navarroa@lilly.com. We thank Lilly Research Laboratories for the use of the laboratories and financial support.

2. Sakai, M.; Hayashi, H.; Miyaura, N. *Organometallics* **1997**, *16*, 4229–4231.

3. The absolute stereochemistry depends on the stereochemistry of the BINAP used as a ligand. Thus, (R)-BINAP gives *(S)-Ethyl 3-(4-bromophenyl)butanoate* and (S)-BINAP gives *(R)-Ethyl 3-(4-bromophenyl)butanoate*.

4. The enantiomeric purity was measured by chiral HPLC by comparing with an authentic sample of the racemate.

5. The corresponding conjugate addition of ethyl crotonate under the influence of (S)-BINAP had been previously reported by Lukin and coworkers to provide ethyl (S)-3-(4-bromophenyl)buteneoate in high ee, although the authors did not specify on what basis they had assigned the absolute configuration (see *J. Org. Chem.* **2009**, *74*, 929–931). When our results appeared to contradict the literature results, the decision was made to carry out the independent proof of configuration described herein.

6. (a) Prelog, V.; Scherrer, H. *Helv. Chim. Acta* **1959**, *42*, 2227–2232. (b) Nicolás, E.; Russell, K. C.; Hruby, V. J. *J. Org. Chem.* **1993**, *58*, 766–770. (c) Kanemasa, S.; Suenaga, H.; Onimura, K. *J. Org. Chem.* **1994**, *59*, 6949–6954.

7. In a typical procedure: (S)-3-Phenylbutyric acid (510 mg, 3.1 mmol), (S)-(+)-2-phenylglycinol (478 mg, 3.48 mmol) and TBTU (1.07 g, 3.27 mmol) are added under air to a flask containing a stir bar. Dimethylformamide (10.0 mL) and triethylamine (1 mL) are added and the solids dissolve yielding a light orange solution. The solution is stirred overnight, and the reaction mixture is added to a separatory funnel with MTBE (100 mL). The mixture is washed with 1 M HCl (2 x 50 mL) followed by 1 M NaOH (2 x 50 mL). The organic layer is dried over anhydrous MgSO$_4$, filtered, and concentrated to furnish 770 mg (87%) of a white, crystalline solid. ^{1}H NMR (399.80 MHz, DMSO-d_6) δ: 1.13 (d, *J* = 6.9 Hz, 3H), 2.42 (d, *J* = 7.6 Hz, 2H), 3.11-3.20 (m, 1H), 3.47 (d, *J* = 6.4 Hz, 2H), 4.78-4.83 (m, 1H), 4.92 (s, 1H), 7.2-7.29 (m, 10H), 8.30 (d, *J* = 8.2 Hz, 1H).

^{13}C NMR (100.53 MHz, DMSO-d_6) δ: 21.4, 36.2, 43.9, 54.9, 64.6, 125.9, 126.6, 126.7, 126.9, 127.9, 128.3, 141.4, 146.5, 170.5.

8. For use of these conditions to debrominate a range of substrates, see *Synth. Commun.* **1986**, *16*, 1067–1072.

9. Stereochemistry confirmation: in order to unequivocally assign the absolute stereochemistry of our compound, a sample of debrominated amide (Scheme 2, 4.4 mg) and a sample of the pure (*S,S*)-amide **4** (4.6 mg) were mixed and dissolved in DMSO-d_6. Analytical data (^{1}H NMR, ^{13}C NMR and LCMS) were consistent with a single isomer, which unequivocally proves that (*R*)-BINAP provided the (*S*) configuration of the product **1** in the RCAA reaction.

10. We have further verified that using (*S*)-BINAP results in the *R* enantiomer with identical yields and enantioselectivity.

Appendix
Chemical Abstracts Nomenclature (Registry Number)

(4-Bromophenyl)boronic acid (5467-74-3)
Isopropyl (*E*)-but-2-enoate (623-70-1)
(*R*)-(+)-2,2'-Bis(diphenylphosphino)-1,1'-binaphthyl (76189-55-4)
Bis(norbornadiene)rhodium(i) tetrafluoroborate (36620-11-8)
(S)-3-Phenylbutyric Acid (772-15-6)
(*S*)-(+)-2-Phenylglycinol (20989-17-7)

Antonio Navarro was born in Valencia, Spain in 1970. He received a B.S. in chemistry in 1993 and completed his Ph.D. in 1999 from University of Valencia, in the laboratories of Prof. Santos Fustero, working on the development of a new methodology to access to α–, and β– amino acids. After completing the first total synthesis of Azinomycin A in the laboratories of Prof. Rob Coleman at The Ohio State University, in Columbus, OH, he joined Eli Lilly as an organic synthetic chemist in Spain. In 2005 he moved to Indianapolis, IN as a Principal Research Scientist in Discovery Chemistry Research & Technologies focusing on developing new chemistry technologies.

Organic
Syntheses

H. George Vandeveer was born in Newport News, VA in 1979. He is currently a chemist working for Albany Molecular Research, Inc. in collaboration with Eli Lilly in Indianapolis. George joined AMRI in August of 2012 at the research facility of Eli Lilly where he worked in a medicinal chemistry group for 7 months, after which he transferred to their scale-up group. Shortly after this transition, he began working as the catalysis liaison between AMRI and Eli Lilly where he has focused his efforts on method development with applications to process chemistry.

J. Craig Ruble was born in Columbus, IN in 1972 and received a B.S. in chemistry from Indiana University in 1994. He completed his Ph.D. from the Massachusetts Institute of Technology in 1999 after working in the laboratories of Professor Gregory C. Fu on planar chiral nucleophilic catalysts. He then joined Pharmacia & Upjohn in Kalamazoo, MI as a medicinal chemist in the area of antibacterials. In 2003, he moved to Eli Lilly and Company where he is currently a Research Advisor in Discovery Chemistry Research & Technologies focused on the application of metal-catalyzed reactions in drug discovery.

Vignesh Palani was born in Tamil Nadu, India in 1994. He received his B.S. degree in 2016 from University of Minnesota-Twin Cities, where he worked with Prof. Thomas R. Hoye. He is currently pursuing his Ph.D. in the laboratories of Prof. Richmond Sarpong at UC Berkeley. His research focuses on total synthesis of natural products.

Preparation of Tributyl(iodomethyl)stannane

Michael U. Luescher, Chalupat Jindakun, and Jeffrey W. Bode[1*]

Laboratorium für Organische Chemie, ETH Zürich, HCI F315, Vladimir-Prelog-Weg 3, CH-8093 Zürich, Switzerland

Checked by Rahul Mondal, Estíbaliz Merino, and Cristina Nevado

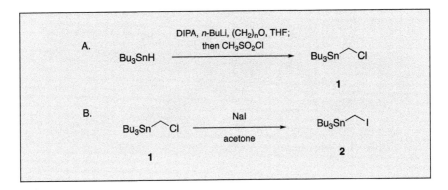

Procedure (Note 1)

A. *Tributyl(chloromethyl)stannane (1)*. An oven-dried 500-mL pear-shaped recovery flask equipped with a 34 x 16 mm, Teflon-coated, oval magnetic stir bar is charged with anhydrous THF (150 mL) (Note 2) and *N,N*-diisopropylamine (10.6 mL, 7.65 g, 75.6 mmol, 1.1 equiv) (Note 3) via a syringe. The flask is fitted with a rubber septum and a nitrogen inlet needle, after which the stir rate is set to ca. 375 rpm (Figure 1A). The reaction is cooled to 0–5 °C using an ice/water bath before the dropwise addition of *n*-BuLi (47.8 mL, 75.6 mmol, 1.1 equiv) (Note 4) via a 20 mL syringe over 25 min (Figure 1B). The resulting clear pale-yellow solution is stirred further at 0–5 °C for 30 min before tributyltin hydride (18.5 mL, 20.0 g, 68.7 mmol, 1.0 equiv) (Note 5) is added dropwise via syringe over 15 min (Figure 1C). The resulting clear yellow solution is stirred further at 0–5 °C for 30 min before paraformaldehyde (2.30 g, 76.6 mmol, 1.1 equiv) (Note 6) is weighed under air into a glass vial and charged in one portion to the reaction flask.

The ice/water bath is removed, and the turbid yellow suspension is stirred at 25–27 °C for 3 h during which a discoloration takes place affording a turbid pale-yellow reaction mixture (Figure 1D).

Figure 1. Color change through the course of the reaction

The resulting turbid reaction solution is cooled in a dry ice-acetone bath for 30 min before methanesulfonyl chloride (6.40 mL, 9.47 g, 82.7 mmol, 1.2 equiv) (Note 7) is added dropwise via a syringe over 10 min (Figure 1E). The cooling bath is removed, and the resulting pale-yellow suspension is stirred at 25–27 °C for 10 h (Note 8) (Figure 1F). At this point water (100 mL) is added in one portion using a graduated cylinder and the mixture is stirred further for 15 min.

The contents of the reaction flask are transferred to a 500-mL separatory funnel containing hexanes (150 mL). The aqueous layer is separated and extracted with hexanes (2 x 60 mL). The combined organic layers are washed with saturated NaCl solution (2 x 50 mL), dried over anhydrous $MgSO_4$ (12 g), and filtered through a 150-mL sintered glass Büchner funnel (medium porosity, 66 mm diameter). The $MgSO_4$ is washed with hexanes (3 x 10 mL) and the combined filtrate is concentrated by rotary evaporation (40 °C, 20 mmHg) to afford ca. 26 g of a pale-yellow oil (Notes 9 and 10). This material is diluted with hexanes (15 mL) and deposited onto a column (90 mm diameter) of 550 g of silica gel (16 cm high) (Note 11) prepared as a slurry in hexanes. Elution is carried out with hexanes collecting 50-mL fractions. The desired product is obtained in fractions 25–46 (Note 12).

Mixed fractions 20–24 are collected separately and concentrated by rotary evaporation (40 °C, 20 mmHg). The resulting colorless oil is diluted with hexanes (10 mL) and loaded onto a column (60 mm diameter) of 300 g of silica gel (8.5 cm high) (Note 11) prepared as a slurry in hexanes. Elution is carried out with hexanes collecting 30-mL fractions. The desired product is obtained in fractions 16–27. All fractions containing pure product according to TLC (Note 9) are combined and concentrated by rotary evaporation (40 °C, 20 mmHg). Further concentration at 23 °C under 0.1 mmHg for 2 h provides 15.8 g (68%) of chloromethyl stannane 1 as a colorless oil (Figure 1G) (Notes 13, 14, 15, and 16).

B. *Tributyl(iodomethyl)stannane (2)*. An oven-dried 500-mL pear-shaped recovery flask equipped with a 34 x 16 mm, Teflon-coated, oval magnetic stir bar is charged with acetone (125 mL) (Note 17) via a graduated cylinder and tributyl(chloromethyl)stannane (1) (10.5 g, 30.9 mmol, 1.0 equiv) (Note 13) from step A via a syringe over 10 min. The flask is fitted with a rubber septum and nitrogen inlet needle, after which the stir rate is set to ca. 375 rmp. Sodium iodide (9.50 g, 63.4 mmol, 2.0 equiv) (Note 18) is weighed in air into a glass vial and added in one portion to the reaction flask (Figure 2A). The colorless suspension is stirred at 25–27 °C for 8 h during which the mixture turns pale yellow (Note 19) (Figure 2B).

Figure 2. Color change through the course of the reaction

The resulting suspension is concentrated by rotary evaporation (40 °C, 40 mmHg) to afford a colorless oil containing excess sodium iodide and formed sodium chloride precipitates (Figure 3).

Figure 3. Crude tributyl(iodomethyl)stannane with sodium chloride and excess sodium iodide

The resulting mixture is suspended in hexanes (30 mL) and filtered over a 45 g silica gel plug (3 cm high) (Note 11) in a 150-mL sintered glass Büchner funnel (medium porosity, 66 mm diameter) rinsing with hexanes (8 x 50 mL). The colorless filtrate is concentrated on a rotavap (40 °C, 20 mmHg) and at 23 °C under 0.1 mmHg for 2 h to afford 12.3 g (92%) of iodomethyl stannane **2** as a colorless oil (Figure 3B) (Notes 16, 20, 21, and 22).

Notes

1. Prior to performing each reaction, a thorough hazard analysis and risk assessment should be carried out with regard to each chemical substance and experimental operation on the scale planned and in the context of the laboratory where the procedures will be carried out. Guidelines for carrying out risk assessments and for analyzing the hazards associated with chemicals can be found in references such as Chapter 4 of "Prudent Practices in the Laboratory" (The National Academies Press, Washington, D.C., 2011; the full text can be accessed free of charge at https://www.nap.edu/catalog/12654/prudent-practices-in-the-laboratory-handling-and-management-of-chemical. See also "Identifying and Evaluating Hazards in Research Laboratories" (American Chemical Society, 2015) which is available via the associated website "Hazard Assessment in Research Laboratories" at

https://www.acs.org/content/acs/en/about/governance/committees/chemicalsafety/hazard-assessment.html. In the case of this procedure, the risk assessment should include (but not necessarily be limited to) an evaluation of the potential hazards associated with, as well as the proper procedures for tetrahydrofuran, *N,N*-diisopropylamine, *n*-butyllithium, tributyltin hydride, paraformaldehyde, methanesulfonyl chloride, hexanes, magnesium sulfate, sodium chloride, silica gel, acetone, sodium iodide, and sodium chloride. Tributyltin hydride and other tributyltin derivatives are moderately toxic. Those reagents should only be handled by individuals trained in their proper and safe use.

2. Tetrahydrofuran (low water inhibitor free HPLC grade) was purchased from Sigma-Aldrich and purified by pressure filtration through activated alumina immediately prior to use.

3. *N,N*-Diisopropylamine (≥99.5%) was obtained from Sigma-Aldrich and distilled from calcium hydride at 83–86 °C (760 mmHg) prior to use.

4. *n*-Butyllithium (1.60 M in hexanes under AcroSeal) was purchased from Sigma-Aldrich and titrated prior to use. The concentration was determined to be 1.58 M according to the reported procedure from Watson, S. C.; Eastham, J. F. *J. Organomet. Chem.* **1967**, *9*, 165–168.

5. Tributyltin hydride (97% containing 0.05% BHT as stabilizer) was purchased from Sigma-Aldrich and used as received.

6. Paraformaldehyde (95%) was purchased from Sigma-Aldrich and used as received.

7. Methanesulfonyl chloride (98%) was purchased from Alfa Aesar and used as received.

8. The reaction progress was not monitored as the intermediate mesylate is rather unstable.

9. TLC analysis (hexanes with KMnO$_4$ stain visualization): Hexabutyldistannane sideproduct (Rf = 0.87) and tributyl(chloromethyl)stannane (**1**) (Rf = 0.74). The tributyl(chloromethyl)stannane (**1**) is not visible using UV 254 nm as the visualization technique.

10. The KMnO$_4$ stain was prepared using 1.5 g of KMnO$_4$ and 10 g of K$_2$CO$_3$ dissolved in 200 mL of water and 1.25 mL of 10% w/v NaOH solution.

11. High-purity silica gel grade (9385), pore size 60 Å, 230–400 mesh particle size purchased from Sigma-Aldrich.

12. Purification is followed using TLC analysis on Silica gel (hexanes with UV 254 nm and KMnO$_4$ stain visualization): Hexabutyldistannane sideproduct (Rf = 0.87) and tributyl(chloromethyl)stannane (**1**) (Rf =

0.74). The tributyl(chloromethyl)stannane (1) is not visible using UV 254 nm as the visualization technique.

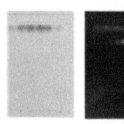

Some fractions contain minor amounts of Bu$_3$SnOSnBu$_3$ and Bu$_3$SnH as impurities; see TLC in Note 12. The second column is not necessary for synthetic purposes as the impurity (ca. 5%) does not affect the follow-up reaction and only slightly diminishes ($\leq$ 5% difference) the yield in following alkylation steps. The second purification was done to meet the purity standards on the publication and improve the yield of pure material.

13. Tributyl(chloromethyl)stannane (1) was prepared according to a modified procedure of Seitz et al.[2]

14. A second reaction on the same scale provided 16.2 g (69%) of the identical product. Tributyl(chloromethyl)stannane (1) decomposes over time when stored neat at ambient temperature. This reagent should be stored as a degassed 1 M solution in hexanes at –10 °C and used within the next few days for best results. The purification of the reaction crude after 7-10 days at – 28 °C, provided 8.40 g (36% yield) of the compound 1.

15. Tributyl(chloromethyl)stannane (1) has the following physical and spectroscopic properties: Rf = 0.74 (hexanes; KMnO$_4$ visualization; Merck Millipore TLC Silica gel 60 F254 plates); ^{1}H NMR (CDCl$_3$, 400 MHz) δ: 0.90 (t, J = 7.3 Hz, 9H), 0.97–1.01 (m, 6H), 1.32 (dq, J = 14.3, 7.2 Hz, 6H), 1.49–1.57 (m, 6H), 3.06 (t, $J(^{117/119}$Sn-1H) = 16.1 Hz, 2H); ^{13}C NMR (CDCl$_3$, 101 MHz) δ: 9.6, 13.8, 24.5, 27.4, 29.0; HRMS (EI) calculated for C$_{12}$H$_{27}$Sn [M – CH$_2$Cl]$^+$ 291.1129, found 291.1132 and calculated for C$_9$H$_{20}$ClSn [M – C$_4$H$_9$]$^+$ 283.0270, found 283.0264; IR (film): 3015, 2969, 2654, 2926, 1738, 1456, 1365, 1228, 1216, 1206, 876, 692, 664, 527, 515 cm^{-1}; Purity of the product was not assessed due to the instability of the product, which was quickly taken on to the second step.

16. ^{1}H NMR chemical shifts are expressed in parts per million (δ) downfield from tetramethylsilane (with the $CHCl_3$ peak at 7.26 ppm used as a standard). ^{13}C NMR chemical shifts are expressed in parts per million (δ) downfield from tetramethylsilane (with the central peak of $CHCl_3$ at 77.00 ppm used as a standard) and $^{117}/^{119}$Sn–^{13}C couplings are not reported.

17. Acetone ($\geq$99.5% analytical reagent grade) was purchased from Fisher Scientific and used as received.

18. Sodium iodide (puriss. p. a. $\geq$99.0%) was purchased from Sigma-Aldrich and used as received.

19. The progress of the reaction was not monitored.

20. Tributyl(iodomethyl)stannane (2) was prepared according to a modified procedure of Seitz et al[2].

21. A second run on the same scale provided 12.6 g (95%) of the identical product. Tributyl(iodomethyl)stannane (2) decomposes over time when stored neat at ambient temperature. This reagent should be stored as a degassed 1 M solution in hexanes at –10 °C and used within a few days for best results.

22. Tributyl(iodomethyl)stannane (2) has the following physical and spectroscopic properties: Rf = 0.76 (hexanes; $KMnO_4$ visualization; Merck Millipore TLC Silica gel 60 F254 plates); ^{1}H NMR ($CDCl_3$, 400 MHz) δ: 0.90 (t, J = 7.3 Hz, 9H), 0.96–1.00 (m, 6H), 1.32 (dq, J = 14.2, 7.2 Hz, 6H), 1.49–1.57 (m, 6H), 1.94 (t, $J(^{117/119}$Sn-1H) = 18.0 Hz, 2H); ^{13}C NMR ($CDCl_3$, 101 MHz) δ: 10.8, 13.8, 27.4, 29.0; HRMS (EI) calculated for $C_{12}H_{27}Sn$ [M – CH_2I]$^{+}$ 291.11292, found 291.11326 and calculated for $C_9H_{20}ISn$ [M – C_4H_9]$^{+}$ 374.96262, found 374.96257; IR (film): 2954, 2924, 2871, 2848, 1739, 1456, 1375, 1365, 1228, 1217, 875, 692, 664, 588, 515, 456 cm^{-1}; Purity was assessed as >95% via Q NMR using 4'–nitroacetophenone as the internal standard.

Working with Hazardous Chemicals

The procedures in *Organic Syntheses* are intended for use only by persons with proper training in experimental organic chemistry. All hazardous materials should be handled using the standard procedures for work with chemicals described in references such as "Prudent Practices in the Laboratory" (The National Academies Press, Washington, D.C., 2011; the full text can be accessed free of charge at http://www.nap.edu/catalog.php?record_id=12654). All chemical waste should be disposed of in accordance with local regulations. For general guidelines for the management of chemical waste, see Chapter 8 of Prudent Practices.

In some articles in *Organic Syntheses*, chemical-specific hazards are highlighted in red "Caution Notes" within a procedure. It is important to recognize that the absence of a caution note does not imply that no significant hazards are associated with the chemicals involved in that procedure. Prior to performing a reaction, a thorough risk assessment should be carried out that includes a review of the potential hazards associated with each chemical and experimental operation on the scale that is planned for the procedure. Guidelines for carrying out a risk assessment and for analyzing the hazards associated with chemicals can be found in Chapter 4 of Prudent Practices.

The procedures described in *Organic Syntheses* are provided as published and are conducted at one's own risk. *Organic Syntheses, Inc.*, its Editors, and its Board of Directors do not warrant or guarantee the safety of individuals using these procedures and hereby disclaim any liability for any injuries or damages claimed to have resulted from or related in any way to the procedures herein.

Discussion

The development of transition metal-catalyzed cross-coupling reactions has greatly influenced the manner in which the synthesis of complex organic molecules is approached. A wide variety of methods are now available for the formation of C(sp2)–C(sp2) bonds, and more recent work has focused on the use of C(sp3) electrophiles and nucleophiles.[3,4] Access to functionalized aliphatic building blocks for such potential cross-coupling efforts are sought after, and organotin reagents remain one of the best approaches. While their applications have been hampered by challenges in removing organotin residues from the final products,[5,6] organotin reagents often allow for mild processes tolerating a wide variety of functional groups.[5,6] Their popularity is also due to their air- and moisture-stable nature,[5,6] and their wide-availability.

Tributyl(iodomethyl)stannane (**2**) has been used as an electrophile in the preparation of α–heteroalkylstannanes nucleophiles such as α–tributylstannylmethyl ethers,[7] amines,[8] or sulfides,[9] as in SnAP reagents for the preparation of functionalized, unprotected N-heterocycles (Scheme 1).[10,11]

Scheme 1. Preparation of an α–tributylstannanemethyl ether and its application in the synthesis of a functionalized morpholines.

Tributyl(iodomethyl)stannane (**2**) is also a useful reagent for the preparation of precursors that generate more reactive nucleophiles, via tin-lithium exchange, used in [2,3]-sigmatropic Wittig rearrangements or react further with other electrophiles such as ketones (Scheme 2).[8,12]

Scheme 2. [2,3]-Sigmatropic Wittig rearrangement

Therefore, tributyl(iodomethyl)stannane (**2**) allows for the facile access of various air- and moisture-stable C(sp3) nucleophiles ready for further transformations.

References

1. Laboratory of Organic Chemistry, Department of Chemistry and Applied Biosciences, ETH Zurich, Vladimir-Prelog-Weg 3, 8093 Zürich, Switzerland. Email: bode@org.chem.ethz.ch. We thank the European Research Council (ERC Starting Grant No. 306793-CASAA) for generous financial support.
2. Seitz, D. E.; Carroll, J. J.; Cartaya M., C. P.; Lee, S.-H.; Zapata, A. *Synth. Commun.* **1983**, *13*, 129–134.
3. Cárdenas, D. J. *Angew. Chem. Int. Ed.* **1999**, *38*, 3018–3020.
4. Cárdenas, D. J. *Angew. Chem. Int. Ed.* **2003**, *42*, 384–387.
5. Cordovilla, C.; Bartolomé, C.; Martínez-Ilarduya, J. M.; Espinet, P. *ACS Catal.* **2015**, *5*, 3040–3053.
6. Le Grognec, E.; Chrétien, J.-M.; Zammattio, F.; Quintard, J.-P. *Chem. Rev.* **2015**, *115*, 10207–10260.
7. (a) Still, W. C.; Mitra, A. *J. Am. Chem. Soc.* **1978**, *100*, 1927–1928. (b) Broka, C. A.; Lee, W. J.; Shen, T. *J. Org. Chem.* **1988**, *53*, 1336–1338. (c) Yoshida, J.-I.; Ishichi, Y.; Isoe, S. *J. Am. Chem. Soc.* **1992**, *114*, 7594–7595.
8. Pearson, W. H.; Lindbeck, A. C.; Kampf, J. W. *J. Am. Chem. Soc.* **1993**, *115*, 2622–2636.
9. Ikeno, T.; Harada, M.; Arai, N.; Narasaka, K. *Chem. Lett.* **1997**, *26*, 169–170.
10. Luescher, M. U.; Geoghegan, K.; Nichols, P. L.; Bode, J. W. *Aldrichimica Acta* **2015**, *48*, 43–48.
11. (a) Vo, C.-V. T.; Mikutis, G.; Bode, J. W. *Angew. Chem. Int. Ed.* **2013**, *52*, 1705–1708. (b) Luescher, M. U.; Vo, C.-V. T.; Bode, J. W. *Org. Lett.* **2014**, *16*, 1236–1239. (c) Vo, C.-V. T.; Luescher, M. U.; Bode, J. W. *Nat. Chem.* **2014**, *6*, 310–314. (d) Luescher, M. U.; Bode, J. W. *Angew. Chem. Int. Ed.* **2015**, *54*, 10884–10888. (e) Siau, W.-Y.; Bode, J. W. *J. Am. Chem. Soc.* **2014**, *136*, 17726–17729. (f) Geoghegan, K.; Bode, J. W. *Org. Lett.* **2015**, *17*, 1934–1937.
12. (a) Paquette, L. A.; Sugimura, T. *J. Am. Chem. Soc.* **1986**, *108*, 3841–3842. (b) Balestra, M.; Kallmerten, J. *Tetrahedron Lett.* **1988**, *29*, 6901–6904. (c) Bol, K. M.; Liskamp, R. M. J. *Tetrahedron* **1992**, *48*, 6425–6438.

Appendix
Chemical Abstracts Nomenclature (Registry Number)

N,N-Diisopropylamine: *N*-(Propan-2-yl)propan-2-amine; (108-18-9)
n-BuLi: *n*-Butyllithium; (109-72-8)
Tributyltin hydride: Tributylstannane; (688-73-3)
Paraformaldehyde: Polyoxymethylene; (30525-89-4)
Methanesulfonyl chloride; (124-63-0)
Sodium iodide; (7681-82-5)

Michael U. Luescher was born in 1985 in Switzerland and was trained as a medicinal chemistry laboratory technician at the Novartis Pharma AG (Switzerland). He then moved on to earn a BSc and MSc degree in chemistry in 2010 and 2012 from the University of Basel under the supervision of Professor Karl Gademann. Afterwards, he joined Professor Jeffrey W. Bode at ETH Zurich (Switzerland) investigating SnAP reagents where he received his Ph.D. in 2017. He is currently a post-doctoral fellow in the laboratory of Professor Emily Balskus at Harvard University.

Chalupat Jindakun received his BSc and MSc in Organic Chemistry from Mahidol University in Bangkok (Thailand). In 2016, he received a Royal Thai Government Scholarship to pursue his doctoral studies in organic chemistry under the guidance of Professor Jeffrey W. Bode at ETH Zurich (Switzerland) where he is currently investigating SnAP reagents for the preparation functionalized, of NH-free N-heterocycles.

Organic
Syntheses

Jeffrey W. Bode studied at Trinity University in San Antonio, TX (USA). Following doctoral studies at the California Institute of Technology (USA) and ETH Zurich and postdoctoral research at the Tokyo Institute of Technology (Japan), he began his independent academic career at UC Santa Barbara (USA) in 2003. He moved to the University of Pennsylvania as an Associate Professor in 2007 and to ETH Zurich as a Full Professor in 2010. Since 2013, he is also a Principal Investigator and Visiting Professor at the Institute of Transformative Biomolecules (WPI-ITbM) at Nagoya University (Japan).

Rahul Mondal completed his master's degree in chemistry in 2016 at the Indian Institute of Technology Kanpur. He then joined Prof. D. Maiti's group at the Indian Institute Technology Bombay to work as Research Assistant. In August 2017, he started his PhD studies in Prof. Cristina Nevado's group at the University of Zurich.

Estíbaliz Merino obtained her Ph.D. degree from the Autónoma University (Madrid-Spain). After a postdoctoral stay with Prof. Magnus Rueping at Goethe University Frankfurt and RWTH-Aachen University in Germany, she worked with Prof. Avelino Corma in Instituto de Tecnología Química-CSIC (Valencia) and Prof. Félix Sánchez in Instituto de Química Orgánica General-CSIC (Madrid) in Spain. At present, she is research associate in Prof. Cristina Nevado's group in University of Zürich. She is interested in the synthesis of natural products using catalytic tools and in the development of new materials with application in heterogeneous catalysis.

Stannylamine Protocol (SnAP) Reagents for the Synthesis of C–Substituted Morpholines from Aldehydes

Michael U. Luescher, Chalupat Jindakun, and Jeffrey W. Bode*[1]

Laboratorium für Organische Chemie, ETH Zürich, HCI F315, Vladimir-Prelog-Weg 3, CH-8093 Zürich, Switzerland

Cedric Hervieu, Estíbaliz Merino, and Cristina Nevado

Procedure (Note 1)

A. *1-((Tributylstannyl)methoxy)propan-2-amine* (**1**). Sodium hydride (1.12 g, 27.9 mmol, 1.1 equiv) (Note 2) is weighed out open to the air and added in one portion to an oven-dried 500-mL, pear-shaped recovery flask equipped with a 40 x 20 mm, Teflon-coated, oval magnetic stir bar, a rubber septum and a nitrogen inlet needle. Pentane (7.5 mL) (Note 3) is added via a syringe through the septum and stirring is started to release the sodium hydride free from the mineral oil. Stirring is stopped after 10 min, the septum is removed, and the supernatant pentane is removed using a glass pipette. The septum is reattached, anhydrous THF (135 mL) (Note 4) is added via a syringe, and stirring with a rate of ca. 400 rpm is started. *N,N*-Dimethylformamide (30 mL) (Note 5) is added after 10 min via a syringe. The grey suspension is cooled to ca. 0–5 °C using an ice/water bath for 30 min before DL-alaninol (2.00 mL, 1.90 g, 25.3 mmol, 1.0 equiv) (Note 6) is added dropwise via syringe over 15 min. The ice/water bath is removed, and the grey suspension is stirred at 25–27 °C for 2 h to afford a yellow suspension that is re-cooled to 0–5 °C using an ice/water bath (Figure 1a and b). At this point, tributyl(iodomethyl)stannane (10.9 g, 25.3 mmol, 1.0 equiv) (Note 7) is added dropwise to the reaction flask over 1.5 h via a syringe pump using a 10-mL plastic syringe. The ice/water bath is removed and stirring at 25–27 °C is continued for 6 h to afford a colorless suspension (Figure 1c) (Notes 8 and 9).

The suspension is re-cooled to 0–5 °C using an ice/water bath and saturated NH₄Cl (25 mL) is added in one portion via a graduated cylinder. The cooling bath is removed and stirring is continued at 25–27 °C. After 10 min, the biphasic mixture is poured into a 500 mL separatory funnel containing ethyl acetate (50 mL) and water (50 mL). The aqueous layer is separated and extracted with ethyl acetate (3 x 25 mL). The combined organic layers are washed with saturated NaCl solution (3 x 25 mL), dried over anhydrous MgSO₄ (12.5 g), and filtered through a 100-mL sintered glass Büchner funnel (medium porosity, 66 mm diameter).

Figure 1. Color change through the course of the reaction

The MgSO₄ is washed with ethyl acetate (3 x 10 mL) and the combined filtrate is concentrated by rotary evaporation (40 °C, 20 mmHg) to afford ca. 11.7 g of a pale yellow oil. This material is diluted with ethyl acetate (2.5 mL) and deposited onto a column (90 mm diameter) of 410 g of silica gel (12 cm high) (Note 10) prepared as a slurry in 10:1 ethyl acetate-MeOH. Elution is carried out with 10:1 ethyl acetate-MeOH collecting 50-mL fractions. The desired product is obtained in fractions 18–34 (Note 11). Mixed fractions 11–17 are collected separately and concentrated by rotary evaporation (40 °C, 20 mmHg) (Note 12). The resulting colorless oil is diluted with ethyl acetate (1.5 mL) and loaded onto a column (60 mm diameter) of 200 g of silica gel (5.5 cm high) (Note 10) prepared as a slurry in 10:1 ethyl acetate-MeOH. Elution is carried out with 10:1 ethyl acetate-MeOH collecting 50-mL fractions. The desired product is obtained in fractions 11–23. All fractions containing pure product according to TLC (Note 11) are combined and concentrated by rotary evaporation (40 °C, 20 mmHg). Further concentration at 25 °C under 0.1 mmHg for 2 h provides 7.5 g (78%) of amino stannane **1** as a colorless oil (Notes 9, 13, 14 and 15).

B. *(±)-cis-3-(2-Chloro-4-fluorophenyl)-5-methylmorpholine* (**2**). An oven-dried 250-mL pear-shaped recovery flask equipped with a 40 x 20 mm, Teflon-coated, oval magnetic stir bar is charged with 4 Å molecular sieves (1.72 g) (Note 16) and acetonitrile (75 mL) (Note 17) via a syringe. The flask is fitted with a rubber septum and nitrogen inlet needle, after which the stir rate is set to ca. 375 rpm. 1-((Tributylstannyl)methoxy)propan-2-amine (**1**) (6.5 g, 17.2 mmol, 1.0 equiv) (Notes 13 and 14) prepared in step A is added in one portion via a syringe followed by 2-chloro-4-fluorobenzaldehyde (2.73 g, 17.2 mmol, 1.0 equiv) (Note 18) that is weighed in air and is added in one portion to the reaction flask. The pale-yellow suspension is stirred at 25 °C for 4 h (Note 19). The resulting yellow suspension is filtered over 2.5 g of Celite in a 30-mL sintered glass funnel (30 mm diameter, medium porosity) into a 250-mL pear-shaped recovery flask. The solid material is rinsed with acetonitrile (3 x 5 mL), and the filtrate is then concentrated by rotary evaporation (40 °C, 20 mmHg) to afford ca. 8.9 g of the imine as a clear yellow oil (Notes 20 and 21).

Separately, an oven-dried 1-L pear-shaped recovery flask equipped with a 40 x 20 mm, Teflon-coated, oval magnetic stir bar, a rubber septum and a nitrogen inlet needle is charged with Cu(OTf)₂ (6.2 g, 17.2 mmol, 1.0 equiv) (Note 22 and 23). Dichloromethane (250 mL) (Note 24) and 1,1,1,3,3,3-hexafluoro-2-propanol (65 mL) (Note 25) are added via a syringe. Stirring with a rate of ca. 400 rpm is started and 2,6-lutidine (1.99 mL, 1.84 g,

17.2 mmol, 1.0 equiv) (Note 26 and 27) is added over 5 min via a syringe to the grey suspension affording a green suspension containing scattered lumps of Cu(OTf)$_2$ (Figure 2a). This suspension is stirred at 25 °C for 1 h affording a more homogeneous dark green suspension (Figure 2b) (Note 28).

Figure 2. Color change through the course of the copper complex formation

A solution of the imine prepared earlier (ca. 8.9 g) (Notes 20 and 21) in dichloromethane (10 mL) (Note 24) is added dropwise via syringe over 5 min affording a brown reaction mixture. The flask that contained the imine is rinsed with dichloromethane (2 x 2.5 mL) (Note 24) that are added to the reaction mixture via syringe in one portion.

The resulting brown reaction mixture is stirred at 25 °C for 12 h affording a green suspension that is quenched with a pre-mixed solution of 1:1 water-NH$_4$OH solution (150 mL) (Note 29), which is added via a graduated cylinder in one portion (Notes 30 and 31). The biphasic mixture is stirred vigorously (ca. 800 rpm) for 30 min before being poured into a 1-L separatory funnel. The blue aqueous layer is separated and extracted with dichloromethane (2 x 25 mL). The combined organic layers are washed with a pre-mixed solution of 1:1 water-NH$_4$OH solution (2 x 40 mL) (Note 29) and saturated NaCl solution (2 x 40 mL), dried over anhydrous MgSO$_4$ (15 g), and filtered through a 150-mL sintered glass Büchner funnel (medium porosity, 66 mm diameter). The MgSO$_4$ is washed with dichloromethane (3 x 15 mL) and the combined filtrate is concentrated by rotary evaporation (40 °C, 20 mmHg). The resulting brown oil is further concentrated at 40 °C and ca. 0.1 mmHg to remove most of the 2,6-lutidine (prior to chromatographic purification) affording ca. 13.0 g of brown oil (Note 32). This material is diluted with dichloromethane (5 mL) and deposited onto a column (90 mm diameter) of 340 g of silica gel (10 cm high) (Note 10) prepared as a slurry in 12:1 dichloromethane-ethyl acetate. Elution is carried out with 12:1 dichloromethane-ethyl acetate (2 L) and then 9:1 dichloromethane-ethyl acetate, collecting 50-mL fractions. The desired product is obtained in fractions 28–65 (Note 33). These fractions are combined, and the solvent is removed by rotary evaporation (40 °C, 20 mmHg). Further concentration at 25 °C under 0.1 mmHg for 2 h provides 2.55 g (65%) of morpholine **2** as a yellow oil (Figure 3) (Notes 9, 15, 34 and 35).

Figure 3. Product of Step B

Notes

1. Prior to performing each reaction, a thorough hazard analysis and risk assessment should be carried out with regard to each chemical substance and experimental operation on the scale planned and in the context of the laboratory where the procedures will be carried out. Guidelines for carrying out risk assessments and for analyzing the hazards associated with chemicals can be found in references such as Chapter 4 of "Prudent Practices in the Laboratory" (The National Academies Press, Washington, D.C., 2011; the full text can be accessed free of charge at https://www.nap.edu/catalog/12654/prudent-practices-in-the-laboratory-handling-and-management-of-chemical. See also "Identifying and Evaluating Hazards in Research Laboratories" (American Chemical Society, 2015) which is available via the associated website "Hazard

Assessment in Research Laboratories" at https://www.acs.org/content/acs/en/about/governance/committees /chemicalsafety/hazard-assessment.html. In the case of this procedure, the risk assessment should include (but not necessarily be limited to) an evaluation of the potential hazards associated with sodium hydride, pentane, mineral oil, tetrahydrofuran, N,N-dimethylformamide, DL-alaninol, tributyl(iodomethyl)stannane, ammonium chloride, ethyl acetate, sodium chloride, magnesium sulfate (anhydrous), methanol, silica gel, molecular sieves, acetonitrile, 2-chloro-4-fluorobenzaldehyde, Celite, copper(II) triflate, dichloromethane, 1,1,1,3,3,3-hexafluoro-2-propanol , 2,6-lutidine, and ammonium hydroxide.

2. Sodium hydride (60% dispersion in mineral oil) was purchased from Sigma-Aldrich and used as received.

3. Pentane (anhydrous, ≥99%) was purchased from Sigma-Aldrich and used as received.

4. Tetrahydrofuran (low water inhibitor free HPLC grade) was purchased from Sigma-Aldrich and purified by pressure filtration through activated alumina immediately prior to use.

5. N,N-Dimethylformamide (low water inhibitor free HPLC grade) was purchased from Sigma-Aldrich and purified by pressure filtration through activated alumina immediately prior to use.

6. DL-Alaninol (98%) was obtained from Sigma-Aldrich and used as received.

7. Tributyl(iodomethyl)stannane was prepared according to a modified procedure of Seitz et al[2] published in *Organic Syntheses* from Bode, et al, *Org. Synth.* **2018**, *95*, 345.

8. TLC analysis indicated that full conversion had occurred, as DL-alaninol was not visible at this point (10:1, ethyl acetate-MeOH with KMnO$_4$ stain visualization): DL-alaninol (Rf = 0.08) and 1-((tributylstannyl)methoxy)propan-2-amine (**1**) (Rf = 0.18).

9. The KMnO$_4$ stain was prepared using 1.5 g of KMnO$_4$ and 10 g of K$_2$CO$_3$ dissolved in 200 mL of water and 1.25 mL of 10% m/v NaOH solution.

10. High-purity Silica gel grade (9385), pore size 60 Å, 230–400 mesh particle size purchased from Sigma-Aldrich.

11. Purification is followed using TLC analysis on Silica gel (10:1, ethyl acetate-MeOH with KMnO$_4$ stain visualization): 1-((tributylstannyl)methoxy)propan-2-amine (**1**) (Rf = 0.18).

12. The impurities in those mixed fractions do not affect the subsequent imine formation and annulation reaction. Mixed fractions that contain mostly product can be combined with the pure material in order to avoid a second column purification and save solvent.

13. A second reaction performed on half-scale provided 3.65 g (78%) of the same colorless oil. 1-((Tributylstannyl)methoxy)propan-2-amine (**1**) decomposes over time when stored neat at ambient temperature. This reagent should be stored as a degassed 1 M solution in dry dichloromethane at –10 °C in which it is stable for months.

14. 1-((Tributylstannyl)methoxy)propan-2-amine (**1**) has the following physical and spectroscopic properties: Rf = 0.18 (10:1 ethyl acetate-MeOH; $KMnO_4$ visualization; Merck Millipore TLC Silica gel 60 F254 plates); ^1H NMR ($CDCl_3$, 400 MHz) δ: 0.88–0.93 (m, 15H), 1.03 (d, J = 6.3 Hz, 3H), 1.26–1.35 (m, 6H), 1.41–1.62 (m, 6H), 1.76 (bs, 2H), 3.01–3.14 (m, 2H), 3.25 (td, J = 3.1, 1.5 Hz, 1H), 3.70 (d, J = 10.3 Hz, 1H), 3.76 (d, J = 10.3 Hz, 1H); ^{13}C ($CDCl_3$, 101 MHz) δ: 9.0, 13.7, 19.6, 27.3, 29.1, 46.5, 62.3, 82.4; HRMS (ESI) calculated for $C_{16}H_{38}NOSn$ [M + H]$^+$ 380.19699, found 380.19654; IR (film): 2955, 2924, 2871, 2853, 1463, 1376, 1086, 864, 726, 688, 664, 594, 504 cm^{-1}. Purity was assessed as 95% by Q NMR using 4′-nitroacetophenone as the internal standard.

15. ^1H NMR chemical shifts are expressed in parts per million (δ) downfield from tetramethylsilane (with the $CHCl_3$ peak at 7.26 ppm used as a standard). ^{13}C NMR chemical shifts are expressed in parts per million (δ) downfield from tetramethylsilane (with the central peak of $CHCl_3$ at 77.00 ppm used as a standard) and $^{117}/^{119}Sn–^{13}C$ couplings are not reported.

16. 4 Å Molecular sieves (powder, activated, 325 mesh particle size) was purchased from Sigma-Aldrich. The sieves are activated at 120 °C and 0.1 mmHg for 12 h.

17. Acetonitrile (≥99.5% HPLC gradient grade) was purchased from Fisher Scientific and purified by pressure filtration through activated alumina immediately prior to use.

18. 2-Chloro-4-fluorobenzaldehyde (98.0%) was purchased from Fluorochem and used as received.

19. A small aliquot was taken and filtered to remove the molecular sieves. Concentration using a rotavap (40 °C, 20 mmHg) afforded an orange oil which upon ^{1}H NMR measurement indicated full conversion.

20. The imine is stable neat at ambient temperature. No special precautions need to be taken.

21. The intermediate imine has the spectroscopic properties: ^{1}H NMR (CDCl$_3$, 400 MHz) δ: 0.84 (t, J = 7.3 Hz, 15H), 1.17–1.29 (m, 9H), 1.37–1.54 (m, 6H), 3.38 (d, J = 3.1 Hz, 1H), 3.40 (d, J = 0.7 Hz, 1H), 3.53–3.65 (m, 1H), 3.67 (d, J = 10.3 Hz, 1H), 3.75 (d, J = 10.3 Hz, 1H), 6.99 (dddd, J = 8.7, 7.9, 2.5, 0.7 Hz, 1H), 7.10 (dd, J = 8.5, 2.5 Hz, 1H), 8.06 (dd, J = 8.8, 6.4 Hz, 1H), 8.61 (s, 1H); ^{13}C (CDCl$_3$, 101 MHz) δ: 8.9, 13.7, 18.7, 27.2, 29.1, 62.3, 65.7, 79.8, 114.5 (d, J_{CF} = 21.4 Hz), 116.8 (d, J_{CF} = 24.8 Hz), 129.9 (d, J_{CF} = 3.5 Hz), 130.1 (d, J_{CF} = 9.1 Hz), 135.8 (d, J_{CF} = 10.6 Hz), 155.8, 163.6 (d, J_{CF} = 253.5 Hz).

22. Cu(OTf)$_2$ (98%) was purchased from Strem Chemical and dried at 115 °C at 0.1 mmHg for 2 h before use. The dried Cu(OTf)$_2$ can be stored in a desiccator and be used for weeks without a negative impact on the reaction outcome.

23. Cu(OTf)$_2$ from other suppliers gave inferior results. A 1:1 Cu(OTf)$_2$ - 2,6-lutidine complex in HFIP (0.1 M) of a suitable copper source affords a green suspension within 0.5–1 h (see picture below - left side) while inferior Cu(OTf)$_2$ sources afford blue or purple suspensions (see picture below - right side).

... with Cu(OTf)$_2$ from Strem!   →   ←   ... with Cu(OTf)$_2$ from other supplier!

24. Dichloromethane (HPLC grade) was purchased from Fisher Scientific and purified by pressure filtration through activated alumina immediately prior to use.

25. 1,1,1,3,3,3-Hexafluoro-2-propanol, also known as 1,1,1,3,3,3-hexafluoroisopropanol or HFIP (99.9%) was purchased from Fluorochem and was used as received.

26. 2,6-Lutidine (ReagentPlus, 98%) was purchased from Sigma-Aldrich and used as received.

27. This is a slightly endothermic reaction.

28. Different colors at this point in time, other than various shades of green, often result in lower yields. This phenomenon was observed using Cu(OTf)$_2$ from suppliers other than Strem Chemicals (Note 23).

29. Ammonium hydroxide solution (ACS reagent, 28.0–30.0% NH$_3$ basis) was purchased from Sigma-Aldrich and used as received.

30. This is a slightly exothermic reaction.

31. The reaction progress was not monitored.

32. TLC analysis (9:1, dichloromethane-ethyl acetate with KMnO$_4$ stain visualization): Side product (Rf = 0.38), (±)-*cis*-3-(2-chloro-4-fluorophenyl)-5-methylmorpholine (**2**) (Rf = 0.31), and 2,6-lutidine (Rf = 0.18). The morpholine **2** is barely visible using UV 254 nm as the visualization technique. In general, product heterocycles are detected by TLC in the unpurified reaction mixture using both, potassium permanganate and ninhydrin stains. The product is visible with both developing agents while 2,6-lutidine is only visible using UV 254 nm and does not stain using KMnO$_4$ or ninhydrin (see pictures below in which 4:1 hexanes-ethyl acetate was used as the eluent for better separation on the TLC plate).

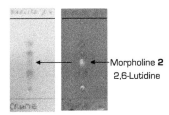

Morpholine **2**
2,6-Lutidine

33. Purification is followed using TLC analysis on silica gel (9:1 dichloromethane-ethyl acetate; KMnO$_4$ visualization; Merck Millipore TLC Silica gel 60 F254 plates): (±)-*cis*-3-(2-Chloro-4-fluorophenyl)-5-methylmorpholine (**2**) (Rf = 0.31). The morpholine **2** is barely visible using UV 254 nm as the visualization technique.

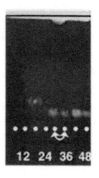

34. (±)-*cis*-3-(2-Chloro-4-fluorophenyl)-5-methylmorpholine (**2**) has the following physical and spectroscopic properties: R*f* = 0.31 (9:1, dichloromethane-ethyl acetate; KMnO$_4$ visualization; Merck Millipore TLC Silica gel 60 F254 plates) and R*f* = 0.45 (4:1 hexanes-ethyl acetate; KMnO$_4$ visualization; Merck Millipore TLC Silica gel 60 F254 plates); ^{1}H NMR (CDCl$_3$, 400 MHz) δ: 1.03 (d, *J* = 5.9 Hz, 3H), 1.72 (bs, 1H), 3.09–3.18 (m, 3H), 3.78 (d, *J* = 8.2 Hz, 1H), 3.90 (ddd, *J* = 10.8, 3.1, 0.7 Hz, 1H), 4.42 (ddd, *J* = 9.9, 5.0, 0.3 Hz, 1H), 6.98 (dddd, *J* = 8.6, 7.9, 2.6, 0.4 Hz, 1H), 7.09 (dd, *J* = 8.5, 2.6 Hz, 1H), 7.67 (dd, *J* = 8.7, 6.3 Hz, 1H); ^{13}C (CDCl$_3$, 101 MHz) δ: 17.8, 50.9, 56.0, 71.2 (d, *J*$_{CF}$ = 1.3 Hz), 73.06, 114.2 (d, *J*$_{CF}$ = 20.6 Hz), 116.6 (d, *J*$_{CF}$ = 24.6 Hz), 129.5 (d, *J*$_{CF}$ = 8.7 Hz), 133.6 (d, *J*$_{CF}$ = 10.1 Hz), 133.7 (d, *J*$_{CF}$ = 3.4 Hz), 161.5 (d, *J*$_{CF}$ = 249.4 Hz); HRMS (ESI) calculated for C$_{11}$H$_{14}$ClFNO [M + H]$^+$ 230.07425, found 230.07385; IR (film): 3015, 2970, 2960, 2949, 1740, 1488, 1436, 1366, 1229, 1216, 1102, 910, 588, 539, 527, 515 cm^{-1}. Purity was assessed as 97% by Q NMR using 4′-nitroacetophenone as the internal standard.

35. A second reaction performed on half-scale provided 1.31 g (66%) of the same yellow oil. In general, the reaction is not very sensitive to oxygen or H$_2$O and can be conducted without extra dry solvents or without pre-dried Cu(OTf)$_2$ with only slightly diminished yields.

Working with Hazardous Chemicals

The procedures in *Organic Syntheses* are intended for use only by persons with proper training in experimental organic chemistry. All hazardous materials should be handled using the standard procedures for work with chemicals described in references such as "Prudent Practices in the Laboratory" (The National Academies Press, Washington, D.C., 2011; the full text can be accessed free of charge at http://www.nap.edu/catalog.php?record_id=12654). All chemical waste should be disposed of in accordance with local regulations. For general guidelines for the management of chemical waste, see Chapter 8 of Prudent Practices.

In some articles in *Organic Syntheses*, chemical-specific hazards are highlighted in red "Caution Notes" within a procedure. It is important to recognize that the absence of a caution note does not imply that no significant hazards are associated with the chemicals involved in that procedure. Prior to performing a reaction, a thorough risk assessment should be carried out that includes a review of the potential hazards associated with each chemical and experimental operation on the scale that is planned for the procedure. Guidelines for carrying out a risk assessment and for analyzing the hazards associated with chemicals can be found in Chapter 4 of Prudent Practices.

The procedures described in *Organic Syntheses* are provided as published and are conducted at one's own risk. *Organic Syntheses, Inc.*, its Editors, and its Board of Directors do not warrant or guarantee the safety of individuals using these procedures and hereby disclaim any liability for any injuries or damages claimed to have resulted from or related in any way to the procedures herein.

Discussion

Functionalized saturated N-heterocycles, such as piperidines, piperazines, or morpholines, can be found with increasing prevalence in small molecule pharmaceuticals, despite their limited commercial abundance and challenging routes for their preparation.[3] Enormous efforts have been made on the synthesis of such cyclic amines, but a direct extension of cross-coupling methods to include saturated N-heterocycles remains elusive. Recent efforts to address this well-known limitation have provided promising methodologies for the derivatization of simple N-heterocycles.[4,5] However, these methods still have considerable shortcomings as a restricted substrate in terms of classes of N-heterocycles, substitution patterns, the requirement of protection-deprotection steps or the limited access of more complex N-heterocycles for further modifications.

As an alternative to traditional cross-coupling approaches, our group developed SnAP (stannyl (Sn) amine protocol) reagents as a versatile, predictable method for the preparation of functionalized, NH-free saturated nitrogen heterocycles, using widely available aromatic, heteroaromatic, and aliphatic aldehydes or ketones to access spirocyclic scaffolds (Scheme 1).[6,7]

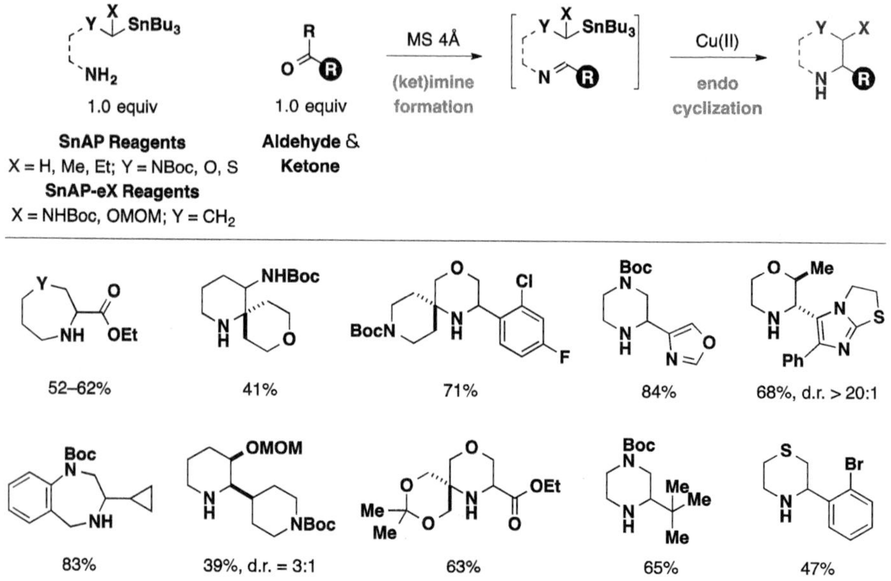

SnAP Reagents
X = H, Me, Et; Y = NBoc, O, S
SnAP-eX Reagents
X = NHBoc, OMOM; Y = CH₂

52–62%   41%   71%   84%   68%, d.r. > 20:1

83%   39%, d.r. = 3:1   63%   65%   47%

Scheme 1. SnAP reagent concept and substrate scope

Following our first report on SnAP reagents for the preparation of thiomorpholines,[7a] we have extended the line of air- and moisture stable SnAP reagents to include ones suitable for the preparation of functionalized morpholines and piperazines,[7b, d–f] pyrrolidines and piperidines,[7f] as well as medium-sized N-heterocycles.[7c] This process has a excellent substrate scope and tolerates electronically and sterically diverse (hetero)aromatic and aliphatic aldehydes and ketones as well as a good functional group tolerance accepting various heterocycles, aryl halides, nitriles, or unprotected phenols. Furthermore, it offers the advantage of affording unprotected products, which obviates the need to cleave the often-difficult-to-remove protecting groups. A further advantage of the SnAP protocol is that the reaction protocol described herein can be used for all SnAP reagents with no substrate specific optimization needed; although substrate-specific optimization might allow to improve isolated yields of the desired N-heterocyclic products. As an example, catalytic amounts of Cu(OTf)$_2$ in combination with a bisoxazoline ligand proved to be beneficial for aldehydes containing proximal heteroatoms (Scheme 2).[7d]

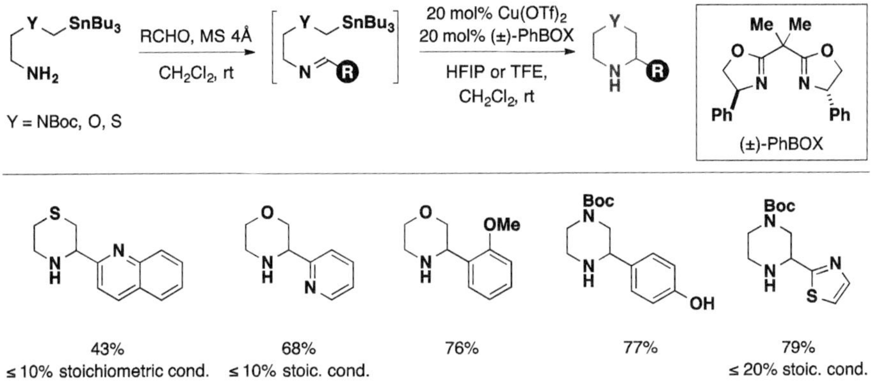

Scheme 2. Catalytic synthesis of 6-membered thiomorpholines, morpholines, and piperazines

The main drawback of the SnAP protocol, however, is its dependence on tin and its suspected toxicity.[8] The large difference in polarity between the NH-free product heterocycles and the tin products, however, simplifies the purification and methods to remove most of the tin species prior to column purification, for example, through extraction with acetonitrile and hexanes further facilitate purification to access products containing trace amounts of tin at most.[8,9] Furthermore, the unprotected N-heterocycles can be converted into their salts to remove last traces of tin impurities,[7c] and we hope that the procedure disclosed herein further simplifies access to various functionalized N-heterocycles currently challenging to prepare, and in great demand in drug development approaches.

References

1. Laboratory of Organic Chemistry, Department of Chemistry and Applied Biosciences, ETH Zurich, Vladimir-Prelog-Weg 3, 8093 Zürich, Switzerland. Email: bode@org.chem.ethz.ch. We thank the European

Research Council (ERC Starting Grant No. 306793-CASAA) for generous financial support.

2. Seitz, D. E.; Carroll, J. J.; Cartaya M., C. P.; Lee, S.-H.; Zapata, A. *Synth. Commun.* **1983**, *13*, 129–134.

3. (a) Vitaku, E.; Smith, D. T.; Njardarson, J. T. *J. Med. Chem.* **2014**, *57*, 10257–10274. (b) Taylor, R. D.; MacCoss, M.; Lawson, A. D. G. *J. Med. Chem.* **2014**, *57*, 5845–5859.

4. For selected recent examples on the functionalization of saturated N-heterocycles, see: (a) Shaw, M. H.; Shurtleff, V. W.; Terrett, J. A.; Cuthbertson, J. D.; MacMillan, D. W. C. *Science* **2016**, *352*, 1304–1308. (b) Jain, P.; Verma, P.; Xia, G.; Yu, J.-Q. *Nat. Chem.* **2016**, *9*, 140–144. (c) Firth, J. D.; O'Brien, P.; Ferris, L. *J. Am. Chem. Soc.* **2016**, *138*, 651–659. (d) Johnston, C. P.; Smith, R. T.; Allmendinger, S.; MacMillan, D. W. C. *Nature* **2016**, *536*, 322–325. (e) Cornella, J.; Li, C.; Malins, L. R.; Edwards, J. T.; Kawamura, S.; Brad, D.; Eastgate, M. D.; Baran, P. S.: A General Alkyl-Alkyl Cross-Coupling Enabled by Redox-Active Esters and Alkylzinc Reagents. *Sciene* **2016**, *352*, 801–805. (f) Wang, J.; Qin, T.; Chen, T.-G.; Wimmer, L.; Edwards, J. T.; Cornella, J.; Vokits, B.; Shaw, S. A.; Baran, P. S. *Angew. Chem. Int. Ed.* **2016**, *55*, 9676–9679. (g) Cornella, J.; Edwards, J. T.; Qin, T.; Kawamura, S.; Wang, J.; Pan, M.; Gianatassio, R.; Schmidt, M. A.; Eastgate, M. D.; Baran, P. S. *J. Am. Chem. Soc.* **2016**, *138*, 2174–2177. (h) Spangler, J. E.; Kobayashi, Y.; Verma, P.; Wang, D. H.; Yu, J.-Q. *J. Am. Chem. Soc.* **2015**, *137*, 11876–11879. (i) Noble, A.; Mccarver, S. J.; MacMillan, D. W. C. *J. Am. Chem. Soc.* **2015**, *137*, 624–627. (j) Zuo, Z.; Ahneman, D. T.; Chu, L.; Terrett, J. A.; Doyle, A. G.; MacMillan, D. W. C.. *Science.* **2014**, *345*, 437–440.

5. For selected recent examples on the preparation of the core scaffolds of saturated N-heterocycles, see: (a) Matlock, J. V.; Svejstrup, T. D.; Songara, P.; Overington, S.; McGarrigle, E. M.; Aggarwal, V. K. *Org. Lett.* **2015**, *17*, 5044–5047. (b) Matlock, J. V.; Fritz, S. P.; Harrison, S. A.; Coe, D. M.; McGarrigle, E. M.; Aggarwal, V. K. *J. Org. Chem.* **2014**, *79*, 10226–10239. (c) Fritz, S. P.; West, T. H.; McGarrigle, E. M.; Aggarwal, V. K. *Org. Lett.* **2012**, *14*, 6370–6373. (d) Yar, M.; McGarrigle, E. M.; Aggarwal, V. K. *Org. Lett.* **2009**, *11*, 257–260. (e) Yar, M.; McGarrigle, E. M.; Aggarwal, V. K. *Angew. Chem. Int. Ed.* **2008**, *47*, 3784–3786. (f) Zhai, H.; Borzenko, A.; Lau, Y. Y.; Ahn, S. H.; Schafer, L. L. *Angew. Chem. Int. Ed.* **2012**, *51*, 12219–12223.

6. Luescher, M. U.; Geoghegan, K.; Nichols, P. L.; Bode, J. W. *Aldrichimica Acta* **2015**, *48*, 43–48.

7. (a) Vo, C.-V. T.; Mikutis, G.; Bode, J. W. *Angew. Chem. Int. Ed.* **2013**, *52*, 1705–1708. (b) Luescher, M. U.; Vo, C.-V. T.; Bode, J. W. *Org. Lett.* **2014**, *16*, 1236–1239. (c) Vo, C.-V. T.; Luescher, M. U.; Bode, J. W. *Nat. Chem.* **2014**, *6*, 310–314. (d) Luescher, M. U.; Bode, J. W. *Angew. Chem. Int. Ed.* **2015**, *54*, 10884–10888. (e) Siau, W.-Y.; Bode, J. W. *J. Am. Chem. Soc.* **2014**, *136*, 17726–17729. (f) Geoghegan, K.; Bode, J. W. *Org. Lett.* **2015**, *17*, 1934–1937. g) Luescher, M. U.; Bode, J. W. *Org. Lett.* **2016**, *18*, 2652–2655.
8. Le Grognec, E.; Chrétien, J.-M.; Zammattio, F.; Quintard, J.-P. *Chem. Rev.* **2015**, *115*, 10207–10260.
9. Berge, J. M.; Roberts, S. M. *Synthesis* **1979**, 471–472.

Appendix
Chemical Abstracts Nomenclature (Registry Number)

Sodium hydride; (7646-69-7)
DL-Alaninol: (±)-2-Amino-1-propanol; (6168-72-5)
Tributyl(iodomethyl)stannane; (66222-29-5)
2-Chloro-4-fluorobenzaldehyde; (84194-36-5)
Cu(OTf)$_2$: Copper(II) trifluoromethanesulfonate (34946-82-2)
HFIP: 1,1,1,3,3,3-Hexafluoro-2-propanol; (920-66-1)
2,6-Lutidine: 2,6-Dimethylpyridine; (108-48-5)
NH$_4$OH: Ammonium hydroxide; (1336-21-6)

Michael U. Luescher was born in 1985 in Switzerland and was trained as a medicinal chemistry laboratory technician at the Novartis Pharma AG (Switzerland). He then moved on to earn a BSc and MSc degree in chemistry in 2010 and 2012 from the University of Basel under the supervision of Professor Karl Gademann. Afterwards, he joined Professor Jeffrey W. Bode at ETH Zurich (Switzerland) investigating SnAP reagents where he received his Ph.D. in 2017. He is currently a post-doctoral fellow in the laboratory of Professor Emily Balskus at Harvard University.

Chalupat Jindakun received his BSc and MSc in Organic Chemistry from Mahidol University in Bangkok (Thailand). In 2016, he received a Royal Thai Government Scholarship to pursue his doctoral studies in organic chemistry under the guidance of Professor Jeffrey W. Bode at ETH Zurich (Switzerland) where he is currently investigating SnAP reagents for the preparation functionalized, of NH-free N-heterocycles.

Jeffrey W. Bode studied at Trinity University in San Antonio, TX (USA). Following doctoral studies at the California Institute of Technology (USA) and ETH Zurich and postdoctoral research at the Tokyo Institute of Technology (Japan), he began his independent academic career at UC Santa Barbara (USA) in 2003. He moved to the University of Pennsylvania as an Associate Professor in 2007 and to ETH Zurich as a Full Professor in 2010. Since 2013, he is also a Principal Investigator and Visiting Professor at the Institute of Transformative Biomolecules (WPI-ITbM) at Nagoya University (Japan).

Cedric Hervieu obtained his master's Degree in chemistry from the Joseph Fourier University (Grenoble-France), in 2017. He joined Prof. Liming Zhang group at the University of California, Santa Barbara in United States to work as a visiting student (2016), followed by a stay at University of Zürich in Prof. Cristina Nevado's group. In August 2017, he started his Ph.D. at the University of Zurich where he is working in Prof. Cristina Nevado's group. His research interests focus on the development and application of new photo redox catalysis based synthetic methods.

Organic
Syntheses

Estíbaliz Merino obtained her Ph.D. degree from the Autónoma University (Madrid-Spain). After a postdoctoral stay with Prof. Magnus Rueping at Goethe University Frankfurt and RWTH-Aachen University in Germany, she worked with Prof. Avelino Corma in Instituto de Tecnología Química-CSIC (Valencia) and Prof. Félix Sánchez in Instituto de Química Orgánica General-CSIC (Madrid) in Spain. At present, she is research associate in Prof. Cristina Nevado´s group in University of Zürich. She is interested in the synthesis of natural products using catalytic tools and in the development of new materials with application in heterogeneous catalysis.

Trimethylsilyldiazo[13C]methane: A Versatile 13C-Labelling Reagent

Chris Nottingham and Guy C. Lloyd-Jones*[1]

EaStChem, University of Edinburgh, Joseph Black Building, David Brewster Road, Edinburgh EH9 3FJ, U.K.

Checked by Bachir Latli and Chris Senanayake

A.

$^{13}CH_3OH$ → (Tosyl Chloride, NaOH, THF:Water, 0 °C to rt) → $^{13}CH_3OTs$ **1**

B.

Ph–C(NH)–Ph → (i) *n*BuLi, THF, -78 °C; (ii) **1**, THF -78 °C to rt → Ph–C(N–$^{13}CH_3$)–Ph **2**

C.

Ph–C(N–$^{13}CH_3$)–Ph **2** → LDA, TMSCl, THF, -78 °C → Ph–C(N–$^{13}CH_2$–SiMe$_3$)–Ph **3**

D.

Ph–C(N–$^{13}CH_2$–SiMe$_3$)–Ph **3** → (i) H$_2$, Pd/C (5 mol% Pd), MeOtBu; (ii) HCl in Et$_2$O → Cl$^-$ H$_3$N$^+$–$^{13}CH_2$–SiMe$_3$ **4**

E.

HO–⋎–OH → (NaNO$_2$, aq HCl) → ONO–⋎–ONO **5**

F.

ONO–⋎–ONO **5**

Cl$^-$ H$_3$N$^+$–$^{13}CH_2$–SiMe$_3$ **4** → (NaOH, Et$_2$O) → H$_2$N–$^{13}CH_2$–SiMe$_3$ → (3-NO$_2$-Phenol (10 mol%), 2-MeTHF/Et$_2$O, rt) → N$_2$=^{13}CH–SiMe$_3$ **6**

Org. Synth. **2018**, *95*, 374-402
DOI: 10.15227/orgsyn.095.0374

Published on the Web 10/15/2018
© 2018 Organic Syntheses, Inc.

Procedure (Note 1)

A. *[¹³C]Methyl-p-toluenesulfonate: (1)*.² A 250-mL, single-necsked round-bottomed flask equipped with a Teflon-coated magnetic stirbar (35 mm x 15 mm, oval) is charged with sodium hydroxide (30.3 g, 758 mmol, 5 equiv) (Note 2). The flask is placed in an ice bath and water (50 mL) (Note 3) is added in one portion with stirring. [¹³C]Methanol (5.00 g, 151.5 mmol, 1 equiv) (Note 4) is weighed in a 12 mL syringe and slowly added to the hydroxide solution at 0 °C. *p*-Toluenesulfonyl chloride (TsCl) (34.7 g, 182 mmol, 1.2 equiv) (Note 5) is weighed into a 250 mL single-necked conical flask equipped with a Teflon-coated magnetic stirbar (30 mm x 15 mm, oval). Tetrahydrofuran (40 mL) (Note 6) is added to this flask and stirred under nitrogen until all the TsCl dissolves. This solution is added over five min via cannulation under nitrogen to the NaOH reaction flask, which is cooled in an ice bath. The conical flask is rinsed with THF (5 mL), which is then added to the round-bottomed flask. The internal sides of the RBF are then rinsed with additional THF (5 mL) via syringe before sealing with a rubber septum and venting with a small needle (Figure 1A). The ice-bath is removed and the mixture is allowed to warm to 25 °C with stirring over 20 h (Note 7). The reaction is neutralized by the slow addition of acetic acid (33 mL, 576 mmol, 3.8 equiv) (Note 8) at 0 °C over 5 min. The reaction mixture is left unstirred for 20 min at 0 °C to induce crystallization of sodium acetate (Note 9). The reaction mixture is then filtered through a sintered glass funnel (100 mm tall, 50 mm wide) to remove solid sodium acetate and the filtrate layers of THF and water are separated in a 250 mL separating funnel. The aqueous phase is extracted with ethyl acetate (2 x 60 mL) (Note 10). The filter cake is dissolved in water (150 mL) along with residue in the reaction flask and extracted with ethyl acetate (2 x 60 mL). The organic phases are combined, transferred to a 500 mL separating funnel, washed with sat. aq. Na₂CO₃ (100 mL) (Note 11) and sat. aq. NaCl (100 mL) (Note 12), dried over Na₂SO₄ (~40 g) for 10 min (Note 13), and filtered through a sintered glass funnel (100 mm tall, 50 mm wide) into a 1 L round-bottomed flask. The sodium sulfate is placed on the same sintered glass funnel and rinsed using additional ethyl acetate (40 mL). The organic solvent is concentrated by rotary evaporation at 40 °C (150 to 7 mmHg). A clear, colorless oil is obtained and transferred to a 250 mL round-bottomed flask with diethyl ether rinsings (~50 mL) (Note 14) and concentrated by rotary evaporation at 40 °C (600 to 7 mmHg). Additional diethyl ether (~50 mL) is added and the concentration repeated to give 27.1 g (96%) of a slightly yellow oil (Figure 1B) (Notes 15 and 16).

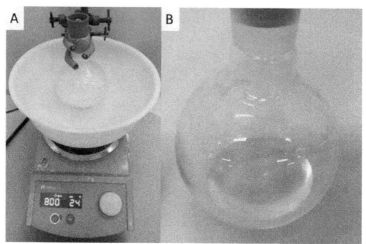

Figure 1. (A) Reaction Assembly for Step A; (B) Product after work-up and concentration (photos provided by submitters)

B.   *N-[¹³C]Methyl benzophenone imine* (**2**). A dry 500 mL, two-necked round-bottomed flask equipped with a Teflon-coated magnetic stirbar (35 mm x 15 mm, oval), a nitrogen inlet and a rubber septum, is filled with a nitrogen atmosphere and maintained this way over the course of the reaction. The flask is charged with anhydrous THF (175 mL) (Note 17) and cooled to –78 °C (Note 18). *n*-Butyllithium (68 mL, 2.45 M in hexanes, 167 mmol, 1.10 equiv) (Notes 19 and 20) is added to the cooled THF with efficient stirring (Note 21). Benzophenone imine (29 mL, 174 mmol, 1.15 equiv) is dissolved in THF (40 mL), and the solution is added via cannulation over a period 10 min to give a blue solution (Note 22). The benzophenone imine flask is rinsed with THF (10 mL) and added to the reaction flask (Figure 2A). [¹³C]Methyl *p*-toluenesulfonate (**1**) is dissolved in THF (40 mL) and added via cannulation using a double-ended needle over 10 min to the cold (–78 °C) solution. The methyl *p*-toluenesulfonate flask is rinsed with THF (10 mL), which is added to the reaction flask. The stir-rate is increased to facilitate a mild vortex (Note 23) and the flask is transferred to a large ice-bath and allowed to warm to 0 °C over 20 min (Note 24). The flask is placed in a large water bath at 23 °C for a further 45 min. The

reaction is quenched with water (5 mL), transferred to a 1 L separatory funnel and partitioned with water (200 mL) and Et₂O (200 mL) (Note 25). The layers are separated and the aqueous layer is back extracted with Et₂O (50 mL). The combined organic layers are washed with sat. aq. Na_2CO_3 (100 mL), sat. aq. NaCl (100 mL), dried with Na_2SO_4 (~40 g) for 10 min, and filtered through a sintered glass funnel (100 mm tall, 50 mm wide) into a 1 L round-bottomed flask. The sodium sulfate is washed in the same sintered glass funnel using additional Et₂O (40 mL). The bright yellow solution is concentrated *in vacuo* (600 to 7 mmHg, 40 °C) to give a yellow oil (35 g). The oil is treated with petroleum ether (250 mL, bp 35–60 °C) (Note 26) to give a cloudy solution, which is placed in a fridge (4 °C) for 20 h. A short pad of Celite™ 545 (3 cm deep) (Note 27), which is first wetted with 50 mL of petroleum ether, is prepared in a 60 mL glass funnel (medium frit). The mixture is then filtered through the Celite™ into a 500 mL round-bottomed flask using petroleum ether (150 mL). The solution is concentrated *in vacuo* (40 °C, 500 to 120 to 7 mmHg) to give a yellow oil (32 g) (Figure 2B) (Note 28). This material is used directly in the next reaction without further purification (Note 29).

Figure 2. (A) Reaction Assembly for Step B; (B) Product after work-up and concentration (photos provided by checkers)

C. *N-[¹³C]Methyl(trimethylsilyl)-benzophenone imine (3).* A dry 1 L, three-necked round-bottomed flask equipped with a Teflon-coated magnetic stirbar (35 mm x 15 mm, oval), a nitrogen inlet and two rubber septa is placed under a nitrogen atmosphere and maintained this way over the course of the reaction (Figure 3A). The flask is charged with anhydrous THF (240 mL) and diisopropylamine (30.3 mL, 216 mmol, 1.5 equiv) (Note 30) before being cooled to –78 °C (Note 31). *n*-Butyllithium (88.1 mL, 2.45 M in hexanes, 216 mmol, 1.5 equiv) is added with stirring to the diisopropylamine solution. After 5 min stirring at –78 °C, chlorotrimethylsilane (29.2 mL, 230 mmol, 1.6 equiv) (Note 32) is added over 3 min followed immediately by *N*-[¹³C]methyl benzophenone imine (**2**) that is dissolved in THF (40 mL), which is added over the course of 5 min. The flask that contained compound **2** is rinsed with THF (10 mL), which is added to the three-necked flask. A dark solution is formed, which can be dark green to dark red depending on slight changes in the amount of unreacted benzophenone imine. Figure 3B shows the dark green color of the reaction mixture. The reaction mixture is allowed to stir at –78 °C for 15 mins, at which time the dry ice/acetone bath is removed and a dry ice/CHCl₃ bath (–60 °C) is added (Note 33). The reaction is maintained at this temperature for 5 h. The reaction is quenched at –60 °C by the slow addition of acetone (6.3 mL, 86 mmol, 0.6 equiv) (Note 34) and the resulting quenched solution left to stir for 10 min at this temperature (Note 35). The reaction mixture is transferred to a large 23 °C water bath and allowed to warm over 15 min to give a yellow-orange solution (Figure 3C). The solution is transferred to a 1 L single-necked, round-bottomed flask, and the three-necked flask is rinsed with petroleum ether (50 mL), which is added to the single-necked, round-bottomed flask. The solution is concentrated by rotary evaporation at 40 °C (500 to 120 to 7 mmHg). The residue is suspended in petroleum ether (150 mL), filtered through a 3 cm plug of Celite™ 545 filter aid (Note 36) into a 1 L round-bottomed flask with petroleum ether (70 mL) and concentrated by rotary evaporation at 40 °C (500 to 120 to 7 mmHg).

The residue is once more dissolved in petroleum ether (150 mL), cooled in an ice-water bath for ~20 min, filtered through a ~3 cm pad of Celite™ 545 with additional petroleum ether (70 mL) into a 500 mL round-bottomed flask and concentrated by rotary evaporation (40 °C, 3 mmHg) to give the crude product (~44 g) as a clear yellow to orange oil (Figure 3D) (Notes 37 and 38). This material is used directly in the next reaction without further purification (Note 39).

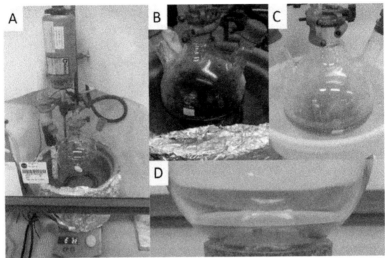

Figure 3. (A) Reaction Assembly for Step C; (B) Dark green reaction mixture; (C) Yellow reaction mixture after acetone quench and warming; (D) Product after work-up and concentration (photos provided by submitters)

D. *Trimethylsilyl[¹³C]methylamine hydrochloride (4)* (Note 40). A solution of *N*-[¹³C]methyl(trimethylsilyl)-benzophenone imine (44 g crude) in MTBE (350 mL) (Note 41) is placed in a hydrogenation apparatus containing 10 g of Pd/C (10%) (Note 42) (Figure 4A). The mixture is stirred under 20 psi of hydrogen for 24 h (Note 43), at which time LCMS show all the starting material is consumed (Note 44).

The mixture is filtered through a short pad of Celite™ 545 and rinsed with MTBE (100 mL). The colorless filtrate solution is stirred for 15 min while bubbling nitrogen slowly into the solution. A solution of 2M HCl in ether (Note 45) is added slowly over 45 min to give a fluffy white solid, and stirring is continued for 2 h (Note 46). The mixture is filtered through a sintered glass funnel (Note 47) to collect the solid trimethylsilylmethylamine hydrochloride salt. After drying under suction for 1 h, the vacuum is removed and the solid is broken up using a spatula. MTBE (50 mL) is added to the solid and stirred using the spatula for 2 min. Vacuum is applied again to remove the solvent and isolate a fluffy white solid (15.5 g) (Note 48). This product is transferred to a 250 mL round bottomed flask and suspended in isopropanol (100 mL) (Note 49). The suspension is heated in an oil bath to 85 °C to give a slight yellow solution. The temperature of the oil bath is reduced to 40 °C and MTBE (300 mL) is added slowly via a funnel. The product begins to precipitate before all MTBE is added. The resulting suspension is allowed to cool to room temperature over 30 min and then placed in the freezer at –20 °C and left overnight.

Filtration through a sintered funnel and rinsing with MTBE (50 mL) gives shiny white fluffy crystals, which are dried under suction using an inverted funnel dispensing nitrogen (Figure 4B) for 2 h to give white crystals (10.56 g, 52%) (Notes 50 and 51) (Figure 4C), which are stored in a screw cap flask.

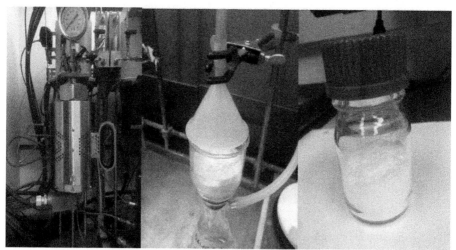

Figure 4. (A) Hydrogenation apparatus used in Step D; (B) TMS^{13}CH$_2$NH$_2$•HCl drying under nitrogen; (C) TMS^{13}CH$_2$NH$_2$•HCl product in a screw cap flask (photos provided by checkers)

E. *2,2-Diethyl-1,3-propanedinitrite (5)*. A 500 mL, single-necked round-bottomed flask equipped with a Teflon-coated magnetic stir bar (35 mm x 15 mm, oval) is charged with sodium nitrite (34.5 g, 99%, Fisher Scientific) and 2,2-diethyl-1,3-propanediol (26.4 g, 99%, Sigma Aldrich) followed by water (130 mL). The resulting suspension is stirred at ~500 rpm at 0–4 °C in an ice-water bath and fitted with a 250 mL pressure-equalizing funnel. The funnel is charged with 6 M aqueous hydrochloric acid (80 mL, 99%, Sigma Aldrich). This HCl solution is added dropwise (~1 drop per 2 seconds) over the course of 30 min at 0 °C, the funnel is removed and the resulting yellow/green solution allowed to stir for a further 30 min at 0 °C (Figure 5). The reaction mixture is transferred to a 500 mL separatory funnel and NaCl (~2 g) is added to assist separation (The mixture should be swirled but not shaken). The aqueous layer is discarded and cold sat. aq. Na₂CO₃ (50 mL)

**Figure 5. Propanedinitrite reaction assembly
(photos provided by checkers)**

is added followed by cold sat. aq. NaCl (100 mL) to assist separation. The product is washed with cold sat. aq. NaCl(100 mL) once more and then collected in a 100 mL beaker with Na₂SO₄ (~20 g). The dried product is transferred to a 100 mL round-bottomed flask under suction filtration through a sintered glass funnel to give the product as a yellow oil (32 g, 84%). Sodium sulfate (5 g) is added and the solution is stored in a sealed container in a fridge (0–4 °C). This reagent is used without further purification (Note 52).

F. *Trimethylsilyldiazo[¹³C]methane* (*6*). Trimethylsilyl[¹³C]-amine hydrochloride (10 g) is transferred to a 125 mL Erlenmeyer flask. Diethyl ether (30 mL) (Note 14) is added, followed by a freshly prepared solution of 2M aq NaOH (40 mL) (Note 2). The biphasic solution is swirled until all solids are dissolved, and the colorless solution is transferred to a 250 mL separatory funnel. The Erlenmeyer flask is rinsed with additional 2M NaOH solution (30 mL) and ether (10 mL), both of which are added to the separatory funnel. The aqueous layer is saturated with NaCl (15.4 g, 0.22 g per mL of 2N NaOH solution). The organic layer is removed and the aqueous layer is extracted with ether (2 x 25 mL). The combined organic extracts are dried over Na_2SO_4 (5 g) for 10 min, then filtered through a sintered funnel to a 250 mL round-bottomed flask. The Na_2SO_4 is rinsed with ether (10 mL). The colorless solution is stirred using a Teflon-coated magnetic stirbar (35 mm x 15 mm, oval). The flask is placed in an oil bath and fitted with a Vigreux column, short condenser and collection flask (Note 53) (Figure 6A and 6B). Diethyl ether is slowly distilled from the colorless solution at 46 °C for 4 h, then at 48 °C for another 4 h. The remaining colorless solution is stored at −10 °C overnight.

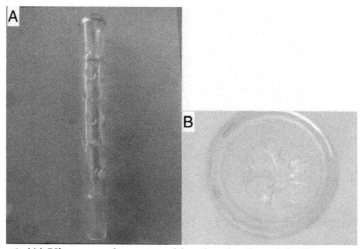

Figure 6. (A) Vigreux column used by the checker; (B) Vigreux column showing indents (photos provided by checkers)

Trimethylsilyl[^{13}C]methylamine as a solution in diethyl ether in a 250 mL round-bottomed flask is placed in a water bath. Anhydrous 2-Me-THF (45 mL) (Note 54) is added followed by 3-nitrophenol (1 g) (Note 55) and 2,2-diethyl-1,3-propanedinitrite (**5**) (20 mL). A yellow solution is obtained. The single-necked round-bottomed flask is fitted with a nitrogen gas adapter and stirred at 25 °C for 1 h. The nitrogen gas adapter is removed and replaced with a Vigreux column (Note 53), which is a topped with a gas adapter connected to a series of two solvent traps via Tygon tubing. The second solvent trap is connected to a digitally controlled (vacuum control V-850) vacuum pump (Buchi V-700) (Note 56) (Figure 7A-B).). The first trap is cooled at –78 °C in a dry-ice/acetone bath, while the second is cooled with liquid nitrogen (Figure 7).

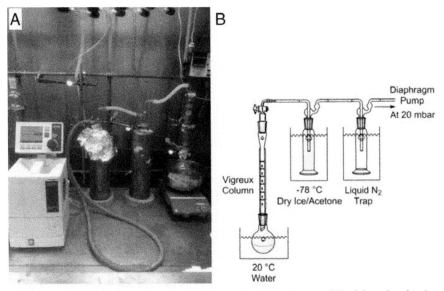

Figure 7. (A) Set-up for distillation step (photo provided by checker); (B) Diagram of set-up for distillation step (diagram from submitter)

The distillation is started at room temperature and the vacuum slowly lowered to 75 mmHg (75 mmHg/5 min) in 45 min. Gas evolution starts around 360 mmHg, and the evolution (bubbling) is rapid at about 80 mmHg. The vacuum is left at 20 mmHg for 30 min, then at 15 mmHg for another 30 min. The bright yellow distillate (Note 57) is warmed to 25 °C and transferred to a 250 mL separatory funnel and rinsed with 2 mL of 2-Me-THF. The solution is washed with sat. aq. NaCl (2 x 10 mL), dried for 10 min over MgSO$_4$ (2 g) (Note 58), gravity filtered through a sintered funnel, and rinsed with 2-MeTHF (3 mL). A total of about 50 mL of the product (Note 59) solution was obtained, and the solution is stored over molecular sieves (Note 60). The concentration of the trimethylsilyl[^{13}C]methane is determined by Q NMR comparison to bibenzyl (Note 61) to be 0.66 M, which indicates the formation of ca.

33 mmol of the product. The reaction flask contains 37 mL of undistilled deep yellow solution (Note 62).

Notes

1. Prior to performing each reaction, a thorough hazard analysis and risk assessment should be carried out with regard to each chemical substance and experimental operation on the scale planned and in the context of the laboratory where the procedures will be carried out. Guidelines for carrying out risk assessments and for analyzing the hazards associated with chemicals can be found in references such as Chapter 4 of "Prudent Practices in the Laboratory" (The National Academies Press, Washington, D.C., 2011; the full text can be accessed free of charge at https://www.nap.edu/catalog/12654/prudent-practices-in-the-laboratory-handling-and-management-of-chemical. See also "Identifying and Evaluating Hazards in Research Laboratories" (American Chemical Society, 2015) which is available via the associated website "Hazard Assessment in Research Laboratories" at https://www.acs.org/content/acs/en/about/governance/committees/chemicalsafety/hazard-assessment.html. In the case of this procedure, the risk assessment should include (but not necessarily be limited to) an evaluation of the potential hazards associated with ([^{13}C]methanol, p-toluenesulfonyl chloride, tetrahydrofuran, acetic acid, ethyl acetate, sodium carbonate, sodium sulfate, diethyl ether, n-butyllithium, benzophenone imine, petroleum ether, diisopropylamine, chlorotrimethylsilane, acetone, methyl *tert*-butyl ether, palladium on carbon, hydrogen gas, 2 M hydrogen chloride (HCl) in diethyl ether, isopropanol, sodium chloride, 2-methyltetrahydrofuran, 3-nitrophenol, sodium nitrite and 2,2-diethyl-1,3-propanediol). Step D involves the use of hydrogen gas, this is highly flammable and explosive, keep away from all sources of heat and potential sources of electrical sparks. Step D also involves the use of palladium on carbon, this can be pyrophoric, especially after use in a hydrogenation reaction. The filtration should be conducted under a flow of nitrogen and the filter cake should never be allowed to become completely dry. Once the filtration is complete, the palladium on carbon residue should be immediately moistened with water to prevent spontaneous ignition. Step E involves the preparation

of 2,2-diethyl-1,3-propanedinitrite, alkyl nitrates are known vasodilators and should not be removed from the fume hood unless stored in a well-sealed container. Step F involves the preparation of trimethylsilyldiazo[^{13}C]methane, this should be regarded as highly toxic and must be handled with all precautions appropriate for work with highly toxic substances. Ensure fume cupboard is working correctly before commencing use/preparation of this reagent and do not remove it from the fume cupboard unless stored in a well-sealed container. An emergency quenching solution of methanol:acetic acid 10:1 should be on hand in the case of a spill.

2. Sodium hydroxide (pellets, 98.9%) was purchased from Fisher Scientific and used as received.

3. Deionized water was used.

4. The checkers used [^{13}C]methanol purchased from Sigma-Aldrich with 100% purity by GC and 99% ^{13}C-labelled. [^{13}C]Methanol was purchased by the submitters from CK-isotopes (98% purity, 99% ^{13}C-labelled) and used as received.

5. The checkers used p-toluenesulfonyl chloride purchased from Sigma-Aldrich (99%). The submitters purchased p-toluenesulfonyl chloride (99%) from Acros Organics and used the material as received.

6. The checkers used tetrahydrofuran purchased from Sigma-Aldrich (>99.9% with 250 ppm BHT). Tetrahydrofuran (>99% with 250 ppm BHT) was purchased by the submitters from Fisher Scientific and used as received.

7. Vigorous stirring with a large stir bar is essential to facilitate efficient mixing. The submitters reported stirring at 800 rpm to achieve a mild vortex.

8. Acetic acid (>99.5%) was purchased from Sigma Aldrich and used as received.

9. The reaction flask may be scratched with a spatula to initiate crystallization.

10. Ethyl acetate (>99%) was purchased from Fisher Scientific and used as received.

11. Sodium carbonate (>99%) was purchased from Fisher Scientific and used as received.

12. Sodium chloride (99%) was purchased from Fisher Scientific and used as received.

13. Sodium sulfate (anhydrous, granular, 99%) was purchased from Fisher Scientific and used as received.

14. Diethyl ether (>99%) was purchased from Fisher Scientific and used as received.

15. Characterization data for [^{13}C]Methyl *p*-toluenesulfonate (**1**): ^{1}H NMR (400 MHz, CDCl$_3$) δ : 2.46 (s, 3H), 3.74 (d, *J* = 150 Hz, 3H), 7.36 (d, *J* = 8.5 Hz, 2H), 7.79 (d, *J* = 8.5 Hz, 2H); ^{13}C {^{1}H} NMR (101 MHz, CDCl$_3$) δ: 21.6, 56.1, 128.1, 129.9, 132.2, 144.9.

16. The weight percent (wt%) purity was determined to be 98.4 wt% by quantitative ^{1}H NMR (Q NMR) using dimethylsulfone (99.96 wt%) purchased from Sigma Aldrich as an internal standard.

17. The checkers used THF (anhydrous, >99.9% stabilized with 250 ppm BHT) purchased from Sigma-Aldrich. Tetrahydrofuran (>99.8%, unstabilized) was purchased by the submitters from Fisher Scientific and dried by passage through an activated alumina column under argon.

18. The submitters used an insulated bucket filled with dry ice and acetone to maintain the temperature at –78 °C.

19. The checkers used *n*-butyllithium (2.5 M in hexanes) from Acros Organics. The certificate of analysis indicated 2.67 M. The *n*-butyllithium solution was used as received and was not titrated. *n*-Butyllithium (2.5 M in hexanes) was purchased by the submitters from Acros Organics and titrated before use (Note 20).

20. Freshly titrated *n*-butyllithium in hexanes (167 mL) should be used, although the volume required will depend on the concentration of the *n*-butyllithium solution. The submitters titrated *n*-butyllithium against diphenylacetic acid (2 mmol) in tetrahydrofuran (15 mL) at room temp (approx. 20 °C in a water bath) until the appearance of a consistent yellow color.[3] The titration was performed in duplicate.

21. The submitters stirred the reaction at 340 rpm, which resulted in a slight vortex being visible.

22. The checkers purchased benzophenone imine (98%) from Oakwood Products, Inc. Benzophenone imine (98%) was purchased by the submitters from Fluorochem and used as received. The submitters report the solution to be a dark red color.

23. The submitters stirred the reaction at 500 rpm, which resulted in a mild vortex being visible.

24. The color of the reaction at this stage can vary from red to blue to black with no noticeable effect on yield or purity after the quench and workup.

25. As the product imine is susceptible to hydrolysis the work-up should be conducted swiftly. An insoluble white precipitate may form at this stage, in such cases the workup should be followed as normal, any precipitate remaining in the organic layer after separation will be removed by filtration.

26. Petroleum ether (bp 35–60, ACS reagent) was purchased from Sigma-Aldrich.

27. The checkers used Celite™ 545 from Fisher, rinsed with petroleum ether before use. Kieselguhr washed with acid was purchased from Fisher Scientific by the submitters and used as received.

28. N-[^{13}C]Methyl benzophenone imine, relevant NMR resonances in crude material: ^{1}H NMR (400 MHz, CDCl$_3$) δ : 3.25 (d, J = 135 Hz, 3H), 7.15–7.19 (m, 2H), 7.29–7.50 (m, 6H), 7.56–7.62 (m, 2H); ^{13}C {^{1}H} NMR (101 MHz, CDCl$_3$) δ : 41.5, 127.8, 128.0, 128.2, 128.3, 128.5, 129.8, 136.5 (d, J = 6.1 Hz), 139.8 (d, J = 7 Hz), 169.6 (d, J= 3.9 Hz).

29. The submitters found this compound to be sensitive to chromatographic media, undergoing hydrolysis to benzophenone, which has an identical R$_f$ to that of the desired product in numerous solvent systems. Distillation is difficult due to the similar boiling points of the starting material and product. If this material were required in pure form the submitters would recommend using equimolar amounts of starting materials rather than the ratios employed here.

30. The checkers used diisopropylamine (redistilled, 99.95%) from Sigma-Aldrich without further purification. Diisopropylamine (>99.5%) was purchased from Sigma Aldrich by the submitters and was purified by distillation over calcium hydride under N$_2$ before use.

31. The checkers used a dry ice/acetone bath (–78 °C), followed by a dry ice/chloroform bath (–60 °C). The submitters utilized an insulated bucket filled with ethanol and cooled with an immersion cooler. An overhead mechanical stirrer with a propeller type paddle (35mm) was used by the submitters to stir the cooling bath.

32. The checkers used TMSCl from Sigma-Aldrich (≥ 99.0% (GC)) without purification. Chlorotrimethylsilane (98%) was purchased from Sigma Aldrich by the submitters and purified by distillation over calcium hydride under N$_2$ before use.

33. At higher temperatures, such as –30 °C, double silylation of the imine methyl group is much more prominent. As such, the reaction was examined at –45 °C and found to provide a product ratio of 0:96:4 (starting material : product : disilylated material), demonstrating some flexibility with temperature control.

34. Acetone (>99%) was purchased from Fisher Scientific and used as received.

35. Quenching with water or an organic alcohol caused protodesilylation; therefore, acetone should be used to quench the reaction.

36. The checker used a 3 cm tall plug of Celite™ 545 filter aid (not acid-washed) powder from Fisher, and the Celite™ was rinsed with petroleum ether (50 mL) before use. The submitters used "Kieselgur washed with acid" for filtration to remove lithium chloride. Other filter aids may provide similar results; however, an aqueous workup should be avoided as protodesilylation can occur.

37. The crude material is a mixture of starting material, product and disilylated material, which contain ^{1}H NMR (CDCl$_3$) resonances at 3.26, 3.31, and 3.04 ppm, respectively. The ratio of products (based solely on these three compounds) can be determined by the following calculation; Product % = $(b/2)*100/((a/3)+(b/2)+(c/1))$. Where a = starting material, b = product and c = disilylated material by integration of their respective ^{1}H NMR peaks.

38. N-[^{13}C]Methyl(trimethylsilyl)benzophenone imine, relevant peaks in crude material: ^{1}H NMR (400 MHz, CDCl$_3$) δ : 0.2 (d, J =1.5 Hz, 9H), 3.27 (d, J = 127.5 Hz, 2H), 7.11 (m, 2H), 7.28–7.30 (m, 2H), 7.36–7.50 (m, 3H), 7.50–7.53 (m, 2H); ^{13}C {^{1}H} NMR (101 MHz, CDCl$_3$) δ : –0.1 (d, J = 3.6 Hz), 130.0, 130.1, 130.2, 130.5, 130.6, 131.3, 138.9 (d, J = 5.7 Hz), 142.7 (d, J = 6.7 Hz), 167.0 (d, J = 4.2 Hz).

39. The submitters found this compound to be sensitive to chromatographic media, undergoing hydrolysis to benzophenone, which has an identical R$_f$ to the desired product in numerous solvent systems. Vacuum distillation caused decomposition of the product to give a black tar.

40. The submitters performed the hydrogenolysis using the following procedure. The 500 mL, single-necked round-bottomed flask containing N-[^{13}C]methyl(trimethylsilyl)-benzophenone imine from the previous step is equipped with a Teflon-coated magnetic stir bar (35 mm x 15 mm, oval), flushed with nitrogen gas and charged with methyl tert-butyl ether (200 mL). Stirring is commenced and palladium on carbon

(7.3 g, 10% wt/wt Pd on carbon, 6.8 mmol, 5 mol% Pd) (Note 42) is added followed by additional methyl *tert*-butyl ether (70 mL) (Note 41) to rinse the sides of the flask. The flask is sealed with a rubber septum and a double walled hydrogen balloon is added (Notes 63 and 64) (Figure 8A and 8B). The flask is flushed with hydrogen by piercing the rubber septum with a small needle (40 mm, gauge 20) as an outlet. After 15 min the outlet needle is removed, the hydrogen balloon is refilled and the reaction left under this pressure at room temperature (23 °C) overnight. The following morning the balloon is replaced with a freshly made hydrogen balloon. This balloon should be refilled once more in the before being left over a second night. The following morning the reaction is checked by TLC (Notes 65, 66, and 67) (Figure 8C and 8D).

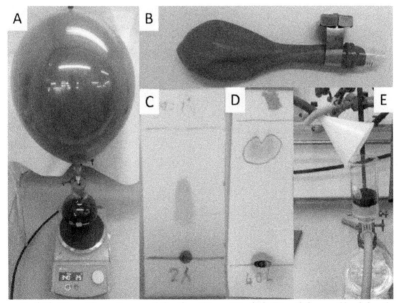

Figure 8. (A) Reaction Assembly for Step D; (B) Double walled hydrogen balloon; (C) TLC of starting material stained with I_2. (D) TLC at end of reaction stained with I_2; (E) Filtration of reaction mixture under a blanket of N_2 gas to give a clear solution (photos provided by the submitters)

Once complete, the remaining hydrogen gas is released slowly in the fume hood, the flask is flushed with nitrogen and the crude mixture filtered through a ~2 cm pad of kieselgur into a 1 L round-bottomed flask under an inverted funnel dispensing nitrogen gas (Figure 8E). The kielselgur is washed with methyl *tert*-butyl ether (MTBE) (2 x 50 mL). Water (~5 mL) is added to the used kieselgur pad, which is then disposed of as heavy metal waste. The clear, near-colorless filtrate obtained is degassed with stirring and gentle nitrogen bubbling over 10 min to remove any ammonia or methylamine side products (Figure 9A). Bubbling is suspended and the 1 L round-bottomed flask is equipped with a Teflon-coated magnetic stirbar (35 mm x 15 mm, oval) and placed in a water bath at room temperature. A 250 mL separating funnel is charged with 2 M HCl in Et₂O (89 mL, 178 mmol, 1.3 equiv) (Note 45). The HCl solution is then slowly added via the 250 mL separating funnel to the reaction mixture over ~15-20 mins with slow stirring (Note 46) (Figure 9B). The resulting suspension is filtered through a sintered glass funnel (Note 47) to collect the solid trimethylsilylmethylamine hydrochloride salt. MTBE (30 mL) is added to the solid and the mixture stirred to form a slurry before removal of the solvent by further vacuum filtration and collection of the product as a fluffy white solid (15 - 17 g) (Note 48).

Figure 9: (A) Degassing of reaction mixture with N₂ bubbling through a needle. (B) Set-up for addition of 2 M HCl in Et₂O to reaction mixture. (C) Suspension after recrystallization step. (D) Crystals collected by filtration. (E) Product in a 100 mL screw cap flask

The product is transferred to a 1 L single-necked round-bottomed flask and suspended in isopropanol (~105 mL, 6 mL/g of crude material)

(Note 49). The suspension is heated to ~80 °C with a heat gun and swirling by hand to give a clear solution with a slight yellow coloration. This hot solution is swirled by hand while warm (~40 °C) MTBE (~313 mL, 18 mL/g of crude material) is added slowly. Material begins to precipitate before the MTBE addition is complete. The resulting suspension is allowed to cool to room temperature over 30 min, placed in a freezer (–20 °C) and left overnight (16 h). The resulting suspension is filtered with suction through a sintered glass funnel to collect the solid trimethylsilylmethylamine hydrochloride salt (Note 47) (Figure 9C-D). The resulting white crystalline solid is washed with MTBE (50 mL) then dried with suction under an inverted funnel dispensing nitrogen gas for 20 min. The white crystalline solid (13.0–14.5 g, 61–68% yield from methanol) is transferred to a 100 mL screw cap flask for storage (Figure 9E) (Notes 50 and 51).

41. The checkers used methyl *tert*-butyl ether purchased from Sigma-Aldrich (ACS reagent, > 99%). Methyl *tert*-butyl ether (99%) was purchased by the submitters from Acros Organics and used as received.

42. Palladium on carbon (10 wt. % loading, matrix activated carbon support) was purchased from Sigma Aldrich and used as received.

43. Hydrogen gas (>99%) was purchased from BOC gases.

44. The checkers used Ultra Performance Liquid Chromatography-Mass Spectrometry (UPLCMS) using a medium polar method: run time 3.0 min, gradient 95% water (0.1% formic acid) and 5% MeCN to 5% water in 2.1 min, hold to 3 min at 5% water, flow 2.5 mL/min; column: BEH C18 (2.1 mm × 50 mm, 1.7 µm), m/z 120-1000, 0.3 µL injection. The starting material showed at 1.03 min, MH^+ = 269. A new peak was observed at 1.85 min corresponding to diphenylmethane (UV active), though no mass peak was observed.

45. 2M HCl in diethyl ether was purchased by the checkers from Sigma-Aldrich. The submitters purchased 2M HCl in diethyl ether from VWR (Alfa Aesar brand) and used the material as received.

46. The submitters stirred the mixture at 360 rpm until the slurry became too thick for magnetic stirring, at which point gentle swirling by hand was sufficient.

47. A fine porosity sintered glass funnel is required to avoid the frit becoming blocked. The submitters used grade 3, (16-40 µm pore size), 65 mm diameter, 60 mm high. The filtration is slow but can be accelerated by gently stirring the product slurry with a spatula.

48. The purity of product (^{1}H NMR and ^{13}C NMR) obtained at this stage varied from 90-95% by QNMR with dimethylsulfone (see Note 51 for Q NMR of purified material); therefore, further purification by a trituration/crystallization is carried out. Alternatively, the crude material can be carried through the next step with no complications, which slightly enhances the overall yield of TMSdiazomethane from methanol by eliminating loss from crystallization. Use of the crude material should only be performed if no methylamine hydrochloride (δ : 2.61 in CD$_3$OD) is detected in the ^{1}H NMR spectrum of the crude material, since methylamine hydrochloride will form diazomethane in the next step.

49. Isopropanol (Chromasolv plus, 99.9%) from Sigma-Aldrich was used.

50. Characterization data for trimethylsilyl[^{13}C]methylamine hydrochloride (**4**): mp 239–242 °C (*i*PrOH:MeO*t*Bu, 1:3); ^{1}H NMR (400 MHz, CD$_3$OD) δ : 0.22 (d, J = 2.5 Hz, 9H), 2.39 (d, J = 131 Hz, 2H). ^{13}C {^{1}H} NMR (101 MHz, CD$_3$OD) δ : −2.8 (d, J = 4.5 Hz), 29.5. IR (ATR) 3200-2800, 2951, 1603, 1503, 1412, 1245 cm^{-1}; HRMS ESI-MS *m/z* calcd for ^{13}C$_1$C$_3$H$_{14}$NSi [M-Cl]$^+$: 105.09236, found: 105.09180.

51. The weight percent (wt%) purity was determined to be 99.4 wt% by quantitative ^{1}H NMR (Q NMR) using dimethylsulfone purchased from Sigma Aldrich as an internal standard (99.96 wt%).

52. Characterization data for 2,2-diethyl-1,3-propanedinitrite (**5**): bp 20 °C (3.5 mmHg); ^{1}H NMR (400 MHz, CDCl$_3$) δ : 0.87 (t, J = 7.5 Hz, 6H), 1.38 (q, J = 7.5 Hz, 4H), 4.57 (s, 4H); ^{13}C {^{1}H} NMR (101 MHz, CDCl$_3$) δ : 7.2, 23.3, 40.9, 69.7.

53. The submitters used a 24 cm tall, B24 Vigreux column (16 cm of effective column, actual height 24 cm) fitted with a condenser and collection flask. The Vigreux column should have deep indents/fingers for efficient separation (Figure 6A-B). The checkers used a slightly different Vigreux column with no deep indents, see photos.

54. 2-Methyltetrahydrofuran (anhydrous, inhibitor free, >99%) was purchased from Sigma Aldrich and used as received.

55. 3-Nitrophenol (99%) was purchased from Sigma Aldrich and used as received.

56. The checker used a digitally controlled (vacuum control V-850) vacuum pump (Büchi V700). The submitters used a vacuubrand MD 4 NT VARIO diaphragm pump with a CVC 3000 vacuum controller. The

vacuum pump exhaust should be vented into a working fume cupboard.

57. The submitters added additional 2-MeTHF (10 mL) to the crude reaction mixture and performed a second distillation prior to washing with sat. aq. NaCl solution.

58. Magnesium sulfate (anhydrous, 99%) was purchased from Fisher Scientific and used as received.

59. Trimethylsilyldiazo[^{13}C]methane, relevant resonances in crude solution, both ^{1}H and ^{13}C NMR spectra are referenced to tetramethylsilane at 0.0 ppm: ^{1}H NMR (400 MHz, 2-MeTHF/Et$_2$O) δ : 0.14 (d, J = 2.8 Hz, 9H), 2.71 (d, J = 171.6 Hz, 1H); ^{13}C {^{1}H} NMR (101 MHz, 2-MeTHF/Et$_2$O) δ : –1.0 (d, J = 5.1 Hz), 21.2; HRMS ESI-MS m/z calcd for C$_3^{13}$C$_1$H$_{10}$N$_2$Si [M]$^+$ 115.16413, found: 115.06501. Slight variation in the reported chemical shifts is observed as a result of different ratios of Et$_2$O:2-MeTHF. CHCl$_3$ may be used as an alternative reference at 7.87 ppm in the ^{1}H NMR spectrum and 79.1 ppm in the ^{13}C NMR spectrum.

60. Molecular sieves (3Å, beads 8-12 mesh) were purchased by the checkers from Aldrich and activated before use. Molecular sieves (3Å, general purpose grade) were purchased by the submitters from Fisher Scientific and stored in an oven at 220 °C for a minimum of 5 days before use.

61. To determine the concentration of trimethylsilyldiazo[^{13}C]methane, bibenzyl (42.3 mg) was weighed in an amber vial, and then 0.7 mL of the trimethylsilyldiazomethane solution was added and swirled to dissolve the bibenzyl. CDCl$_3$ (Sigma-Aldrich, 99.8 atom% D) was added and ^{1}H NMR spectrum was acquired. The concentration of TMSdiazomethane was then calculated using the following calculation; C=(4*m*b)/(M*V*a) where M = molecular weight of bibenzyl, V = volume of trimethylsilyldiazo[^{13}C]methane solution, m = mass of bibenzyl, a = integral value of the methylene protons (δ 2.89, (s)) of bibenzyl and b = integral value of the methine proton (δ 2.71 (d)) of trimethylsilyldiazo[^{13}C]methane.

62. No attempts were made to continue the distillation and improve the yield.

63. Thick latex balloons (0.015 inch (15 mil)) rated for 12 L gas volume were purchased from Sigma Aldrich.

64. The submitters used two 0.015 inch (15 mil) thick latex balloons attached to a 5 mL disposable syringe barrel with electrical tape and a

small metal hose clamp (Figure 8B). A needle (40 mm, 20 gauge) is attached and used to pierce the rubber septum.

65. Glass-backed TLC plates (Al$_2$O$_3$) were purchased from Sigma Aldrich.

66. Using petroleum ether:Et$_2$O (19:1) as the eluent, the reaction is deemed complete when only diphenylmethane (R_f = 0.9) and a baseline spot are visible. The spots can be viewed by fluorescence quenching on suitable TLC plates at 254 nm or by I$_2$ staining. (Figure 8C-D).

67. The time required for the hydrogenation reaction varies with scale. When the reaction is performed on smaller scales (ca. 30 mmol), the reaction was complete overnight (16 h). On full scale the reaction will theoretically consume 7.8 L of H$_2$ gas. Replacing the balloons to ensure an excess of H$_2$ is essential.

Working with Hazardous Chemicals

The procedures in *Organic Syntheses* are intended for use only by persons with proper training in experimental organic chemistry. All hazardous materials should be handled using the standard procedures for work with chemicals described in references such as "Prudent Practices in the Laboratory" (The National Academies Press, Washington, D.C., 2011; the full text can be accessed free of charge at http://www.nap.edu/catalog.php?record_id=12654). All chemical waste should be disposed of in accordance with local regulations. For general guidelines for the management of chemical waste, see Chapter 8 of Prudent Practices.

In some articles in *Organic Syntheses*, chemical-specific hazards are highlighted in red "Caution Notes" within a procedure. It is important to recognize that the absence of a caution note does not imply that no significant hazards are associated with the chemicals involved in that procedure. Prior to performing a reaction, a thorough risk assessment should be carried out that includes a review of the potential hazards associated with each chemical and experimental operation on the scale that is planned for the procedure. Guidelines for carrying out a risk assessment and for analyzing the hazards associated with chemicals can be found in Chapter 4 of Prudent Practices.

The procedures described in *Organic Syntheses* are provided as published and are conducted at one's own risk. *Organic Syntheses, Inc.,* its

Editors, and its Board of Directors do not warrant or guarantee the safety of individuals using these procedures and hereby disclaim any liability for any injuries or damages claimed to have resulted from or related in any way to the procedures herein.

Discussion

Trimethylsilyldiazomethane (TMSDAM) is a highly versatile reagent for organic synthesis. It exhibits reactivity as a nucleophile under basic conditions, an electrophile under acidic conditions, a 1,3-dipole under neutral/basic conditions and can even function as a source of diazomethane.[4-7] This ambivalent reactivity allows it to participate in a plethora of different transformations, from simple S_N2 reactions to cycloadditions, metal mediated insertions, carbene chemistry and a variety of rearrangements (Schemes 1-3).[8-10] Its reactivity is well known and has

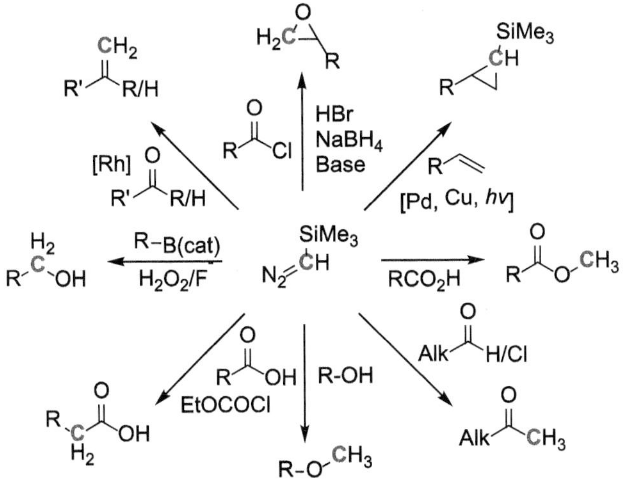

Scheme 1. Selected examples of TMSDAM as a C1 synthon[8]

been extensively documented (over 10,000 recorded uses on Reaxys) while the reagent itself has a reliable preparatory procedure[11] and is commercially available.[12]

Scheme 2. Selected examples of Li-TMSDAM as a C1 synthon[9]

Scheme 3. Selected examples of Li/TMSDAM as a CNN synthon[10]

Based on this, we propose a ^{13}C-isotopologue of TMSDAM to be a highly valuable reagent for the predictable and facile synthesis of ^{13}C-labelled compounds for both chemical and biological studies.[8-10] Indeed, the potential application of ^{13}C-labelled-TMSDAM as a derivatization agent for mass spectrometry has recently formed part of a patent claim.[13] However, to date no synthesis or application of such a reagent has been published in either the academic or patent literature. Aware that a practical synthesis enhances the application opportunities of any new reagent, we have developed a high-yielding, chromatography free synthesis of ^{13}C-labelled-TMSDAM starting from [13]CH₃OH as a readily available and cheap source of [13]C.[14]

The final step in this synthesis involves the diazotization of trimethylsilylmethylamine which was recently reported by Lebel and co-workers in 60-68% conversion.[10f] This provides an ethereal solution of TMSDAM which can be utilized without isolation for reactions which tolerate the presence of organic nitrite residues and water. A purification procedure was also reported to remove the nitrite residues and dry the solution. While we found the conversions to be reproducible the yield of TMSDAM after purification was between 40-47% in our hands. Seeking to improve this for a ^{13}C-labelled synthesis we noted that the acid catalyst employed (1-adamantanecarboxylic acid, 15 mol%) consumed 15% of the TMSDAM by methylation. We also noted that the multi-step washing procedure employed before distillation was required to remove the nitrite reagent (1,3-propanedinitrite). Finally, the diazotization was conducted by addition of the nitrite reagent at reflux, causing a powerful exothermic reaction which we sought to avoid for safety reasons.

Our studies revealed that 3-NO₂-phenol (10 mol%) functioned as an excellent acid catalyst. It allowed for a slightly lower catalyst loading, a homogeneous reaction mixture, a much lower reaction temperature (room temp, ~20 °C) and the catalyst was much more resistant to methylation by TMSDAM (compared with 1-adamantane carboxylic acid). We also found that by replacing the nitrite reagent with a higher boiling analogue (2,2-diethyl-1,3-propanedinitrite) we could eliminate the nitrite washes from the purification procedure and simply carry out a vacuum distillation. Combined, this allowed for higher in-situ conversions of 80-88% (compared to 60-68%) and isolated yields of 57-67%.

The preparation of the diazotization precursor trimethylsilyl-[^{13}C]methylamine hydrochloride was optimized with the goal of avoiding time-consuming purification procedures. The resulting route facilitates the synthesis of pure trimethylsilyl[^{13}C]methylamine hydrochloride in 61-68% yield from [^{13}C]MeOH with only one purification step (trituration/recrystallization). This material can be utilized for in-situ diazotization and subsequent reactions as reported by Lebel[10f] or purified from the nitrite residues and dried as described in the above procedure.

References

1. Current address: University of Edinburgh, EaStChem, Joseph Black Building, David Brewster Road, Edinburgh, EH9 3FJ, U.K. Email: guy.lloyd-jones@ed.ac.uk. The research leading to these results has received funding from the European Research Council under the European Union's Seventh Framework Programme (FP7/2007-2013) / ERC grant agreement n° [340163].

2. Based on a procedure for the preparation of CD$_3$OTs with slight modification: Weidong F.; Xiaoyong, G.; Xiaojun, D. U.S. Pat. Appl. 2013/0060044, A1, March 7, 2013.

3. Kofron, W. G.; Baclawski, L. M. *J. Org. Chem.* **1976**, 41, 1879–1880.

4. Shioiri, T.; Aoyama, T. *J. Synth. Org. Chem. Jpn.* **1996**, 54, 918–928.

5. Podlech, J. *J. Prakt. Chem.* **1998**, 340, 679–682.

6. Kühnel, E.; Laffan, D. D. P.; Lloyd-Jones, G. C.; Martínez del Campo, T.; Shepperson, I. R.; Slaughter, J. L. *Angew. Chem. Int. Ed.*, **2007**, 46, 7075–7078.

7. Yoshiyuki, H.; Takayuki, S.; Toyohiko, A. *J. Synth. Org. Chem. Jpn.* **2009** 67, 357–368 (Japanese only).

8. (i) Epoxide formation: (a) Galley, G.; Norcross, R.; Pflieger, P. WO2012/126922, A1, 27/09/2012.
   (ii) Silylcyclopropanation: (b) Haszeldine, R. N.; Scott, D. L.; Tipping, A. E. *J. Chem. Soc., Perkin Trans. 1*, **1974**, 1440–1443. (c) Aoyama, T.; Iwamoto, Y.; Nishigaki, S.; Shioiri, T. *Chem. Pharm. Bull.* **1989**, 37, 253–256. (d) France, M. B.; Milojevich, A. K.; Stitt, T. A.; Kim, A. J. *Tetrahedron Lett.* **2003**, 44, 9287–9290. (e) Crowley, B. M.; Campbell, B. T.; Duffy, J. L.; Greshock, T. J.; Guiadeen, D. G.; Harvey, A. J.; Huff, B. C.; Leavitt, K. J.; Rada, V. L.; Sanders, J. M.; Shipe, W. D.; Suen, L. M.; Bell, I. M. US2017/275260, A1, 28/09/2017.

(iii) Methyl esterification: (f) Seyferth, D.; Menzel, H.; Dow, A.W.; Flood, T. C. *J. Am. Chem. Soc.* **1968**, 90, 1080–1082. (g) Hashimoto, N.; Aoyama, T.; Shioiri, T. *Chem. Pharm. Bull.* **1981**, 29, 1475–1478. (h) For a study on the reaction mechanism see ref. 6 above.

(iv) Methyl ketone synthesis: (i) Aoyama, T.; Shioiri, T. *Synthesis* **1988**(3), 228–229. (j) Giovannini, R.; Bertani, B.; Ferrara, M.; Lingard, I.; Mazzaferro, R.; Rosenbrock, H. US2013/197011, 2013, A1. 01/08/2013.

(v) Methyl etherification: (k) Hashimoto, N.; Aoyama, T.; Shioiri, T. *Chem. Pharm. Bull.* **1981**, 29, 1475–1478. (l) Aoyama, T.; Terasawa, S.; Sudo, K.; Shioiri, T. *Chem. Pharm. Bull.* **1984**, 32, 3759–3760. (m) Aoyama, T.; Shioiri, T.; *Tetrahedron Lett.* **1990**, 31, 5507–5508.

(vi) Arndt-Eistert Homologation: (n) Aoyama, T.; Shioiri, T. *Chem. Pharm. Bull.* **1981**, 11, 3249–3255. (o) Cesar, J.; Sollner Dolenc, M. *Tetrahedron Lett.* **2001**, 42, 7099–7102.

(vii) Alcohol synthesis: (p) Goddard, J. P.; Le Gall, T.; Mioskowski., C. *Org. Lett.* **2000**, (10), 1455–1456.

(viii) Terminal alkene synthesis: (q) Lebel, H.; Paquet, V.; Proulx, C.; *Angew. Chem. Int. Ed.* **2001**, 40, 2887–2890. (r) Lebel, H.; Paquet, V. *J. Am. Chem. Soc.* **2004**, 126, 320–328. (s) Paquet, V.; Lebel, H. *Synthesis* **2005**, 11, 1901–1905.

9. (i) Silyl enol ether preparation: (a) Aggarwal, V. K.; Sheldon, C. G.; Macdonald, G. J.; Martin, W. P. *J. Am. Chem. Soc.* **2002**, 124 (35), 10300–10301.

(ii) Thiophene synthesis: (b) Miyabe, R.; Shioiri, T.; Aoyama, T.; *Heterocycles* **2002**, 57, 1313–1318.

(iii) Ketone Homologation: (c) Liu, H.; Sun, C.; Lee, N.-K.; Lee, D. *Chem. Eur. J.* **2012**, 18, 11889–11893.

(iv) Synthesis of aldehydes: (d) Miwa, K.; Aoyama, T.; Shioiri, T. *Synlett* **1994**, (2), 109.

(v) Synthesis of alkynes: (e) Miwa, K.; Aoyama, T.; Shioiri, T. *Synlett* **1994**, (2), 107–108.

(vi) Synthesis of 5-trimethylsilyl-2,3-dihydrofurans: (f) Miwa, K.; Aoyama, T.; Shioiri, T. *Synlett* **1994** (6), 461–462.

(vii) Pyrrole synthesis: (g) Ogawa, H.; Aoyama, T.; Shioiri, T. *Heterocycles* **1996**, 42, 75–82.

10. (i) 1,2,3-Thiadiazoles: (a) Shioiri, T.; Iwamoto, Y.; Aoyama, T. *Heterocycles* **1987**, 26, 1467–1470.

(ii) Pyrazolines from α,β–unsaturated esters: (b) O'Connor, M.; Sun, C.; Lee, D. *Angew. Chem. Int. Ed.* **2015**, 54, 9963–9966.

(iii) Pyrazoles: (c) Amegadzie, A. K.; Gardinier, M. K.; Hembre, E. J.; Hong, J. E.; Muehl, B. S.; Robertson, M. A.; Savin, K. A.; Remick, D. M.; Jungheim, L. N. WO2003/91226, 2003, A1, 6/11/2003. (d) Zrinski, I.; Juribasic, M.; Eckert-Maksic, M. *Heterocycles* **2006**, 68, 1961–1967. (e) Vuluga, D. ; Legros, J.; Crousse, B. ; Bonnet-Delpon, D. *Green Chem.* **2009**, 11, 156–159. (f) Audubert, C.; Gamboa Marin, O. J.; Lebel, H. *Angew. Chem. Int. Ed.* **2017**, 56, 6294–6297.

(iv) Triazoles: (g) Aoyama, T.; Sudo, K.; Shioiri, T. *Chem. Pharm. Bull.* **1982**, 30, 3849–3851. (h) Armer, R. E.; Akzo, N.; Dutton, C. J.; Gibson, S. P.; Verrier, K.; Tommasini, I.; Gethin, D. M.; Critcher, D. J. EP1072601, 2001, A2, 31/01/2001.

(v) Indazole: (i) Liu, Z.; Shi, F.; Martinez, P. D. G.; Raminelli, C.; Larock, R. C. *J. Org. Chem.* **2008**, 73, 219–226.

(vi) Cyano-pyrazoline: (j) Ahn, J. H.; Kim, H. M.; Jung, S. H.; Kang, S. K.; Kim, K. W.; Rhee, S. D.; Yang, S. D.; Cheon, H. G.; Kim, S. S. *Bioorg. Med. Chem. Lett.* **2004**, 14, 4461–4465.

11. Shioiri, T.; Aoyama, T.; Mori, S. *Org. Synth.* **1990**, 68, 1.
12. www.sigmaaldrich.com, CAS number: 18107-18-1 checked on 21/11/2017.
13. Kurland, I. J.,. Int. Pat. Appl. WO 2017/079102, A1, May 11, 2017.
14. Commercially available from a range of suppliers including Sigma-Aldrich, CKisotopes and Goss Scientific. We received quotes ranging from £1.98 - £3.46 per mmol.

Appendix
Chemical Abstracts Nomenclature (Registry Number)

[^{13}C]Methanol: Methanol-^{13}C; (14742-26-8)
p-Toluenesulfonyl chloride: Benzenesulfonyl chloride, 4-methyl-; (98-59-9)
Tetrahydrofuran: Furan, tetrahydro-; (109-99-9)
Acetic acid: Acetic acid; (64-19-7)
Sodium carbonate: Carbonic acid disodium salt; (497-19-8)
Sodium sulfate: Sulfuric acid disodium salt; (7757-82-6)
n-Butyllithium: Lithium, butyl-; (109-72-8)
Diphenylacetic acid: Benzeneacetic acid, α-phenyl-; (117-34-0)
Benzophenone imine: (1013-88-3)
Diisopropylamine: 2-Propanamine, *N*-(1-methylethyl)-; (108-18-9)
Chlorotrimethylsilane: Silane, chlorotrimethyl-; (75-77-4)

Methyl *tert*-butyl ether: Propane, 2-methoxy-2-methyl-; (1634-04-4)
2 M HCl in diethyl ether: Hydrochloric acid; (7647-01-0)
Dimethylsulfone: Methane, sulfonylbis-; (67-71-0)
Isopropanol: 2-Propanol; (67-63-0)
2-Methyltetrahydrofuran: Furan, tetrahydro-2-methyl-; (96-47-9)
3-Nitrophenol: Phenol, 3-nitro-; (554-84-7)
Sodium nitrite: Nitrous acid, sodium salt; (7632-00-0)
2,2-Diethyl-1,3-propanediol: (115-76-4)
Hydrochloric acid: Hydrochloric acid; (7647-01-0)
Magnesium sulfate: Sulfuric acid magnesium salt (1:1); (7487-88-9)
Bibenzyl: Benzene, 1,1'-(1,2-ethanediyl)bis-; (103-29-7)

Guy Lloyd-Jones FRS studied Chemical Technology at Huddersfield, obtained his doctorate at Oxford with John Brown FRS, and did postdoctoral research with Andreas Pfaltz at Basel. He began his independent career in 1996 at the University of Bristol, building a research group specializing in kinetics, NMR and isotopic labelling. In 2013 he moved to take up The Forbes Chair at the University of Edinburgh and in the same year was elected to the UK National Academy of Science (FRS).

Chris Nottingham completed a B.Sc. in Chemical and Pharmaceutical Science at the Galway-Mayo Institute of Technology in 2012 before obtaining a Ph.D. in organic chemistry under the supervision of Prof. Patrick Guiry MRIA at University College Dublin in 2016. He is currently performing postdoctoral research at the University of Edinburgh under the direction of Prof. Guy C. Lloyd-Jones FRS.

Organic Syntheses

Bachir Latli obtained his Ph.D. in organic chemistry in 1991 at Stony Brook University under the supervision of Prof. Glenn Prestwich. He then joined the laboratory of the late Prof. John Casida at the University of California in Berkeley. In 1998 he moved to Boehringer Ingelheim Pharmaceuticals in Ridgefield, Connecticut, where he is a Senior Research Fellow.

Stereoretentive Iron-catalyzed Cross-coupling of an Enol Tosylate with MeMgBr

Takeshi Tsutsumi, Yuichiro Ashida, Hiroshi Nishikado, and Yoo Tanabe*[1]

Department of Chemistry, School of Science and Technology, Kwansei Gakuin University, 2-1 Gakuen, Sanda, Hyogo 669-1337, Japan

Checked by Zhaobin Han and Kuiling Ding

Procedure (Note 1)

A. *Methyl (Z)-2-phenyl-3-(tosyloxy)acrylate [(Z)-1)].* A 300-mL, three-necked, round-bottomed flask is attached to a CaCl$_2$ drying tube, which is capped with a glass stopper, and fitted with a thermometer and a Teflon-coated magnetic stir bar (Note 2). Methyl 2-phenylacetate (10.54 g, 70 mmol) (Note 3), methyl formate (6.52 mL, 105 mmol) (Note 4), and THF (70 mL) (Note 5) are added and a septum is added to the middle neck. The vigorously stirred colorless solution is immersed in an ice-cooling bath, and sodium *tert*-butoxide (NaO*t*-Bu) (10.12 g, 105 mmol) (Note 6) is added in 5 portions over 5 ~ 10 min after temporarily removing the septum while maintaining the inner temperature below 10 °C (Figure 1). The suspension becomes a well-mixed pale yellow slurry (Figure 2) ca. 5 min after the addition of NaO*t*-Bu.

Figure 1. Addition of sodium *tert*-butoxide

Figure 2. Reaction mixture after the addition of NaO*t*-Bu

The reaction mixture is stirred at 0 ~ 5 °C for 1 h. *p*-Toluenesulfonyl chloride (TsCl) (20.03 g, 105 mmol) (Note 7) is added in 5 portions over 5 ~ 10 min after temporarily removing the septum while maintaining the inner temperature below 10 °C (Note 8) (Figure 3). The septum is replaced and the solution is stirred at 0 ~ 5 °C for 1 h. Water (70 mL) is added to the resulting mixture over the course of 1 min to maintain the inner temperature below 20 °C. The suspension immediately develops two transparent phases.

Figure 3. Addition of TsCl

The mixture is transferred to a 500-mL separatory funnel and the flask is rinsed with ethyl acetate (2 x 10 mL), which is added to the separatory funnel. The organic phase is separated and the aqueous phase is re-extracted with ethyl acetate (2 x 50 mL). The combined organic phases are washed with H_2O (70 mL) and brine (70 mL), dried over Na_2SO_4 (70 g), filtered, and concentrated under reduced pressure using a rotary evaporator (15 mmHg, 40 °C). The flask containing the residue was attached to a vacuum line with gentle heating using a dryer to completely remove the ethyl acetate.

The slightly pale yellow oil solidifies after a few minutes in a 100-mL round-bottomed flask (Note 9) (Figure 4). The crude solid (38.34 g) is crushed into particles, and the flask is equipped with a reflux condenser (10 cm height) and a Teflon-coated magnetic stir bar (Note 10). 2-Propanol (30 mL) (Note 11) is poured into the flask, which is then immersed in a water bath heated to 70 °C. The solution is stirred for ca. 5 min and the solid completely dissolves (Figure 5). After removal of the water bath, the solution is allowed to cool gradually to room temperature (20 ~ 25 °C). Over the course of 3 h colorless crystals appear. Using a glass filter (G3, 70 mm diameter), the first crop is filtered, collected, and washed twice with 2-propanol (2 x 20 mL) to yield 18.76 g (80%, E/Z = 1:99) of the desired product (Z)-**1** as colorless crystals (Note 12) (Figure 6).

Figure 4. The crude product before recrystallization

Figure 5. Recrystallization of the crude product

Figure 6. Product (Z)-1

B. *Methyl (Z)-2-phenyl-2-butenoate [(Z)-2]*. A 300-mL, three-necked, round-bottomed flask equipped with a 100-mL dropping funnel with rubber septum and inert gas inlet, capped with a septum stopper, a thermometer, and a Teflon-coated magnetic stirring bar (Note 13) is charged with methyl (Z)-2-phenyl-3-(tosyloxy)acrylate [(Z)-1)] (13.30 g, 40 mmol), tris(2,4-pentanedionato)iron(III) [Fe(acac)$_3$] (706 mg, 2 mmol) (Note 14), ethyl acetate (80 mL), and TMEDA (6.0 mL, 40 mmol) (Note 15) with gentle stirring.

The flask, which contains an orange-colored reaction mixture, is immersed in a temperature-controlled water bath. Methylmagnesium bromide (MeMgBr) (60.0 mL, 60 mmol) (Note 16) is added to the vigorously stirred solution mixture over the course of 30 ~ 40 min using a 100 mL-dropping funnel, while maintaining the inner temperature below 25 °C (Figure 7). The reaction mixture is stirred vigorously while maintaining the temperature between 20 ~ 25 °C (internal temperature) for 1 h (Figure 8).

The reaction mixture is poured onto sat. aqueous NH$_4$Cl (30 mL) and ice (30 g) with vigorous stirring over a period of ca. 1 min, after which 1 M aqueous HCl (ca. 100 mL) is added to the stirred mixture (Note 17) (Figure 9). The mixture is transferred into a 500-mL separatory funnel and the flask is rinsed with ethyl acetate (2 x 10 mL), which is added to the separatory funnel. The organic phase is separated and the aqueous phase is re-extracted with ethyl acetate (50 mL). The combined organic phases are washed with water (80 mL) and brine (80 mL), dried over Na$_2$SO$_4$ (60 g), filtered, and concentrated under reduced pressure using a rotary evaporator (15 mmHg, 40 °C).

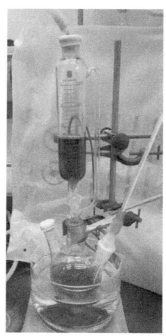

Figure 7. Apparatus assembly for Step B

Figure 8. Reaction mixture after addition of MeMgBr

Figure 9. Reaction mixture after quenched by addition of HCl

The black-colored oil (6.81 g) is moved into a 20 mL round-bottomed flask with a Teflon-coated magnetic stirring bar (Note 18). Distillation while the flask is immersed in a temperature-controlled oil bath and distilled under reduced pressure using a vacuum pump (62.0 ~ 65.0 °C/0.20 mmHg) to give the desired product (Z)-**2** (6.01 g, 85% yield, E/Z = 3:97) as a pale yellow oil (Note 19) (Figure 10). Product with similar purity can be isolated via chromatography (Note 20).

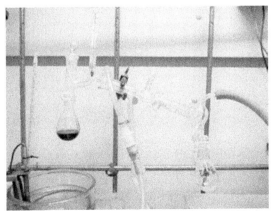

Figure 10. Apparatus assembly for distillation

Notes

1. Prior to performing each reaction, a thorough hazard analysis and risk assessment should be carried out with regard to each chemical substance and experimental operation on the scale planned and in the context of the laboratory where the procedures will be carried out. Guidelines for carrying out risk assessments and for analyzing the hazards associated with chemicals can be found in references such as Chapter 4 of "Prudent Practices in the Laboratory" (The National Academies Press, Washington, D.C., 2011; the full text can be accessed free of charge at https://www.nap.edu/catalog/12654/prudent-practices-in-the-laboratory-handling-and-management-of-chemical. See also "Identifying and Evaluating Hazards in Research Laboratories" (American Chemical Society, 2015) which is available via the associated website "Hazard Assessment in Research Laboratories" at https://www.acs.org/content/acs/en/about/governance/committees/chemicalsafety/hazard-assessment.html. In the case of this procedure, the risk assessment should include (but not necessarily be limited to) an evaluation of the potential hazards associated with Calcium chloride, methyl 2-phenylacetate, methyl formate, sodium *tert*-butoxide, *p*-toluenesulfonyl chloride, ethyl acetate, sodium sulfate, 2-propanol, ethylene carbonate, tris(2,4-pentanedionato)iron(III), tetramethyl-ethylenediamine, methylmagnesium bromide, ammonium chloride, hydrochloric acid, hexanes, silica gel, and chloroform.
2. An egg-shaped stir bar (50 mm length x 20 mm diameter) is used, since the reaction mixture produces a large quantity of salts after addition of TsCl. The yellow slurry is smoothly stirred throughout the reaction.
3. Methyl 2-phenylacetate (GC purity 99.0%) was purchased from Tokyo Chemical Industry Co., Ltd. and used as received.
4. Methyl formate (95.0%) was purchased from Tokyo Chemical Industry Co., Ltd. and used as received.
5. Tetrahydrofuran (THF), stabilizer free was purchased from Kanto Chemical Co., Inc. and used as received. The checkers purchased THF (99.5%) from Acros Organics and used the material as received.
6. Sodium *tert*-butoxide (NaO*t*-Bu) (98.0%) was purchased from Tokyo Chemical Industry Co., Ltd. and used as received.

7. *p*-Toluenesulfonyl (tosyl) chloride (TsCl) (>99.0%) was purchased from Tokyo Chemical Industry Co., Ltd. and used as received; a fresh lot was used. Once the container is opened, the cap should be replaced securely.

8. Addition of *p*-tosyl chloride is an exothermic reaction that results in the production of salts.

9. The compound usually solidifies immediately upon sitting.

10. An egg-shaped stir bar (35 mm length x 15 mm diameter) is used.

11. 2-Propanol (>99.7%, GLC) was purchased from Wako Pure Chemical Industries, Ltd. and used as received. The checkers purchased 2-propanol (99.5%) from Acros Organics and used it as received.

12. A second reaction on identical scale provided 18.66 g (80%) of the product (*Z*)-**1**. Stable solids could be stored in a brown colored bottle at room temperature over a period of months. Physical and spectroscopic properties of (*Z*)-**1**: colorless crystals; mp 92.0–93.0 °C; ^{1}H NMR (400 MHz, CDCl$_3$) δ: 2.47 (s, 3H), 3.74 (s, 3H), 7.06 (s, 1H), 7.22–7.41 (m, 7H), 7.84 (d, *J* = 8.0 Hz, 2H); ^{13}C NMR (100 MHz, CDCl$_3$) δ: 21.7, 52.2, 123.3, 127.8, 128.0, 128.57, 128.60, 130.0, 132.1, 132.3, 138.9, 145.9, 164.9. ν$_{μαξ}$ (film) 3072, 2960, 1727, 1072, 883, 764, 668 cm^{-1} . HRMS (ESI, [M+H]$^+$) *m/z* calcd for C$_{17}$H$_{17}$O$_5$S: 333.0791, Found: 333.0794; Anal. Calcd for C$_{17}$H$_{16}$O$_5$S: C, 61.43; H, 4.85; found: C, 61.44; H, 4.95. The purity of the material was determined to be >97% purity by Q NMR using ethylene carbonate as the internal standard. A second crop of crystals does not appear from the 2-propanol. Concentration of 2-propanol of the mother and the washing liquors by vacuum pump gave a yellow oil (8.59 g) [Molar ratio; (*Z*)-**1**, (*E*)-**1**, and unreacted TsCl = c.a. 1 : 0.4 : 0.6]. Rf value (hexane/EtOAc = 5:1): (*Z*)-**1**: 0.53 and (*E*)-**1**: 0.48.

13. An egg-shaped stir bar (50 mm length x 20 mm diameter) is used.

14. Tris(2,4-pentanedionato)iron(III) [Fe(acac)$_3$] (>98.0%) was purchased from Tokyo Chemical Industry Co., Ltd. and used as received.

15. *N*,*N*,*N′*,*N′*-Tetramethylethylenediamine (TMEDA) (>98.0%) was purchased from Tokyo Chemical Industry Co., Ltd. and used as received.

16. Methylmagnesium bromide (MeMgBr) in tetrahydrofuran (0.95 M) was purchased from Kanto Chemical Co., Inc. and used as received. The checkers purchased methylmagnesium bromide (MeMgBr) in tetrahydrofran (1.0 M) from Acros Organics and used as received.

17. The 1 M aqueous HCl solution was added until the color of aqueous phase changed to black (pH=ca. 3.0). Caution; direct quench using 1 M

aqueous HCl (ca. 100 mL) caused considerable isomerization from Z to E ($Z : E$ = ca. 4:1 ~ 3:1).

18. An egg-shaped stir bar (15 mm length x 7 mm diameter) is used.

19. A second reaction on identical scale provided 6.06 g (86%) of the product (Z)-**2**. Physical and spectroscopic properties of (Z)-**2**: 62.0–65.0 °C/0.20 mmHg. pale yellow oil; ^{1}H NMR (400 MHz, CDCl$_3$) δ: 2.06 (d, J = 7.2 Hz, 3H), 3.81 (s, 3H), 6.29 (q, J = 7.2 Hz, 1H), 7.23-7.37 (m, 5H); ^{13}C NMR (100 MHz, CDCl$_3$) δ: 16.0, 51.5, 127.2, 127.4, 128.2, 135.3, 135.6, 138.0, 168.4; IR (neat) vmax 3069, 2950, 1716, 1200, 1000, 753, 696 cm^{-1} ; HRMS (ESI) calcd for C$_{11}$H$_{13}$O$_2$ (M+H)$^+$ 177.0910, found 177.0912. R$_f$ (hexane/EtOAc = 20:1): (Z)-**2**: 0.61. Isomerization from Z to E did not occur during the distillation. The purity of (Z)-**2** was estimated at least >97% by a quantitative ^{1}H NMR measurement.

20. As an alternative purification procedure, the crude oil (1.90 g) is subjected to flash column chromatography. Silica-gel (11 g) (Note 21) is loaded into a 15 mm-diameter column. The column is slurry packed and the sample is loaded with hexane and then eluted with hexanes/ethyl acetate = 5/1. Fractions are collected (10 mL) at a flow rate of 1.2 mL/s. Overall, 10 fractions are collected and product is observed at fractions 2-4. These fractions are concentrated in vacuo to furnish 1.51 g of (Z)-**2** (95% yield, purity, E/Z = 2:98) as a colorless oil.

21. Type of 60 (spherical), 63-210 mesh, was purchased from Kanto Chemical Co., Inc. and used as received.

Working with Hazardous Chemicals

The procedures in *Organic Syntheses* are intended for use only by persons with proper training in experimental organic chemistry. All hazardous materials should be handled using the standard procedures for work with chemicals described in references such as "Prudent Practices in the Laboratory" (The National Academies Press, Washington, D.C., 2011; the full text can be accessed free of charge at http://www.nap.edu/catalog.php?record_id=12654). All chemical waste should be disposed of in accordance with local regulations. For general guidelines for the management of chemical waste, see Chapter 8 of Prudent Practices.

In some articles in *Organic Syntheses*, chemical-specific hazards are highlighted in red "Caution Notes" within a procedure. It is important to recognize that the absence of a caution note does not imply that no significant hazards are associated with the chemicals involved in that procedure. Prior to performing a reaction, a thorough risk assessment should be carried out that includes a review of the potential hazards associated with each chemical and experimental operation on the scale that is planned for the procedure. Guidelines for carrying out a risk assessment and for analyzing the hazards associated with chemicals can be found in Chapter 4 of Prudent Practices.

The procedures described in *Organic Syntheses* are provided as published and are conducted at one's own risk. *Organic Syntheses, Inc.*, its Editors, and its Board of Directors do not warrant or guarantee the safety of individuals using these procedures and hereby disclaim any liability for any injuries or damages claimed to have resulted from or related in any way to the procedures herein.

Discussion

Regio- and stereo-controlled syntheses of (*E*)- and (*Z*)-stereodefined α, δ - unsaturated esters are pivotal in organic chemistry because of their wide distribution as key structural building blocks in natural products, pharmaceuticals, and supramolecules. They also serve as useful structural scaffolds for various (*E*)- and (*Z*)-stereodefined olefins, conjugate (Michael) addition acceptors, and catalytic asymmetric hydrogenation and oxidation substrates.

Methyl (*Z*)-2-phenyl-2-butenoate [(*Z*)-**2**] and its aryl analogues have a simple common structure, and they are fundamentally useful and promising synthetic building blocks for various stereodefined alkenes. Despite a high demand, (*Z*)-stereoselective synthetic methods are quite limited compared with those for the formation of (*E*)-isomers, due to the inherent (*E*)-stable nature of these esters. Here we present a practical, accessible, and robust synthesis of (*Z*)-**2** and the substrate-generality for synthesis of a variety of analogues of (*Z*)-**1** and (*Z*)-**2** including the complementary (*E*)-**1** and (*E*)-**2** stereoisomers.

Four approaches relevant to the synthesis of (*Z*)-**2** have been reported, as follows.

intermediate is smoothly converted to (Z)-**2** in 90% with excellent (Z)-stereoselectivity by using triflic anhydride and pyridine.[8] This approach is considered to be the most practical since Stille-type cross-couplings can be avoided. This method was applied successfully for the large-scale synthesis of cholecystokinin receptor antagonist drug.[9] Three major drawbacks remain, however; (i) the dianion formation procedure under −78 °C conditions is not economic, (ii) Tf$_2$O is ca. 15-30 times more expensive than TsCl, and (iii) Tf$_2$O is highly toxic and hazardous with a low boiling point (81–83 °C), and reacts violently with water.

OH
Ph⌒CO$_2$Me

→ (2 LHMDS, EtBr / THF, −78 °C) →

Me
OH
Ph⌒CO$_2$Me

→ (Tf$_2$O, Pyridine / CH$_2$Cl$_2$, 0 °C to 20 °C, ca. 10 h) →

Me
Ph⌒CO$_2$Me
(Z)-**2**
90%
E / Z = 1 / 40

Scheme 4. Stereoselective dehydration of methyl 2-ethyl mandelate derived from methyl mandelate

This background led us to investigate a more accessible and robust method. Our synthetic procedure comprises two steps: (i) A one-pot α-formylation and Z-selective enol tosylation of methyl phenylacetate using HCO$_2$Me/tBuONa reagents (Step A) and (ii) A mild iron-catalyzed methylation using MeMgI/Fe(acac)$_3$/TMEDA reagent (Kochi-Fürstner reaction) (Step B).[10-12] The present sequence involves a couple of carbonhomologation (α-formylation and Fe-catalyzed cross-coupling). Both steps are performed using readily accessible substrates, reagents, and solvents under user-friendly conditions (ambient temperature, short reaction and work-up periods). Two purification procedures were carried out using recrystallization (Step A) and simple distillation or short column chromatography (Step B). The most expensive substrate or reagent required throughout the process, is MeMgI/THF.

The original report[13] addressed the isolation of α-formylated intermediates obtained by Ti-Claisen condensation,[14,15] followed by stereocomplementary enol tosylation to give (E)-**1** and (Z)-**1** (Scheme 4).[13] Cross-coupling partner (Z)-**1** was a novel compound when the original report was published. Notably, the present revised procedure involves the obvious advantages that α-formylation and subsequent enol tosylation are performed in "a one-pot procedure" using bench-top handling reagents to produce shelf-stable solid enol tosylate (Z)-**1** in good yield with nearly perfect Z-selectivity. The Fe(acac)$_3$-catalyzed methylation cross-coupling using (Z)-**1** proceeds smoothly to produce (Z)-**2** in good yield with excellent Z-stereoretentivity under mild and accessible conditions.[16] The absence of the Fe(acac)$_3$ catalyst resulted in no reaction. TMEDA was used instead of conventional NMP for enhancing Z-selectivity. The use of EtOAc main-solvent for such Grignard-type reactions is unusual, but effective in this case. These improvements significantly increased the scalability with accessible reaction temperatures (0–25 °C), short reaction periods (total 3 h), and easy operations in all of the procedures.

Scheme 5 outlines the preparation of enol tosylate intermediates (Z)-**1**, and related analogues (E)-**3** and (Z)-**3**. Due to the higher acidity, aryl esters can be formylated by using HCO$_2$Me/tBuONa reagent, followed by Z–selective tosylation in a one-pot manner to produce tosylate (Z)-**1** (Scheme 5, Method A). The Ti-Claisen condensation method covers not only aryl esters but also less accessible alkyl α-formylesters (Method B).[13,15] Successive E- and Z-stereocomplementary enol tosylations and enol phosphorylations were successfully performed using not only α-formylesters[13,16] but also β-ketoesters[17-19] under the substrate-dependent fine-tuned conditions.

Method A (one-pot)

HCO₂Me + Ar⌒CO₂Me

NaOt-Bu (1.5 eq.)
———————————→
/ THF,
0 – 10 °C, 1 h

$$\left[\begin{array}{c} \text{O–Na} \\ \text{Ar} \quad \text{O} \\ \text{OMe} \end{array} \right]$$

TsCl (1.5 eq.)
———————————
THF,
0 – 10 °C
1 h

OTs
Ar⌒CO₂Me

(Z)-**1**

Method B (stepwise)

HCO₂Me + R¹⌒CO₂R²

(R¹ = alkyl, aryl)

TiCl₄ - Et₃N
———————————→
/ CH₂Cl₂ (or toluene)
0 °C, 1 h and rt, 1 h

$$\begin{array}{c} \text{O} \\ \text{H} \quad \text{CO₂R²} \\ \text{R¹} \end{array}$$

TsCl - N-methylimidazole -
Et₃N, Toluene, 0 - 5 °C, 1 h
———————————→

TsO
R¹⌒CO₂R²   (E)-**3**

TsCl - N-methylimidazole -
LiOH, Toluene, 0 - 5 °C, 1 h
———————————→

OTs
R¹⌒CO₂R²   (Z)-**3**

Scheme 5. α-Formylation of esters and (E)- and (Z)- stereocomplementary enol tosylations

Scheme 6 shows the substrate-scope for stereoretentive iron-catalyzed *alkylation* cross-couplings using enol (E)- and (Z)-tosylates derived from α-formylesters. In addition to (Z)-**4a** and (Z)-**4b**, five sets of (E)- and (Z)-**4c–4g** (≥98% ds) are described. Notably, the reactive p-Br atom in (Z)-**4b** was compatible during the course of reaction.

In addition, Scheme 7 lists iron-catalyzed *methylation* cross-couplings using enol (E)- and (Z)-tosylates **5** derived from β-keto esters.

Recently, Franz's group disclosed related iron-catalyzed stereoselective cross-coupling using enol (E)- and (Z)-carbamates.[20]

Scheme 6. Stereoretentive *alkyl*-cross-couplings using enol (*E*)- and (*Z*)-tosylates *derived from* **α**-formylesters

Scheme 7. Stereoretentive *methyl* cross-couplings using enol (*E*)- and (*Z*)-tosylates *derived from* **β**-ketoesters

Iron-catalyzed *arylation* cross-coupling causes small amounts of undesired $Z \rightarrow E$ isomerization.[16] Notably, Suzuki-Miyaura cross-coupling solved the problem; excellent stereoretentive *arylations*[13] were performed as depicted in

Scheme 8. Recently improved conditions [ArB(OH)₂ (1.5 equiv), Pd(OAc)₂ (0.05 equiv), SPhos (0.05 equiv), K₂CO₃ (3.0 equiv) / toluene–H₂O, 80 C°, 2 h] are milder than those shown in Scheme 8.[21]

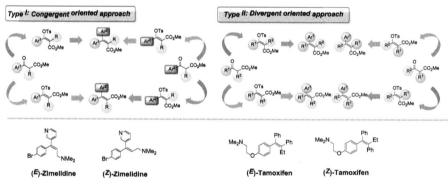

Scheme 8. Stereoretentive Suzuki-Miyaura *aryl*-cross-couplings using (*E*)- and (*Z*)-stereodefined enol tosylates *derived from α-formylesters*

Utilizing these protocols, the first *E*- and *Z*-stereocomplementary and parallel syntheses (convergent and divergent approaches) of Zimeridine and Tamoxifen were successfully performed as illustrated in Scheme 9)[17b-d]

Scheme 9. *E*- and *Z*-stereocomplementary and parallel syntheses: convergent and divergent approaches

(*E*)- and (*E*)-rich compounds **2** and their analogues are representative probes for asymmetric hydrogenation,[22] asymmetric dihydroxylation (Sharpless AD),[23] asymmetric oxohydroxylation,[24] asymmetric Michael-type addition,[25] hydrosilylation,[26] hydrostannylation,[27] etc. However, the corresponding reactions using (*Z*)-**2** and its analogues have been limited to only a handful of reports, likely due to the lack of a practical supply of these precursors. The present method with a broad substrate-scope provides the synthetic accessibility to this type of both *E*- and *Z*-stereodefined α,β–unsaturated esters.

In conclusion, the straightforward and user-friendly synthetic protocol for the formation of (*Z*)-**2** and the related analogues as useful synthetic building blocks provides a new promising avenue for synthetic organic chemistry. This strategy will contribute to the construction of a library for (*E*)-and (*Z*)-stereodefined α,β–unsaturated esters.

References

1. Department of Chemistry, School of Science and Technology, Kwansei Gakuin University, 2-1 Gakuen, Sanda, Hyogo, 669-1337, Japan. Email: tanabe@kwansei.ac.jp. This research was partially supported by Grant-in-Aids for Scientific Research on Basic Areas (B) "18350056" and (C) 15K05508, Priority Areas (A) "17035087" and "18037068", and Exploratory Research "17655045" from the Ministry of Education, Culture, Sports, Science and Technology (MEXT).

2. (a) Schwenker, G.; Gerber, R. *Chem. Ber.* **1967**, *100*, 2460–2461. (b) Lawson, J. A.; Cheng, A.; DeGraw, J.; Frenking, G.; Uyeno, E.; Toll L.; Loew, G. H. *J. Med. Chem.* **1988**, *31*, 2015–2021.

3. (a) Reich, H. J.; Reich, I. L.; Renga, J. M. *J. Am. Chem. Soc.* **1973**, *95*, 5813–5815. (b) Kaur, K.; Adediran, S. A.; Lan, M. J. K.; Pratt, R. F. *Biochemistry* **2003**, *42*, 1529–1536.

4. Bellina, F.; Carpita, A.; De Santis, M. Rossi, R. *Tetrahedron* **1994**, *50*, 12029–12046.

5. Palmer, F. N.; Lach F.; Poriel, C.; Pepper, A. G.; Bagley, M. C.; Slawin, A. M. Z.; Moody, J. C. *Org. Biomol. Chem.* **2005**, *3*, 3805–3811.

6. Consiglio, B. C.; Gaggini, F.; Mordini, A.; Reginato, G. *Amino Acids* **2010**, *39*, 175–180.

7. Brenna, E.; Gatti, F. G.; Manfredi, A.; Monti, D.; Parmeggiani, F. *Org. Process Res. Dev.* **2012**, *16*, 262–268.

8. Mani, N. S.; Mapes, C. M.; Wu, J.; Deng, X.; Jones, T. K. *J. Org. Chem.* **2006**, *71*, 5039–5042.

9. Deng, X.; Mani, N.; Mapes, C. M. USP 2006/0004195 A1.

10. (a) Tamura, M.; Kochi, J. K. *J. Am. Chem. Soc.* **1971**, *93*, 1487–1489. (b) Tamura, M.; Kochi, J. *Synthesis* **1971**, 303–305. (c) Neumann, S. M.; Koch, J. *J. Org. Chem.* **1976**, *41*, 502–509.

11. Representatives: (a) Fürstner, A.; Leitner, A. *Angew. Chem. Int. Ed.* **2002**, *41*, 609–612. (b) Fürstner, A.; Leitner, A.; Mendez, M.; Krause, H. *J. Am. Chem. Soc.* **2002**, *124*, 13856–13863. Review: (c) Fürstner, A.; Martin, R. *Chem. Lett.* **2005**, *34*, 624–629.

12. Fürstner, A.; Leitner, A.; Seidel, G. *Org. Synth.* **2005**, *81*, 33–41.

13. Nakatsuji, H.; Nishikado, H.; Ueno, K.; Tanabe, Y. *Org. Lett.* **2009**, *11*, 4258–4261.

14. Representatives: (a) Tanabe, Y. *Bull. Chem. Soc. Jpn.* **1989**, *62*, 1917–1924. (b) Misaki, T.; Nagase, R.; Matsumoto, K.; Tanabe, Y. *J. Am. Chem. Soc.* **2005**, *127*, 2854–2855. (c) Ref. 13.

15. A mini-review: Ashida, Y.; Kajimoto, S.; Nakatsuji, H.; Tanabe, Y. *Org. Synth.* **2016**, *93*, 286–305.

16. Nishikado, H.; Nakatsuji, H.; Ueno, K.; Nagase, R.; Tanabe, Y. *Synlett* **2010**, 2087–2092.

17. Tosylations: (a) Nakatsuji, H.; Ueno, K.; Misaki, T.; Tanabe, Y. *Org. Lett.* **2008**, *10*, 2131–2134. (b) Ashida, Y.; Sato, Y.; Suzuki, T.; Ueno, K.; Kai, K.; Nakatsuji, H. *Chem. Eur. J.* **2015**, *21*, 5934–5945. (c) Ashida, Y.; Honda, A.; Sato, Y.; Nakatsuji, H.; Tanabe, Y. *ChemistryOpen* **2017**, *6*, 73–89. (d) Ashida, Y.; Sato, Y.; Honda, A.; Nakatsuji, H.; Tanabe, Y. *Synform* **2017**, A38–A43.

18. Phosphorylations: Nakatsuji, H.; Ashida, Y.; Hori, H.; Sato, Y.; Honda, A.; Taira, M.; Tanabe, Y. *Org. Biomol. Chem.* **2015**, *13*, 8205–8210.

19. A mini-review: Ashida, Y.; Nakatsuji, H.; Tanabe, Y. *Org. Synth.* **2017**, *94*, 93–108.

20. Rivera, A. C. P.; Still, R.; Franz, D. E. *Angew. Chem. Int. Ed.* **2016**, *55*, 6689–6693.

21. Sato, Y.; Ashida, Y.; Yoshitake, D.; Hoshino, M.; Takemoto, T.; Tanabe, Y. *Synthesis* in press. DOI: 10.1055/s-0037-1610652

22. Yang, S.; Che, W.; Wu, H.-L.; Zhu, S.-F.; Zhou, Q.-L. *Chem. Sci.* **2017**, *8*, 1977–1980.

23. Kolb, H. C.; VanNiewenhze, M. S.; Sharpless, K. B. *Chem. Rev.* **1994**, *94*, 2483–2547.

24. Wang, C.; Zong, L.; Tan, C-H. *J. Am. Chem. Soc.* **2015**, *137*, 10677–10682.

25. Nishimura, K.; Tomioka, K. *J. Org. Chem.* **2002**, *67*, 431–434.

26. (a) Crump, R. A. N. C.; Fleming, I.; Hill, J. H. M.; Parker, D.; Reddy, N. L. Waterson, D. *J. Chem. Soc., Perkin Trans. 1* **1992**, 3277–3294. (b) Masterson, D. S.; Porter, N. A. *J. Org. Chem.* **2004**, *69*, 3693–3700.

27. Chopa, A. B.; Koll, L. C.; Savini, M. C.; Podestá, J. C. *Organometallics*, **1985**, *4*, 1036–1041.

Appendix
Chemical Abstracts Nomenclature (Registry Number)

Methyl 2-phenylacetate; (101-41-7)
Methyl formate; (107-31-3)
Sodium *tert*-butoxide; (865-48-5)
p-Toluenesulfonyl chloride; (98-59-9)
Methyl (Z)-2-phenyl-3-(tosyloxy)acrylate; (1188274-63-6)
Ethylene carbonate; (96-49-1)
Tris(2,4-pentanedionato)iron(III) ; (14024-18-1)
Tetramethylethylenediamine; (110-18-9)
Methylmagnesium bromide; (75-16-1)
Methyl (Z)-2-phenyl-2-butenoate; (50415-84-4)

Takeshi Tsutsumi was born in Hyogo, Japan, in 1993. He received his B. S. degree from Kwansei Gakuin University under the direction of Professor Yoo Tanabe (2017). Presently, he is the M.S. student and continues his studies on the development of practical syntheses of (E)-, (Z)-stereocomplementary synthesis of α,β,δ-unsaturated esters and asymmetric total syntheses of natural products containing pyron structure, which are directed for process chemistry.

Yuichiro Ashida was born in Kyoto, Japan, in 1989. He received his B. S. degree (2012), M.S. degree (2014), and Ph.D. degree (2017) on the development of divergent synthetic methods for (E)- and (Z)-stereodefined multi-substituted alkene scaffolds from Kwansei Gakuin University under the supervision of Prof. Yoo Tanabe. Presently, he continues his studies, which focuses on (E)-, (Z)-stereocomplementary parallel synthesis of α,β,δ-unsaturated esters utilizing (E)-, (Z)-stereodefined enol tosylates and the related Ti-Claisen condensation, as a researcher in Kwansei Gakuin University.

Hiroshi Nishikado was born in Hyogo, Japan in 1986. He received his B.S. degree (2009) and M.S. degree (2011) on the Iron-catalyzed cross-coupling of (E)-, (Z)-stereodefined enol tosylates and Ti-Claisen condensation from Kwansei Gakuin University under the direction of Prof. Yoo Tanabe. Since 2011, he has been working as a process chemist at Kongo Chemical Co., Ltd.

Yoo Tanabe received his B.S. degree at Tokyo (Professor Kenji Mori). He received his Ph.D. at Tokyo under the direction of Professor Teruaki Mukaiyama on the development of practical acylation reactions. After leaving Sumitomo Chemical Co. Ltd, Dr. Tanabe moved to Kwansei Gakuin University in 1991 as Associate Professor and promoted to Full Professor (1997). In 1996-1997, he studied at University of Groningen with Professors Richard M. Kellogg and Ben L. Feringa. His research focuses on the exploitation of useful synthetic reactions directed for process chemistry: concise synthesis of useful fine chemicals and of total synthesis of biologically active natural products.

Dr. Zhaobin Han graduated from Department of Chemistry, Nanjing University in 2003. He received his Ph.D. degree from Shanghai Institute of Organic Chemistry under the supervision of Prof. Kuiling Ding and Prof. Xumu Zhang in 2009, working on development of novel chiral ligands for asymmetric catalysis. He is currently an associate professor in the same institute and his research interests focus on the development of efficient catalytic methods for organic synthesis based on homogeneous catalysis.

Synthesis of Methyl 2-Bromo-3-oxocyclopent-1-ene-1-carboxylate

Rama Rao Tata and Michael Harmata[*1]

Department of Chemistry, University of Missouri-Columbia, MO 65211

Checked by Leonardo J. Nannini and Erick M. Carreira

A.
1
tBuOOH, 4 Å sieves
Mn(OAc)$_3$ · 2H$_2$O
EtOAc, O$_2$
rt
CO_2Me
2

B.
2
Br$_2$, Et$_3$N
CH$_2$Cl$_2$
0 °C - rt
Br
CO_2Me
3

Procedure (Note 1)

A. *Methyl 3-oxocyclopent-1-ene-1-carboxylate* (**2**).[2] A 1000-mL, one-necked, oven-dried, round-bottomed flask equipped with a teflon-coated magnetic stir bar (25 mm x 10 mm) is fitted with a septum and an argon balloon, and then cooled to ambient temperature (21–23 °C). The flask is charged with methyl 1-cyclopentene-1-carboxylate (**1**) (Note 2) (7.0 g, 55 mmol, 1.0 equiv). Ethyl acetate (280 mL) (Note 3) is added, followed by 4 Å molecular sieves (7.0 g) (Note 4). The septum is replaced with a pressure-equalizing addition funnel that is capped with a septum. A *tert*-butylhydroperoxide solution in decane (5.5 M) (41 mL, 221 mmol, 4.0 equiv) is added via the addition funnel over 15 min (Figure 1) (Note 5). The

Figure 1. Addition of *t*-BuOOH

reaction mixture is stirred for 30 min. Manganese (III) acetate dihydrate (1.487 g, 5.5 mmol, 0.1 equiv) (Figure 2) (Note 6) is transferred to a vial in a glove bag under a nitrogen atmosphere and the solid is added quickly to the reaction mixture. The argon balloon is removed and the reaction flask is

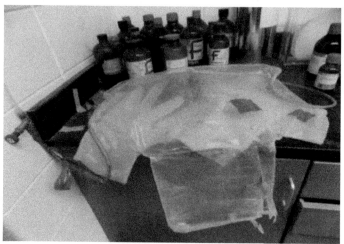

Figure 2. Glove bag

flushed with oxygen using a balloon for 10 min (Figure 3). The reaction mixture is stirred under an oxygen atmosphere at 23 °C for 48 h using a balloon filled with oxygen (Figure 4) (Note 7). The reaction is monitored by TLC on silica using 30% EtOAc-hexane as the eluent (Note 8). After completion of the reaction, the reaction mixture is filtered through a Celite pad (Note 9) that is then washed with diethyl ether (40 mL). The filtrate is diluted with water (150 mL) and transferred to 1 L separatory funnel. The aqueous phase is separated and the organic layer is washed again with water (150 mL). The organic extract is tested for peroxides with KI starch paper and no color is observed (Note 10). The organic extract is dried over anhydrous sodium sulfate for 10 min and concentrated using a rotary evaporator (40 °C, 10 mmHg). A chromatography column (6.4 cm x 45 cm) is loaded with 150 g of silica gel (Note 11) and wetted with pentane (200 mL). Sand is placed on top of the silica gel. The crude compound is loaded on the column. Elution begins with 100% pentane (100 mL) and continues with diethyl ether in pentane 10% (100 mL), 20% (100 mL), and 30% (200 mL), which are collected in 250 mL Erlenmeyer flasks. The desired compound elutes with 40% diethyl ether/pentane (700 mL), which is collected in 10 mL test tubes. The collected fractions (5-75) are concentrated by rotary evaporation (40 °C, 10 mmHg). The compound is obtained as colorless oil (3.0 g, 40% yield) (Notes 12 and 13). The purity of the compound is determined to be 94.5 % by qNMR (Note 14).

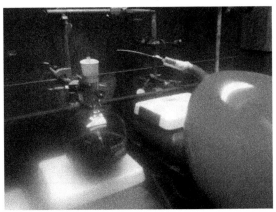

Fig 3. Purging with oxygen

Fig 4. Oxidation under oxygen

B. *Methyl 2-bromo-3-oxocyclopent-1-ene-1-carboxylate* (**3**).[3] A 1000-mL, three-necked, oven-dried, round-bottomed flask is equipped with a teflon-coated magnetic stir bar (25 mm x 10 mm). The middle neck is fitted with a pressure equalizing addition funnel (150 mL) equipped with a rubber septum and an argon balloon. One side neck is equipped with a septum pierced by a thermometer and the remaining neck is sealed with a rubber septum. The flask is charged with methyl 3-oxocyclopent-1-enecarboxylate (**2**) (3.0 g, 21.4 mmol, 1.0 equiv), followed by the addition of dichloromethane (214 mL, 0.1 M) by syringe. The reaction flask is cooled in an ice bath with stirring for 10 min (Note 15). A solution of bromine (1.62 mL, 32.1 mmol, 1.5 equiv) (Note 16) in dichloromethane (40 mL) is charged into the addition funnel and is added dropwise to the reaction mixture over 30 min. The addition funnel is rinsed with dichloromethane (4 mL). The initially colorless solution turned red over the course of this addition (Figure 5A). After complete addition of the bromine solution, the reaction mixture is warmed to 23 °C and stirred for 1 h (Figure 5B). The reaction mixture is again cooled in an ice bath and a solution of triethylamine (5.98 mL, 42.8 mmol, 2.0 equiv) (Note 17) in dichloromethane (40 mL) (Note 18) is carefully added dropwise through the addition funnel to the reaction mixture over 30 min (Note 19). The red-colored solution turns yellow during the addition of triethylamine (Figure 6A) (Note 20). After complete addition, the reaction mixture is again warmed to 23 °C and stirred for 8 h, during which time the yellow color turns red (Figure 6B). The reaction is monitored by TLC on silica using 30% EtOAc-hexane as the eluent (Note 21). The reaction mixture is diluted with dichloromethane (200 mL) and washed with 1M HCl (2 x 250 mL) followed by saturated $Na_2S_2O_3$ solution (1 x 200 mL) in a 1 L separatory funnel. The organic extract is dried over anhydrous Na_2SO_4 and concentrated by rotary evaporator (40 °C, 10 mmHg). Dry silica gel (100 g) (Note 11) is added to a chromatography column (6.4 cm x 45 cm) and pentane (200 mL) is used to wet the column. Sand is added on the top of the silica gel, and then the crude compound is loaded onto the sand at the top of the column. Elution is started with 100% pentane (100 mL) and continues with 5% diethyl ether in pentane (100 mL), 10% (100 mL), 20% (100 mL), 30% (100 mL), 40% (100 mL) and 50% (300 mL). Fractions are collected in 10 mL test tubes beginning with 40% diethyl ether in pentane. Fractions containing the desired compound (13-49) are concentrated by rotary evaporator (40 °C, 10 mmHg). The compound is obtained as a off-white solid (3.0 g, 64%) (Notes 22 and 23), and the purity of the product is 95% as determined by qNMR (Note 24).

Figure 5. A) Addition of bromine; B) After addition of bromine

Figure 6. A) Addition of triethylamine; B) After addition of triethylamine

Notes

1. Prior to performing each reaction, a thorough hazard analysis and risk assessment should be carried out with regard to each chemical substance and experimental operation on the scale planned and in the context of the laboratory where the procedures will be carried out. Guidelines for carrying out risk assessments and for analyzing the hazards associated with chemicals can be found in references such as Chapter 4 of "Prudent Practices in the Laboratory" (The National Academies Press, Washington, D.C., 2011; the full text can be accessed free of charge at https://www.nap.edu/catalog/12654/prudent-practices-in-the-laboratory-handling-and-management-of-chemical. See also "Identifying and Evaluating Hazards in Research Laboratories" (American Chemical Society, 2015) which is available via the associated website "Hazard Assessment in Research Laboratories" at https://www.acs.org/content/acs/en/about/governance/committees/chemicalsafety/hazard-assessment.html. In the case of this procedure, the risk assessment should include (but not necessarily be limited to) an evaluation of the potential hazards associated with methyl 1-cyclopentene-1-carboxylate, molecular sieves, *tert*-butylhydroperoxide, decane, manganese (III) acetate dihydrate, oxygen, nitrogen, argon,

ethyl acetate, hexane, celite, sodium sulfate, silica gel, pentane, diethyl ether, dichloromethane, bromine, triethylamine, iodine, and sodium thiosulfate.

2. Methyl 1-cyclopentene-1-carboxylate was purchased from Ark Pharm and used as received.

3. Ethyl acetate was purchased from Fisher Scientific and distilled over CaH_2.

4. Molecular sieves (4Å) were activated in an oven at 100 °C heat for more than 2 weeks.

5. *tert*-Butyl hydroperoxide solution (5.5 M in decane) was purchased from Sigma Aldrich and used as received.

6. Manganese(III) acetate dihydrate (97%) was purchased from Sigma-Aldrich and used as received. This catalyst was weighed in a glove bag under nitrogen atmosphere. The checkers compared the reaction with and without glove bag, observing no change in the yield.

7. Two oxygen balloons are used over the course of the two days. One balloon is used each day.

8. TLC was performed using 30% EtOAc-hexane as eluent. Starting material and product can be visualized under UV lamp and with an iodine stain. The starting material has $R_f = 0.6$ (pink) and the product has $R_f = 0.3$ (pink).

9. Celite was purchased from Fisher and used as received.

10. Starch paper was purchased from Fisher Scientific.

11. Silica gel (F60, particle size 230-400 mesh) was purchased from Silicycle and used as received.

12. The product has been characterized as follows: ^{1}H NMR (500 MHz, CDCl$_3$) δ: 2.53–2.55 (m, 2H), 2.85–2.87 (m, 2H), 3.87 (s, 3H), 6.76 (dd, $J = 3.0, 2.0$ Hz, 1H); ^{13}C NMR (125 MHz) δ: 27.6, 35.7, 52.7, 138.4, 164.0, 164.9, 209.3.

13. Other reactions performed on half and full scale provided yields of 1.6 g (42%) and 2.7 g (36%), respectively.

14. Checkers determined the purity by qNMR using 8.8 mg of compound **2** and 3.3 mg of standard 1,2,4,5-tetrachlorobenzene (purity 88.1%). The purity was determined using the following equation.

$$Px = \frac{I\,x}{I\,std} \times \frac{N\,std}{N\,x} \times \frac{m\,std}{m\,x} \times \frac{M\,x}{M\,std} \times P\,std$$

P = Purity, x = product, std = standard, Ix = Integration of the methyl peak, Istd = Integration of the standard, Nstd = Number of protons for the standard, Nx = Number of protons for methyl signal, m = Prepared weight, M = Molecular weight.

15. The internal temperature of the reaction mixture was measured to be between 1 °C and 3 °C during the reaction.

16. Reagent grade bromine was purchased from Sigma-Aldrich and used as received.

17. Triethylamine was purchased from Sigma-Aldrich and used as received.

18. Dichloromethane was purchased from Fisher and distilled over CaH$_2$.

19. White fumes were formed during the addition of triethylamine. Triethylamine was carefully added dropwise.

20. The internal temperature of the reaction mixture was measured to be between 1 °C and 3 °C during the reaction.

21. TLC was performed using 30% EtOAc-hexane as eluent. Starting material and product were visualized under a UV lamp. Both starting material and product have same R_f = 0.3 (pink).

22. The product has been characterized as follows: mp 66–68 °C; ^{1}H NMR (500 MHz, CDCl$_3$) δ: 2.65 (m, 2H), 2.89 (m, 2H), 3.91 (s, 3H); ^{13}C NMR (125 MHz) δ: 28.4, 32.7, 52.8, 131.1, 157.7, 163.9, 201.6; FTIR (cm^{-1}) 2954, 1725, 1436, 1282, 1202, 1178; ESI [M + H] m/z calcd for C$_7$H$_8$BrO$_3$: 218.9651. Found: 218.9651.

23. A second run on half scale provided 0.9 g (41%) of compound **3** as an off-white solid.

24. Checkers determined the purity by qNMR using 4.3 mg of compound **3** and 4 mg of standard 1,2,4,5-tetrachlorobenzene (purity 99.8%). Purity was calculated using the equation in Note 13.

Working with Hazardous Chemicals

The procedures in *Organic Syntheses* are intended for use only by persons with proper training in experimental organic chemistry. All hazardous materials should be handled using the standard procedures for work with chemicals described in references such as "Prudent Practices in the Laboratory" (The National Academies Press, Washington, D.C., 2011; the full text can be accessed free of charge at http://www.nap.edu/catalog.php?record_id=12654). All chemical waste should be disposed of in accordance with local regulations. For general guidelines for the management of chemical waste, see Chapter 8 of Prudent Practices.

In some articles in *Organic Syntheses*, chemical-specific hazards are highlighted in red "Caution Notes" within a procedure. It is important to recognize that the absence of a caution note does not imply that no significant hazards are associated with the chemicals involved in that procedure. Prior to performing a reaction, a thorough risk assessment should be carried out that includes a review of the potential hazards associated with each chemical and experimental operation on the scale that is planned for the procedure. Guidelines for carrying out a risk assessment and for analyzing the hazards associated with chemicals can be found in Chapter 4 of Prudent Practices.

The procedures described in *Organic Syntheses* are provided as published and are conducted at one's own risk. *Organic Syntheses, Inc.*, its Editors, and its Board of Directors do not warrant or guarantee the safety of individuals using these procedures and hereby disclaim any liability for any injuries or damages claimed to have resulted from or related in any way to the procedures herein.

Discussion

Cyclopentadienones can be used as dienophiles in [4+2] cycloaddition reactions,[4,5] but they are highly reactive and dimerize rapidly.[4] Treatment of 2-bromocyclopentenone derivative **3** with triethylamine in refluxing toluene for 1 h affords the indanone **6** in 71% yield. The reaction presumably proceeds via formation of intermediate cyclopentadienone **4**, which undergoes dimerization and subsequent decarbonylation to generate the product **6** (Scheme 1).

Scheme 1. Generation of a cyclopentadienone 4 from 3

Recently, our group developed conditions suitable for [4+2] cycloadditions using 3. To optimize the reaction conditions, we treated 3 and 2,3-dimethylbutadiene (7) with various bases and in various solvents. Toluene and tetrahydrofuran (THF) showed promising results. The starting material decomposed when treated with strong bases such as DBU, and the use of hindered base 2,2,6,6-tetramethylpiperidine (TMP) in THF solvent produced cycloadducts in only moderate yield. The reaction proceeded very well with triethylamine (Scheme 2).

Scheme 2. [4+2] Cycloaddition reaction of 3 with 7

With the optimized conditions in hand, several dienes were tested in the cycloaddition. The diene 8 produced the cycloaddition product 8a in toluene in very good yield (82%). Cyclopentadiene (9) afforded the corresponding cycloadduct in THF in 89% yield. Cyclohexadiene (10) afforded 10a in only 17% yield under these conditions. However, diene 10 gave a better yield (57%) using toluene as solvent. Interestingly, the diene 11 produced the corresponding cylcoadduct 11a as single regio- and stereoisomer in 80% yield. The product 11a has the carbocylic skeleton of a steroid, so this methodology could be used in various syntheses of steroid-like molecules.

Table 1. [4+2] Cycloaddition reactions of cyclopentadienone 6

Et₃N (3 equiv), diene
solvent, reflux

3 → **8a - 17a**

Substrate	Diene	Time	Solvent	Product	Yield (%)
1	**8**	1.25	Toluene	**8a**	82
2	**9**	2.75	THF	**9a**	89
3	**10**	4	THF	**10a**	17
		1.5	Toluene		57
4	**11**	1.5	Toluene	**11a**	80
5	**12**	1.25	Toluene	**12a**	66
6	**13**	6	THF	**13a**	52
7	**14**	1.75	Toluene	**14a**	72

Table 1 (cont.)

Substrate	Diene	Time	Solvent	Product	Yield (%)
8	TMSO OMe **15**	5	Benzene	CO$_2$Me ÖMe **15a**	53
9	nBu N Ac **16**	1.75	Toluene	CO$_2$Me NAc nBu **16a**	84
10	SPh **17**	1.25	Toluene	SPh CO$_2$Me **17a**	69

To ascertain the electronic effects on the cycloaddition, various experiments were conducted with the 1-substituted dienes (Table 1, entries 5-7). The dienes 1-phenylbutadiene **12** and 1-methoxybutadiene **13** afforded the corresponding Diels-Alder adducts **12a** and **13a** in yields of 66% and 52%, respectively. However, under the same conditions the diene **14** (2 equiv) gave **14a** in only 30% yield. This might be due to the lower reactivity of diene **14**. Increasing the equivalents of diene to five increased the cycloadduct yield to 72%. Danishefsky's diene **15** produced the cycloadduct **15a** in 53% yield under standard conditions but benefited from the presence of 20 mol% triethylamine hydrobromide. The nitrogen and sulfur substituted dienes **16** and **17** also produced the corresponding cycloadducts in good yields.

In conclusion, we have developed an efficient generalized procedure for the synthesis of methyl 2-bromo-3-oxocyclopent-1-ene-1-carboxylate **3**. This substrate can be used in [4+2] cycloaddition reactions and these reactions proceed with excellent regioselectivity and diasteroselectivity.

References

1. Department of Chemistry, University of Missouri-Columbia, 125 Chemistry Building, Columbia, MO 65211. Email: HarmataM@missouri.edu. Eurofins Biopharm Product Testing, 7200 East ABC Lane, Columbia, MO 65202. This work was supported by the National Science Foundation.
2. (a) Turrini, N. G.; Cioc, R. C.; Van der niet, D. J. H.; Ruijter, E.; Orru, R. V. A.; Hall, M.; Faber, K. *Green. Chem.* **2017**, *19*, 511–518. (b) Nicolaou, K. C.; Pulukuri, K. K.; Yu, R.; Rigol, S.; Heretsch, P.; Grove, C.; Hale, C. R. H.; Elmarrouni, A. *Chem. Eur. J.* **2016**, *22*, 8559–8570. (c) Vogensen, S. B.; Marek, A.; Bay, T.; Wellendorph, P.; Kehler, J.; Bundgaard, C.; Frolund, B.; Pedersen, M. H. F.; Clausen, R. P. *J. Med. Chem.* **2013**, *56*, 8201–8205. (d) Aye, Y.; Davies, S. G.; Garner, A. C.; Roberts, P. M.; Smith, A. D.; Thomson, J. E. *Org. Biomol. Chem.* **2008**, *6*, 2195–2203. (e) Na, S. J.; Lee, B. Y.; Bui, N.; Mho, S.; Jang, H. *J. Organomet. Chem.* **2007**, *692*, 5523–5527. (f) Catino, A. J.; Forslund, R. E.; Doyle, M. P. *J. Am. Chem. Soc.* **2004**, *126*, 13622–13623. (g) Yu, J.; Corey, E. J. *J. Am. Chem. Soc.* **2003**, *125*, 3232–3233. (h) Lange, G. L.; Decicco, C. P.; Willson, J.; Strickland, L. A. *J. Org. Chem.* **1989**, *54*, 1805–1810.
3. (a) Harmata, M.; Gomes, M. G. *Eur. J. Org. Chem.* **2006**, *10*, 2273–2277. (b) Gyorkos, A. C.; Stille, J. K.; Hegedus, L. S. *J. Am. Chem. Soc.* **1990**, *112*, 8465–8472.
4. Harmata, M.; Barnes, C. L.; Brackley, J.; Bohnert, G.; Kirchoefer, P.; Kürti, L.; Rashtasakhon, P. *J. Org. Chem.* **2001**, *66*, 5232–5236.
5. (a) Pearson, A. J.; Kim, J. B. *Tetrahedron Lett.* **2003**, *44*, 8525–8527. (b) Imbriglio, J. E.; Rainier, J. D. *Tetrahedron Lett.* **2001**, *42*, 6987–6990. (c) Rainier, J. D.; Imbriglio, J. E. *Org. Lett.* **1999**, *1*, 2037–2039.

Appendix
Chemical Abstracts Nomenclature (Registry Number)

Methyl 1-cyclopentene-1-carboxylate (25662-28-6)
tert-Butyl hydroperoxide solution (75-91-2)
Manganese(III) acetate dihydrate (19513-05-4)
Bromine (7726-95-6)
Triethylamine (121-44-8)

Rama Rao Tata was born in India. He received his Master degree in organic chemistry in 2007 from the Pondicherry University. He received his doctoral degree in 2015 from the University of Missouri-Columbia under the supervision of Prof. Michael Harmata. He performed his postdoctoral work in University of Missouri-Columbia under the supervision of Prof. Bret Ulery and worked on the synthesis of biodegradable polymers and their applications in bone and spinal cord regeneration. Currently, he is employed as a staff scientist II at Eurofina Biopharma Product Testing in Columbia, MO, working on radiolabeled pharmaceuti-cals and 14C-labeled pesticides.

Michael Harmata was born in Chicago, Illinois and lived on the south side of Chicago for the first 20 years of his life. He received a bachelor's degree from the University of Illinois-Chicago and earned his Ph.D. with Scott Denmark at UIUC. He then headed west to do an NIH postdoctoral fellowship with Paul Wender at Stanford University. He started doing his own thing in 1986 at the University of Missouri-Columbia, where he is now the Norman Rabjohn Distinguished Professor of Chemistry.

Leonardo J. Nannini was born in Bellinzona (Switzerland), where he lived for the first 20 years of his life. He then moved to Zurich, where he earned the bachelor's and master degree in interdisciplinary science at ETH. He has been a Ph.D. candidate under the supervision of Erick M. Carreira since 2014.

Discussion Addendum for:
Preparation of (S)-tert-ButylPHOX and (S)-2-Allyl-2-Methylcyclohexanone

Alexander W. Sun and Brian M. Stoltz[*1]

Warren and Katharine Schlinger Laboratory for Chemistry and Chemical Engineering, Division of Chemistry and Chemical Engineering, California Institute of Technology, Pasadena, California 91125, United States

Original Articles: Krout, M. R.; Mohr, J. T.; Stoltz, B. M. Org. Synth. **2009**, *86, 181–193; Krout, M. R.; Mohr, J. T.; Stoltz, B. M. Org. Synth.* **2009**, *86, 194–211.*

A. $\quad$ 1. allyl alcohol, *p*-TsOH
toluene, reflux

2. NaH, THF, 40 °C
then CH_3I

B. $\quad$ $Pd_2(dba)_3$
(*S*)-*tert*-ButylPHOX

THF, 30 °C

C. $\quad$ 1. semicarbazide · HCl
NaOAc, H_2O

2. recrystallization
3. 3 N HCl (aq), Et_2O

Phosphinooxazoline (PHOX) ligands, originally developed by Pfaltz, Helmchen, and Williams,[2-4] comprise a privileged class of P,N type ligands with extensive applications in various transition-metal catalyzed processes, including allylic alkylations, Heck-type reactions, and catalytic hydrogenation.[5,6] The PHOX ligand class is highly modular and has grown to encompass numerous structural variations, of which only a few major classes are shown in Figure 1. Every component of the ligand scaffold is modifiable as exemplified by R^1 variations on the oxazoline ring (**1, 2**), substitution of the aryl rings with perfluoroalkyl groups (**2**) or ferrocenyl systems (**3**), and various alkyl and spirocyclic⁷ linkers joining the oxazoline ring to the phosphine backbone (**4a-b**).⁶

R¹ = *i*-Pr, *t*-Bu, Ph, Bn

Figure 1. Phosphinooxazoline (PHOX) ligands are highly modular

The substituted tri-aryl PHOX ligands **1** are of particular interest to our laboratory. Since our 2009 *Organic Syntheses* articles, which described an optimized route to the valuable PHOX ligand (*S*)-*t*-BuPHOX **1a** and its use in a Pd-catalyzed decarboxylative allylic alkylation reaction toward the synthesis of (*S*)-2-allyl-2-methylcyclohexanone,⁸,⁹ electronic and steric modifications to the scaffold of **1a** have found increasing traction in the literature.¹⁰⁻¹² In particular, electron-deficient counterparts such as (*S*)-(CF₃)₃-*t*-BuPHOX (**5**), (*S*)-(CF₃)₄-*t*-BuPHOX (**6**), and (*S*)-(CF₃)₄(F)-*t*-BuPHOX (**7**), which contain trifluoromethyl groups at strategic positions on the aryl rings, have been uniquely effective and in many cases superior to **1a** in various highly enantioselective metal-catalyzed reactions.¹³,¹⁴ These ligands may be synthesized through modified procedures based on our original *Organic Syntheses* article.⁸,¹⁰,¹³ This Discussion Addendum highlights recent

1a: (*S*)-*t*-BuPHOX   **5**: (*S*)-(CF₃)₃-*t*-BuPHOX   **6**: R³ = H; (*S*)-(CF₃)₄-*t*-BuPHOX
7: R³ = F; (*S*)-(CF₃)₄(F)-*t*-BuPHOX

Figure 2. Electron-deficient variants of (*S*)-*t*-BuPHOX
PHOX Ligands in Reaction Development

applications of **1a** and its electron-deficient variants **5-7** in total synthesis and reaction development, with an emphasis on decarboxylative asymmetric allylic alkylation reactions.

Aside from their common use in various asymmetric allylic allylation-type reactions,[15,16,17] PHOX ligands **1a** and **5-7** have recently been utilized in diverse transformations including sigmatropic rearrangements,[18] hydroarylation reactions,[19] carboaminations,[20] asymmetric alkylations and protonations,[21,22] and cascade reactions.[23] The chemical motifs produced by these PHOX-catalyzed reactions are often found in biologically active small molecules and can also serve as synthetically valuable intermediates. For instance in 2012, Kozlowski presented a rare example of a catalytic enantioselective Saucy-Marbet Claisen rearrangement using **1a** to effect the transformation of propargyl ethers **8** into allenyl oxindoles **9** (Scheme 1).[18] However, **1a** was effective only for a subset of aryl-substituted alkyne substrates.

R
CO₂R ‖‖‖
[structure 8]   Pd/(S)-t-BuPHOX (1a)   RO₂C R
                 ─────────────────────→   [structure 9]
                 CH₂Cl₂, 0 °C

8   **9**

Scheme 1. Synthesis of quaternary allenyl indoles using an enantioselective Saucy-Marbet Claisen rearrangement

Later in 2015, **1a** was also utilized by Wolfe in a Pd-catalyzed alkene carboamination reaction involving allylphenyltriflates **10** and aliphatic amines to generate chiral aminoindanes **11**, a motif that appears in a variety of pharmacologically active molecules (Scheme 2).[20] Interestingly, the crucial aminopalladation step consists of an intermolecular reaction.

OTf
R─[structure 10]   + amine   Pd/(S)-t-BuPHOX (1a)   R─[structure 11]─NR₂
                               ─────────────────────→
                               LiO-t-Bu, PhMe, 95 °C

10   **11**
                                                       up to 98% ee
                                                       up to 98% yield

Scheme 2. Synthesis of aminoindanes via alkene carboamination

In 2017, Zhu used **1a** and a stoichiometric amount of tetrahydroxy diboron-water as the hydride donor to effect an asymmetric intramolecular reductive Heck reaction of *N*-aryl acrylamides **12** to give 3,3-disubstituted oxindoles **13** (Scheme 3).[19] Notably, the choice of ligand affected the reaction pathway: whereas PPh₃ led to carboboration products, **1a** gave the desired hydroarylation product **13**.

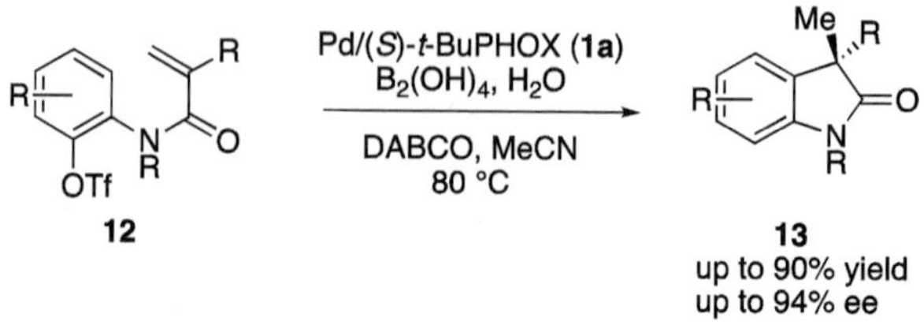

Pd/(*S*)-*t*-BuPHOX (**1a**)
B₂(OH)₄, H₂O
———————————
DABCO, MeCN
80 °C

12

13
up to 90% yield
up to 94% ee

Scheme 3. Synthesis of quaternary oxindoles via asymmetric intramolecular reductive Heck reaction

Recently, in an elegant extension of our work on asymmetric alkylation of 3-halooxindoles,[24] Bisai and co-workers disclosed the use of a Cu-**1a** catalyst to effect malonate addition onto 3-hydroxy 3-indolyl-2-oxindoles **14** (Scheme 4).[21] It is thought that the copper catalyst facilitates a stereoablative elimination of water, followed by asymmetric addition of the malonate.

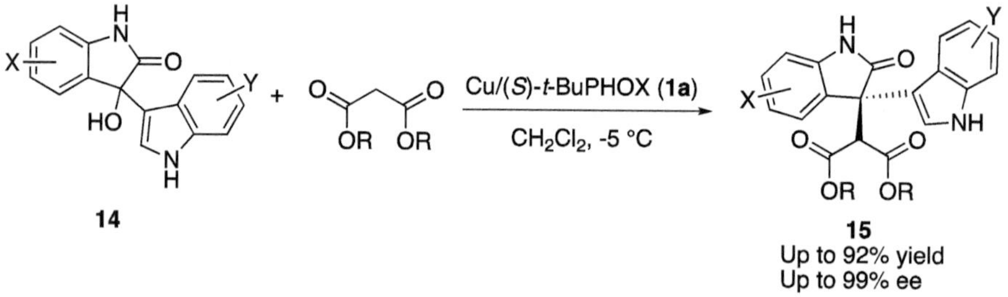

Cu/(*S*)-*t*-BuPHOX (**1a**)
————————————
CH₂Cl₂, -5 °C

14

15
Up to 92% yield
Up to 99% ee

Scheme 4. Synthesis of quaternary dimeric oxindoles by malonate addition onto 3-hydroxy-2-oxindoles

Lam discovered that a Nickel-PHOX catalyst comprised of **1a** or (*S*)-PhPHOX effectively catalyzed the coupling of alkynyl malonate esters **16** with arylboronic esters in a desymmetrizing arylative cyclization (Scheme 5).[23] The resulting cyclopentenone products **17** contain a synthetically challenging fully substituted olefin and a chiral quaternary center.

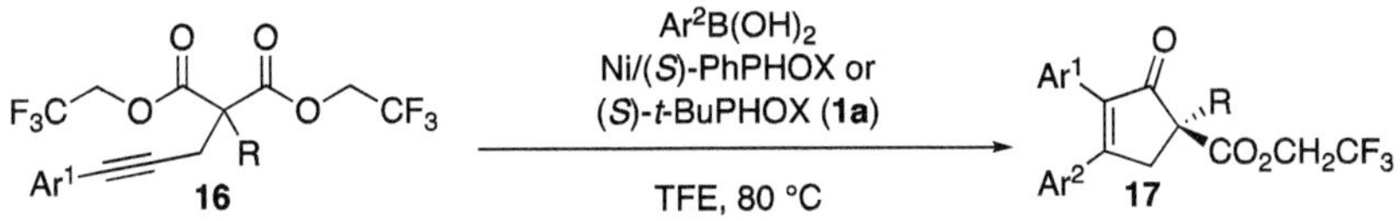

Scheme 5. Synthesis of chiral cyclopentenones via nickel-catalyzed desymmetrization of alkynyl malonate esters with arylboronic acids

With regard to allylic alkylation-type reactions, **1a** and its electron deficient counterparts **5-7** have found extensive applications: After our initial reports on asymmetric decarboxylative protonation,[25,26] the Guiry group has continued to expand the scope and understanding of this reaction. In one example in 2017, they utilized (S)-$(CF_3)_3$-t-BuPHOX (**5**) to enable the enantioselective synthesis of tertiary α-arylated indanones **19** (Scheme 6).[27]

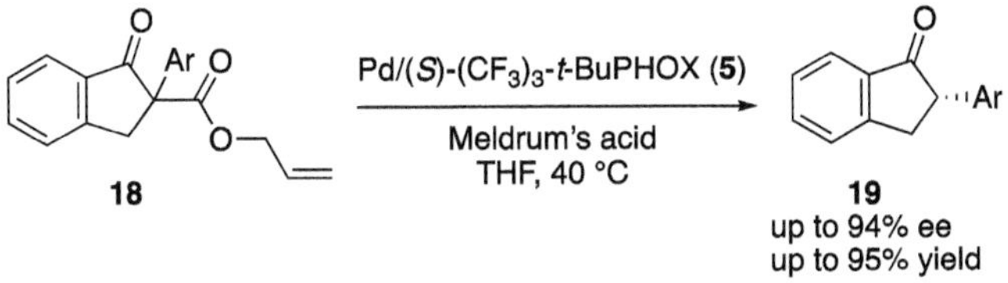

Scheme 6. Synthesis of chiral tertiary α-aryl indanones using decarboxylative protonation

In 2017, Malcolmson introduced a method to synthesize chiral allylic amines **21** through the intermolecular addition of aliphatic amines to acyclic 1,3-dienes **20** (Scheme 7).[15] Here, the electron deficient PHOX ligand **6** was critical in achieving high regioselectivity for the desired 1,2-hydroamination product.

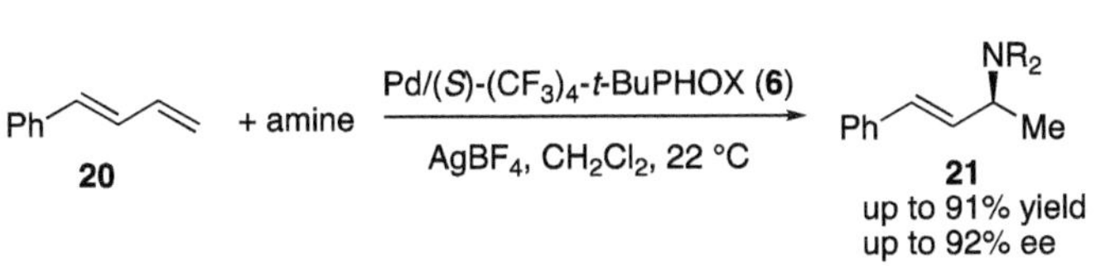

Scheme 7. Synthesis of allylic amines via diene hydroamination

Recently, You employed a Pd-**7** catalyst in a dearomative formal [3+2] cycloaddition of nitroindoles **22** and epoxybutenes **23** to synthesize chiral quaternary tetrahydrofuroindoles **24** (Scheme 8).[14] Notably, the diastereoselectivity of this reaction was affected by the choice of solvent, with toluene and acetonitrile providing differing diastereomers.

Scheme 8. Synthesis of tetrahydrofuroindoles via dearomative formal [3+2] cycloaddition of nitroindoles and epoxybutenes

Along with the Trost laboratory and others, our group has pioneered the development of transition metal-catalyzed decarboxylative asymmetric allylic alkylation reactions to generate quaternary stereocenter-containing compounds. Following our initial efforts on cycloalkanone systems using (S)-t-BuPHOX,[28,29] we have extended the reaction to a wide range of cyclic substrates important in pharmaceuticals and natural product synthesis (Table 1).[30–35] This versatile methodology now enables the synthesis of chiral disubstituted carbocycles of ring size four to eight, "Mannich" adducts, and heterocycles. Especially in the case of lactam systems **29-36**, the electron deficient ligand **5** was hypothesized to generate a more reactive palladium catalyst and was required to achieve high levels of asymmetry.

Table 1. Chiral α-disubstituted carbocycles and lactams accessible by decarboxylative allylic alkylation

25[a]
reference 26, 27

26
reference 28

27
reference 29

28[a]
reference 30

29
reference 31

30
reference 31

31
reference 31

32
reference 31

33
reference 33

34
reference 32

35
reference 32

36
reference 31

[a] (S)-t-BuPHOX (**1a**) instead of (S)-(CF₃)₃-t-BuPHOX (**5**)

Furthermore, in an effort to reduce catalyst loadings to facilitate industrial scale applications, we developed a low-catalyst loading method employing Pd(OAc)₂ instead of the usual zero-valent palladium source (Scheme 9).[36] With this protocol, catalyst loadings as low as 0.075 mol %, corresponding with turnover numbers (TON) of up to 1320, could be used while still providing yields and ee's up to 99%.

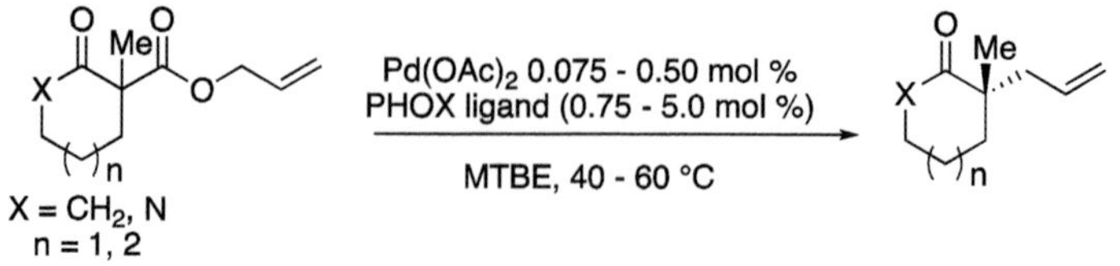

Scheme 9. Low-palladium loading protocol for decarboxylative allylic alkylation

Most recently in 2018, we extended the scope of decarboxylative allylic alkylation toward challenging acyclic substrates by reporting the first asymmetric decarboxylative alkylation of fully substituted acyclic enol carbonates to provide linear α-quaternary ketones **38**, again employing electronic deficient PHOX ligand **5** (Scheme 10).[37,38] Of particular interest, the same enantiomer of product was obtained with comparably high *ee* regardless of the E/Z ratio of the starting material, suggesting a possible dynamic kinetic enolate equilibration during the reaction.

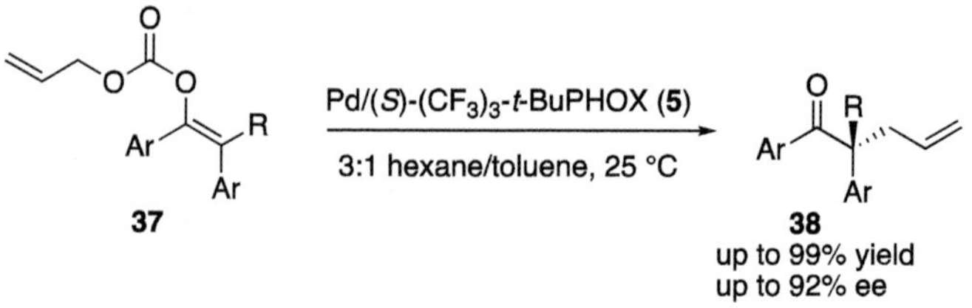

Scheme 10. Synthesis of acyclic α-quaternary ketones using decarboxylative alkylation

PHOX ligands in Natural Product Synthesis

As a testament to the versatility of allylic alkylation-type reactions, recent total syntheses that utilize PHOX ligands do so heavily in the context of transition-metal catalyzed allylic alkylations and protonations. For instance, in Takemoto's 2014 synthesis of the tricyclic alkaloid (−)-aurantioclavine (**39**), **1a** was used in an intramolecular allylic amination of allyl carbonate **37** to generate an unusual seven-membered azepane **38** (Scheme 11).[39]

Scheme 11. Total synthesis of (–)-aurantioclavine via allylic amination

Similarly, in 2012 and 2014, we used a series of powerful allylic alkylations to intercept chiral *gem*-disubstituted cyclic intermediates **41**, **44**, and **47** en route to the formal syntheses of seven natural products including (–)-thujopsene (**42**), (–)-quinic acid (**45**), and (+)-rhazinilam (**48**) (Scheme 12).[33,40]

Scheme 12. Formal syntheses of diverse natural products via allylic alkylation

Decarboxylative protonation has also found utility in natural product synthesis. In route to a 2014 catalytic enantioselective formal synthesis of the hexacyclic caged indole alkaloid (–)-aspidofractinine (**51**), Shao used **1a** in an enantioselective decarboxylative protonation reaction to forge C3-monosubstitutions on medicinally important carbazolone heterocycles **50** (Scheme 13).[41] Later in 2017, Robinson and co-workers also used **1a** in a decarboxylative protonation protocol to make α-substituted ketone **53**, which was then divergently advanced toward the spirocyclic marine alkaloids (–)-fasicularin (**54**) and (–)-lepadiformine (**55**).[42]

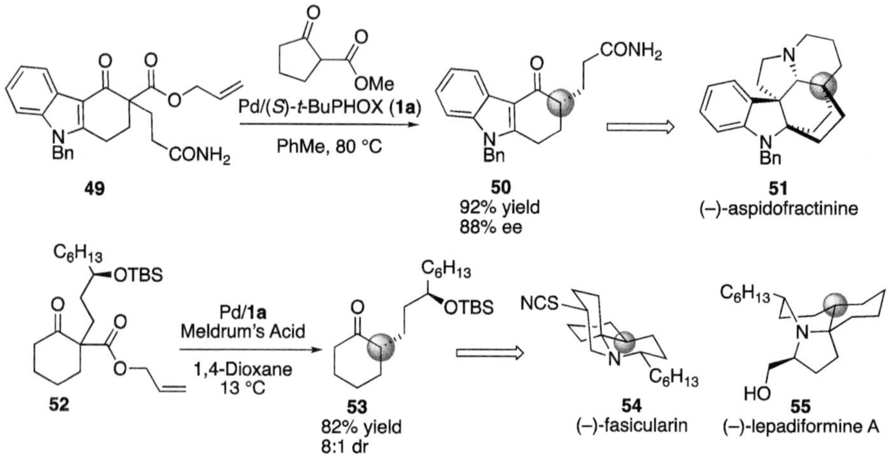

Scheme 13. Total syntheses of (–)-aspidofractinine, (–)-lepadiformine, and (–)-fasicularin using decarboxylative protonation

Most of the remaining recent examples of PHOX ligands **1a** and **5-7** in total synthesis have been found in the context of decarboxylative allylic alkylation toward the synthesis of indole alkaloid natural products (Scheme 14).[43] For example, vinylogous ether **56**, prepared from a (*S*)-*t*-BuPHOX (**1a**)-catalyzed allylic alkylation, can be advanced to (–)-aspidospermidine (**57**) in a formal total synthesis.[40] Similarly, the familiar *gem*-disubstituted lactam **47** serves as an intermediate in the formal syntheses of (+)-quebrachamine (**60**) and (–)-vincadifformine (**61**).[40] In 2014, Mukai synthesized the pentacyclic indole (+)-kopsihainanine A (**59**) utilizing (*S*)-(CF₃)₃-*t*-BuPHOX (**5**) in a decarboxylative alkylation to generate lactam intermediate **58**.[44] Notably, (+)-kopsihainanine A was previously independently synthesized by both Shao and Lupton in 2013, who also also utilized decarboxylative alkylation to synthesize a structurally distinct carbazolone intermediate.[45,46] In 2016,

Scheme 14. Decarboxylative allylic alkylation in the total synthesis of indole alkaloid natural products

Zhu accomplished the syntheses of indole alkaloids (–)-rhazinilam (**63**), (–)-leucomidine B (**64**), and (+)-leuconodine F (**65**) via a divergent route from quaternary cyclopentanone **62**, which was synthesized via allylic alkylation using ligand **1a**.[47] Ligand **5** was again utilized to obtain the dihydropyrido[1,2-a]indolone (DHPI) scaffold **66**, which was crucial in our 2016 and 2017 total and formal syntheses of five indole alkaloids, (+)-

limaspermidine (**67**), (+)-kopsihainanine (**59**), (–)-quebrachamine (**68**), (+)-aspidospermidine (**69**), and (-)-goniomitine (**70**).[48,49] Most recently, Qin in 2017 leveraged ligand **1a** to synthesize the quaternary indole **71**, paving the way to a divergent synthesis of five *Kopsia* indole alkaloids: (–)-fruticosine (**72**), (+)-methyl chanofruticosinate (**73**), (–)-isokopsine (**74**), (–)-kopsine (**75**), and (–)-kopsanone (**76**).[50]

Finally, in 2018 Li and co-workers reported an asymmetric synthesis of the pentacyclic anti-tumor alkaloid (–)-cephalotaxine (**79**) using decarboxylative alkylation on allyl enol carbonate precursor **77**.[51] Here, they utilized ligand **5** to attain **78** with high enantioselectivity; in contrast, (*S*)-*t*-BuPHOX (**1a**) provided the desired product in only 80% ee.

Pd/(*S*)-(CF₃)₃-*t*-BuPHOX (**5**)

MTBE, 40 °C

77

78
82% yield
98% ee

79
(–)-cephalotaxine

Scheme 15. Total synthesis of (–)-cephalotaxine

These aforementioned natural product syntheses highlight the impressive synthetic versatility of enantioselective allylic alkylation reactions facilitated by PHOX ligands. Especially in the cases of indole alkaloid syntheses, decarboxylative alkylation was used to synthesize a range of structurally diverse quaternary stereocenter-bearing intermediates, which were advanced toward the desired natural products. Additionally, electron-deficient variants of (*S*)-*t*-BuPHOX (**1a**) such as (*S*)-(CF₃)₃-*t*-BuPHOX (**5**) have demonstrated superior efficacy in many instances, and are likely to find even wider use in the future, especially toward the synthesis of novel all-carbon quaternary stereocenter-bearing scaffolds using decarboxylative asymmetric allylic alkylation methodologies. The PHOX ligand class has shown extensive versatility in various transition-metal catalyzed asymmetric transformations, thus enabling access to novel and biologically important chemical space. Their modular nature ensures the continual development of new electronically and sterically modified versions with even greater catalytic potential.

References

1. Division of Chemistry and Chemical Engineering, California Institute of Technology, Pasadena, California 91125, United States. email: stoltz@caltech.edu. Our research program is supported by the NIH-NIGMS (Grants R01GM080269 to B.M.S., F30GM120836 and T32GM008042 to A.W.S.), and the UCLA-Caltech Medical Scientist Training Program (A.W.S.).
2. Matt, P. von; Pfaltz, A. *Angew. Chem. Int. Ed.* **1993**, *32*, 566–568.
3. Dawson, G. J.; Frost, C. G.; Williams, J. M. J.; Coote, S. J.; *Tetrahedron Lett.* **1993**, *34*, 3149–3150.
4. Sprinz, J.; Helmchen, G. *Tetrahedron Lett.* **1993**, *34*, 1769–1772.
5. Helmchen, G.; Pfaltz, A. *Acc. Chem. Res.* **2000**, *33*, 336–345.
6. Carroll, M. P.; Guiry, P. J. *Chem. Soc. Rev.* **2014**, *43*, 819–833.
7. Zhu, S.-F.; Xie, J.-B.; Zhang, Y.-Z.; Li, S.; Zhou, Q.-L. *J. Am. Chem. Soc.* **2006**, *128*, 12886–12891.
8. Krout, M. R.; Mohr, J. T.; Stoltz, B. M. *Org. Synth.* **2009**, *86*, 181–193.
9. Mohr, J. T.; Krout, M. R.; Stoltz, B. M. *Org. Synth.* **2009**, *86*, 194–211.
10. Kadunce, N. T.; Reisman, S. E. *J. Am. Chem. Soc.* **2015**, *137*, 10480–10483.
11. Chen, W.; Meng, D.; N'Zemba, B.; Morris, W. J. *Org. Lett.* **2018**, *20*, 1265–1268.
12. Cochran, B.; Henderson, D.; Thullen, S.; Rovis, T. *Synlett* **2018**, *29*, 306–309.
13. McDougal, N. T.; Streuff, J.; Mukherjee, H.; Virgil, S. C.; Stoltz, B. M. *Tetrahedron Lett.* **2010**, *51*, 5550–5554.
14. Cheng, Q.; Zhang, F.; Cai, Y.; Guo, Y.-L.; You, S.-L. *Angew. Chem. Int. Ed.* **2018**, *57*, 2134–2138.
15. Adamson, N. J.; Hull, E.; Malcolmson, S. J. *J. Am. Chem. Soc.* **2017**, *139*, 7180–7183.
16. Adamson, N. J.; Wilbur, K. C. E.; Malcolmson, S. J. *J. Am. Chem. Soc.* **2018**, *140*, 2761–2764.
17. Balaraman, K.; Wolf, C. *Angew. Chem. Int. Ed.* **2017**, *56*, 1390–1395.
18. Cao, T.; Deitch, J.; Linton, E. C.; Kozlowski, M. C. *Angew. Chem. Int. Ed.* **2012**, *51*, 2448–2451.
19. Kong, W.; Wang, Q.; Zhu, J. *Angew. Chem. Int. Ed.* **2017**, *56*, 3987–3991.
20. White, D. R.; Hutt, J. T.; Wolfe, J. P. *J. Am. Chem. Soc.* **2015**, *137*, 11246–11249.

21. Babu, K. N.; Kinthada, L. K.; Das, P. P.; Bisai, A. *Chem. Commun.* **2018**, *54*, 7963–7966.
22. Doran, R.; Guiry, P. J. *J. Org. Chem.* **2014**, *79*, 9112–9124.
23. Karad, S. N.; Panchal, H.; Clarke, C.; Lewis, W.; Lam, H. W. *Angew. Chem. Int. Ed.* **2018**, *57*, 9122–9125.
24. Ma, S.; Han, X.; Krishnan, S.; Virgil, S. C.; Stoltz, B. M. *Angew. Chem. Int. Ed.* **2009**, *48*, 8037–8041.
25. Mohr, J. T.; Nishimata, T.; Behenna, D. C.; Stoltz, B. M. *J. Am. Chem. Soc.* **2006**, *128*, 11348–11349.
26. Marinescu, S. C.; Nishimata, T.; Mohr, J. T.; Stoltz, B. M. *Org. Lett.* **2008**, *10*, 1039–1042.
27. Kingston, C.; Guiry, P. J. *J. Org. Chem.* **2017**, *82*, 3806–3819.
28. Behenna, D. C.; Stoltz, B. M. *J. Am. Chem. Soc.* **2004**, *126*, 15044–15045.
29. Mohr, J. T.; Behenna, D. C.; Harned, A. M.; Stoltz, B. M. *Angew. Chem. Int. Ed.* **2005**, *44*, 6924–6927.
30. Reeves, C. M.; Eidamshaus, C.; Kim, J.; Stoltz, B. M. *Angew. Chem. Int. Ed.* **2013**, *52*, 6718–6721.
31. Numajiri, Y.; Pritchett, B. P.; Chiyoda, K.; Stoltz, B. M. *J. Am. Chem. Soc.* **2015**, *137*, 1040–1043.
32. Hong, A. Y.; Krout, M. R.; Jensen, T.; Bennett, N. B.; Harned, A. M.; Stoltz, B. M. *Angew. Chem. Int. Ed.* **2011**, *50*, 2756–2760.
33. Behenna, D. C.; Liu, Y.; Yurino, T.; Kim, J.; White, D. E.; Virgil, S. C.; Stoltz, B. M. *Nat. Chem.* **2012**, *4*, 130–133.
34. Numajiri, Y.; Jiménez-Osés, G.; Wang, B.; Houk, K. N.; Stoltz, B. M. *Org. Lett.* **2015**, *17*, 1082–1085.
35. Korch, K. M.; Eidamshaus, C.; Behenna, D. C.; Nam, S.; Horne, D.; Stoltz, B. M. *Angew. Chem. Int. Ed.* **2015**, *54*, 179–183.
36. Marziale, A. N.; Duquette, D. C.; Craig, R. A.; Kim, K. E.; Liniger, M.; Numajiri, Y.; Stoltz, B. M. *Adv. Synth. Catal.* **2015**, *357*, 2238–2245.
37. Alexy, E. J.; Zhang, H.; Stoltz, B. M. *J. Am. Chem. Soc.* **2018**, *140*, 10109–10112.
38. Starkov, P.; Moore, J. T.; Duquette, D. C.; Stoltz, B. M.; Marek, I. *J. Am. Chem. Soc.* **2017**, *139*, 9615–9620.
39. Suetsugu, S.; Nishiguchi, H.; Tsukano, C.; Takemoto, Y. *Org. Lett.* **2014**, *16*, 996–999.
40. Liu, Y.; Liniger, M.; McFadden, R. M.; Roizen, J. L.; Malette, J.; Reeves, C. M.; Behenna, D. C.; Seto, M.; Kim, J.; Mohr, J. T.; Virgil, S. C.; Stoltz, B. M. *Beilstein J. Org. Chem.* **2014**, *10*, 2501–2512.

DOI: 10.15227/orgsyn.095.0439

41. Zhao, R.; Sun, Z.; Mo, M.; Peng, F.; Shao, Z. *Org. Lett.* **2014**, *16*, 4178–4181.
42. Burnley, J.; Wang, Z. J.; Jackson, W. R.; Robinson, A. J. *J. Org. Chem.* **2017**, *82*, 8497–8505.
43. Pritchett, B. P.; Stoltz, B. M. *Nat. Prod. Rep.* **2018**, *35*, 559–574.
44. Mizutani, M.; Yasuda, S.; Mukai, C. *Chem. Commun.* **2014**, *50*, 5782–5785.
45. Gartshore, C. J.; Lupton, D. W. *Angew. Chem. Int. Ed.* **2013**, *52*, 4113–4116.
46. Li, Z.; Zhang, S.; Wu, S.; Shen, X.; Zou, L.; Wang, F.; Li, X.; Peng, F.; Zhang, H.; Shao, Z. *Angew. Chem. Int. Ed.* **2013**, *52*, 4117–4121.
47. Dagoneau, D.; Xu, Z.; Wang, Q.; Zhu, J. *Angew. Chem. Int. Ed.* **2016**, *55*, 760–763.
48. Pritchett, B. P.; Kikuchi, J.; Numajiri, Y.; Stoltz, B. M. *Angew. Chem. Int. Ed.* **2016**, *55*, 13529–13532.
49. Pritchett, B. P.; Donckele, E. J.; Stoltz, B. M. *Angew. Chem. Int. Ed.* **2017**, *56*, 12624–12627.
50. Leng, L.; Zhou, X.; Liao, Q.; Wang, F.; Song, H.; Zhang, D.; Liu, X.-Y.; Qin, Y. *Angew. Chem. Int. Ed.* **2017**, *56*, 3703–3707.
51. Zhang, Z.-W.; Wang, C.-C.; Xue, H.; Dong, Y.; Yang, J.-H.; Liu, S.; Liu, W.-Q.; Li, W.-D. Z. *Org. Lett.* **2018**, *20*, 1050–1053.

Alexander W. Sun was born in Eugene, OR in 1990 and obtained his B.A. and M.Sc. in Chemistry in 2012 from the University of Pennsylvania, having worked in the laboratories of Amos B. Smith, III and David R. Tyler. He then matriculated into the UCLA-Caltech MD/PhD program, where he is currently a 5th year Ph.D. candidate in the laboratory of Brian M. Stoltz. His research focuses on asymmetric catalysis and medicinal chemistry.

Brian M. Stoltz was born in Philadelphia, PA in 1970 and obtained his B.S. degree from the Indiana University of Pennsylvania in Indiana, PA. After graduate work at Yale University in the labs of John L. Wood and an NIH postdoctoral fellowship at Harvard with E. J. Corey, he took a position at the California Institute of Technology. A member of the Caltech faculty since 2000, he is currently Professor of Chemistry. His research interests lie in the development of new methodology for general applications in synthetic chemistry.

Synthesis of Acyl Derivatives of Cotarnine

Laxmidhar Rout,[1#*] Bibhuti Bhusan Parida,[#] Ganngum Phaomei,[#]
Bertounesque Emmanuel,[€] and Akhila Kumar Sahoo[&]

[#]Department of Chemistry, Berhampur University, Ganjam, Odisha-760007,
India; [€]Institute Curie, Centre de Recherche, Pavillon Trouillet-Rossignol, 4e
étage, 26 rue d'Ulm, 75248 PARIS Cedex 05, France; [&]Dept. of Chemistry,
University of Hydrabad, Prof. C. R. Rao Road, Gachibowli, Hyderabad,
500046, India.

Checked by Lee Mains, Lauren Markham, Austin Medley, and John L. Wood

Procedure (Note 1)

A. *Cotarnine (1).* A 250-mL, round-bottomed, three-necked flask is
equipped with a magnetic stir bar (elliptical, 32 x 16 mm, Teflon-coated).
The flask is dried under vacuum (1.8 mmHg) with a propane torch and then
allowed to cool to 25 °C. The dried flask is then charged with nitric acid
(150 mL, 18% (v/v) in H_2O) and fitted with a reflux condenser complete
with a nitrogen inlet, a thermometer, and the final neck is sealed with a
rubber septum (Figure 1) (Notes 2, and 3). Slow stirring of the solution
(150 rpm) is commenced. The rubber septum is briefly removed, the solid

(S,R)-Noscapine (20 g, 48.4 mmol, 1.0 equiv) is added to the flask in a single portion at 25 °C, after which the rubber septum is replaced (Notes 4 and 5). The mixture is observed to turn from a white to slightly yellow turbid solution. (Figure 1).

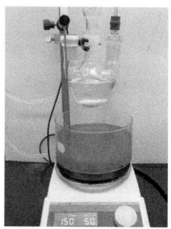

Figure 1. Experimental set-up and addition of (2S,3R)-noscapine

The flask is then lowered into a pre-heated oil bath maintained at 50 °C (Note 6) with the internal temperature monitored throughout the reaction *via* the thermometer. The mixture forms a yellow mixture upon heating and dissolves completely at 38 °C. During the reaction, a series of color changes are observed: the solution changes color from yellow to brown, then brown to red, and finally red to yellow (Note 7). Upon completion of the reaction, the mixture consists of a yellow solution and a white precipitate, which gradually forms during the course of the reaction (Note 8) (Figure 2).

Figure 2. Yellow solution and white precipitate upon completion of the reaction

No more starting material is observed by TLC (Note 9) after 1.5 h, and the reaction flask is removed from oil bath and cooled to 25 °C. A 1-L filter flask is equipped with a sintered glass Büchner funnel (150 mL, 6.7 cm diameter, coarse porosity) and connected to a vacuum source (5 mmHg, 24 °C) for filtration. The mixture is poured through the filtration set-up to remove the precipitate. The collected precipitate is washed with distilled water (3 x 5 mL) and eventually discarded. The Büchner funnel is then removed and the 1-L filter flask containing the yellow filtrate is submerged in an ice-water bath (0 °C). A large magnetic stir bar (polygon, 51 x 8 mm Teflon-coated) is added to the filter flask along with a thermometer to monitor the internal reaction temperature, and continuous stirring is initiated (200 rpm). A 150-mL addition funnel containing an 40% (wt/wt) aqueous solution of KOH (140 mL) (Note 10) is positioned over the filter flask and the KOH solution is added dropwise over 25 min (Figure 3).

Figure 3. KOH addition to the filtrate

An exotherm is observed upon addition of the KOH solution, thus the ice-water bath is employed to maintain the temperature ≤ 20 °C. A precipitate (the desired product) is formed over the course of the reaction. Upon reaching a *p*H = 11, the reaction is deemed complete as no additional precipitation is observed (Note 11). A sintered glass Büchner funnel (150 mL, 6.8 cm diameter, coarse porosity) is fitted to a 1-L filter flask and equipped for vacuum filtration. The reaction mixture is poured through the filtration set-up to collect the crude yellow precipitate. The crude yellow solid is washed with cold (0 °C) distilled water (4 x 5 mL), followed by additional washes with water (4 x 10-mL (60 mL total). The yellow microcrystalline product is dried under high vacuum (1.8 mmHg, 25 °C) for 3 h to afford 9.85 g (87 %) of cotarnine (1) (Figure 4). The obtained product is highly pure (>99%) as established by qNMR and is used directly for Step B (Notes 12 and 13).

Figure 4. Isolated cotarnine (1)

B. *2-(4-Methoxy-6-methyl-5,6,7,8-tetrahydro-[1,3] dioxolo[4,5-g]iso quinol in-5-yl)-1-phenylethan-1-one (3a).* A 50-mL round-bottomed flask is equipped with a magnetic stir bar (elliptical, 22 x 12 mm, Teflon-coated). The flask is dried under vacuum (1.8 mmHg) with a propane torch and then allowed to cool to 25 °C. The dried flask is charged with cotarnine (1) (4.74 g, 20 mmol, 1.0 equiv) and dissolved in distilled methanol (10 mL) under nitrogen. The resultant mixture is stirred (300 rpm) for 5 min to allow for homogeneity resulting in an opaque, orange solution. (Figure 5).

Figure 5. Cotarnine dissolved in MeOH

Acetophenone (2.48 mL, 2.4 g, 20 mmol, 1.0 equiv) is added dropwise *via* syringe to the solution over the course of 2–5 min, and formation of a white precipitate is observed after *ca.* 10 min (Note 14). Stirring is continued for an additional 15 min to ensure completion of the reaction (25 min total) (Figure 6).

Figure 6. Precipitate formation 10 minutes after addition of the acetophenone

The white precipitate is isolated by gravity filtration through a Büchner funnel (115 mL, 6.6 cm diameter) containing filter paper (Whatman 40, 70 mm diameter). The crude solid is washed with cold (0 °C) methanol (10 x 2 mL) and dried under vacuum (25 °C) on the Büchner funnel, then transferred to a 50 mL round-bottomed flask and dried under high vacuum (1.8 mmHg, 25 °C) for 1 h to afford 5.85 g (86%) of **3a** as a white microcrystalline solid (Figure 7). Purity of the product is assessed at >98% by qNMR analysis (Notes 15 and 16).

Figure 7. Isolated product 3a

Notes

1. Prior to performing each reaction, a thorough hazard analysis and risk assessment should be carried out with regard to each chemical substance and experimental operation on the scale planned and in the context of the laboratory where the procedures will be carried out. Guidelines for carrying out risk assessments and for analyzing the hazards associated with chemicals can be found in references such as Chapter 4 of "Prudent Practices in the Laboratory" (The National Academies Press, Washington, D.C., 2011; the full text can be accessed free of charge at https://www.nap.edu/catalog/12654/prudent-practices-in-the-laboratory-handling-and-management-of-chemical. See also "Identifying and Evaluating Hazards in Research Laboratories" (American Chemical Society, 2015) which is available via the associated website "Hazard Assessment in Research Laboratories" at https://www.acs.org/content/acs/en/about/governance/committees

/chemicalsafety/hazard-assessment.html. In the case of this procedure, the risk assessment should include (but not necessarily be limited to) an evaluation of the potential hazards associated with (*(2S, 3R)-Noscapine, concentrated nitric acid, cotarnine, acetophenone, methanol, potassium hydroxide*, acetone, ethyl acetate, 1,4-diiodobenzene and formic acid as well as the proper procedures for *flame drying glassware and vacuum filtration*.

2. (*S,R*)-Noscapine (97%), concentrated nitric acid (69% (wt/wt); KOH pellets (EMPLURA brand by EMD Millipore), acetophenone (99%, ACS reagent grade); 1,4-diiodobenzene (99%), deuterated chloroform (99.8 atom% deuterated); and glass-backed, extra-hard layer TLC plates (60 Å, 250 μm thickness containing F-254 indicator by EMD Millipore) were all purchased from Sigma-Aldrich and were used as received.

3. The 18% (v/v) aqueous nitric acid stock solution is obtained from concentrated nitric acid (69% (wt/wt) in H_2O) by dilution ($N_1V_1=N_2V_2$). To a 250 mL volumetric flask, 170 mL of distilled water is added, followed by 65.2 mL of 69% concentrated nitric acid and an additional 14.8 mL of distilled water to bring the meniscus of the solution to the appropriate line on the volumetric flask.

4. Noscapine is not fully soluble in the reaction medium at room temperature, but upon heating and reaching an internal temperature of 38 °C (*ca.* 5-10 min after immersion in the oil bath) full dissolution occurs.

5. To avoid evaporation of nitric acid solution (which may change the HNO_3 concentration), a condenser with a nitrogen inlet and cooled water circulation is employed until the reaction is complete.

6. The temperature is maintained below 50 °C, and the submitters state that care should be taken to maintain the reaction temperature between 50–55 °C. The submitters reported that the reaction temperature increases to 80 °C upon the formation of the precipitate; however, the checkers did not observe this temperature increase, as the maximum internal temperature reached was 48 °C.

7. These color changes occur over a 10–30 minute time period and a white precipitate (by-product) is observed after 30–35 minutes of heating (Figure 8).

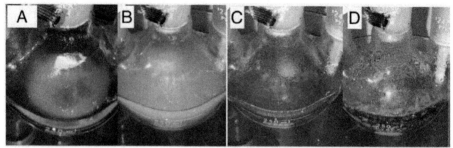

**Figure 8. Color change throughout heating the Noscapine in HNO₃,
A. Before heating, B. After 10 minutes, C. After 15 minutes, D. Color
change and initial precipitate formation after 30 minutes**

8. The precipitate begins to form *ca.* 20 min into the reaction.

9. The yellow color of the solution is observed at the extreme end of the reaction, at which point, no more precipitate is observed to form. The TLC analysis was performed on an acetone solution of the starting material and an aliquot (~2 drops) of the reaction mixture using a solution of EtOAc/MeOH (1:3) containing 1% formic acid as eluent and visualized with ceric ammonium molybdate (CAM) stain. Under these conditions, the product was observed to have an R*f*=0.28 (Figure 9).

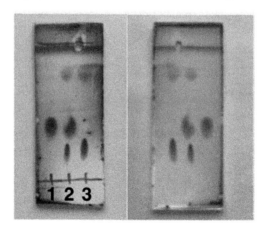

**Figure 9. TLC of the noscapine with nitric acid and stained with CAM
stain (Left: Front of TLC plate, Right: Back of TLC plate); Lane 1: starting
material; Lane 3: reaction mixture; Lane 2: co-spot**

10. Preparation of 40% (wt/wt) KOH solution: In a 200 mL volumetric flask that contains 100 mL of distilled water, 80 g of solid KOH pellets

was added in two 40 g portions. After dissolution, additional distilled water is added to bsring the meniscus of the solution to the appropriate line on the volumetric flask.

11. A precipitate is formed after addition of potassium hydroxide.

Figure 10. Precipitate crashing out of solution at pH 11 upon completion of reaction

12. Yellow crystalline solid. mp 126–128 °C; ^{1}H NMR (600 MHz, CDCl$_3$) δ: 2.33–2.36 (m, 1H), 2.59 (s, 3H), 2.60–2.64 (m, 1H), 2.84–2.91 (m, 1H), 3.06 (td, J =11.5, 4.2 Hz, 1H), 4.04 (s, 3H), 5.41 (s, 1H), 5.88 (s, 2H), 6.31 (s, 1H); ^{13}C NMR (150 MHz, CDCl$_3$) δ: 29.0, 41.1, 43.6, 59.8, 79.2, 100.9, 102.7, 122.5, 129.0, 134.2, 140.6, 149.1; HRMS (ESI) C$_{12}$H$_{14}$NO$_3$ [M-wOH]$^+$: calc. 220.0968, found 220.0967; FTIR (neat, ATR) 3082, 2953, 2870, 1617, 1477, 1445, 1261, 1091, 1036, 1091, 1036, 982, 970, 932, 791 cm $^{-1}$; Rf=0.23, eluent: EtOAc: MeOH (3:7) with 1% formic acid. The purity of the compound was calculated by qNMR with a delay of relaxation of 30 seconds using 11.4 mg of 1,4-diiodobenzene (purity 99%) and 8.2 mg of cotarnine (**1**) and was found to be of 99.4% purity.

13. A reaction performed on half scale provided 5.02 g (88%) of the identical product.

14. In most of the cases, the product precipitated during 10 min of stirring. If a precipitate does not appear in 5–10 min, precipitation can be induced *via* addition of a "seed crystal". This material is prepared in a separate 5-mL vial by reacting 0.5 mmol of cotarnine and 0.5 mmol of acetophenone in 0.5 mL methanol. A crystal is removed from the solution by use of a spatula and placed into the reaction mixture to facilitate crystallization.

15. White crystalline solid. mp 124–125 °C; ^{1}H NMR (600 MHz, CDCl$_3$) δ: 2.37 (s, 3H), 2.41–2.47 (m, 1H), 2.75–2.80 (m, 1H), 2.85–2.92 (m, 1H), 3.10–3.27 (m, 3H), 3.94 (s, 3H), 4.47 (dd, J = 8.6, 3.9 Hz, 1H), 5.86 (s, 2H), 6.32 (s, 1H), 7.46 (t, J = 7.5 Hz, 2H), 7.54 (t, J = 6.9 Hz, 1H), 8.01 (d, J = 8 Hz, 2H); ^{13}C NMR (150 MHz, CDCl$_3$) δ: 23.9, 42.0, 43.7, 44.6, 55.2, 59.3, 100.6, 102.7, 122.4, 128.2, 128.4, 128.6, 132.7, 134.0, 137.4, 140.3, 147.9, 199.0; HRMS (ESI) C$_{20}$H$_{22}$NO$_4$, [M+H]$^+$: calc. 340.1549, found 340.1546. FTIR (neat, ATR) 2915, 2838, 2794, 1678, 1619, 1597, 1498, 1318, 1283, 1257, 1062, 1041, 766, 696 cm^{-1}; Rf=0.84, eluent: EtOAc/MeOH (3:1). The purity of the compound was calculated by qNMR with a delay of relaxation of 30 seconds using 12.5 mg of 1,4-diiodobenzene (purity 99%) and 10 mg of product **3a** and was found to be of 98.1% purity.

16. A reaction performed on half scale provided 2.78 g (82%) of the identical product.

Working with Hazardous Chemicals

The procedures in *Organic Syntheses* are intended for use only by persons with proper training in experimental organic chemistry. All hazardous materials should be handled using the standard procedures for work with chemicals described in references such as "Prudent Practices in the Laboratory" (The National Academies Press, Washington, D.C., 2011; the full text can be accessed free of charge at http://www.nap.edu/catalog.php?record_id=12654). All chemical waste should be disposed of in accordance with local regulations. For general guidelines for the management of chemical waste, see Chapter 8 of Prudent Practices.

In some articles in *Organic Syntheses*, chemical-specific hazards are highlighted in red "Caution Notes" within a procedure. It is important to recognize that the absence of a caution note does not imply that no significant hazards are associated with the chemicals involved in that procedure. Prior to performing a reaction, a thorough risk assessment should be carried out that includes a review of the potential hazards associated with each chemical and experimental operation on the scale that is planned for the procedure. Guidelines for carrying out a risk assessment and for analyzing the hazards associated with chemicals can be found in Chapter 4 of Prudent Practices.

The procedures described in *Organic Syntheses* are provided as published and are conducted at one's own risk. *Organic Syntheses, Inc.*, its Editors, and its Board of Directors do not warrant or guarantee the safety of individuals using these procedures and hereby disclaim any liability for any injuries or damages claimed to have resulted from or related in any way to the procedures herein.

Discussion

The tetrahydroisoquinoline family of alkaloids exhibit a broad range of biological properties such as anti-tumor and anti-microbial activities.[2] Within this family, the anti-tussive drug *(S,R)*-noscapine and its analogues display anti-tumor activity by impairing tubulin polymerization without severe side effects.[2,3] In addition, noscapine causes mitotic arrest of tumor cells, induces apoptosis of tumor cells *in vivo*, and is in phase I/phase II clinical trials for multiple myeloma.[4]

Cotarnine is the central core of noscapine and is documented to have hemostatic activity.[5] Further, cotarnine is the key component of tritoqualin (inhibostamin®) which is used as an anti-allergic drug,[6,7] and has been shown to have a preventive effect on liver injury in rats induced by treatment with CCl_4.[8]

It is significant that the synthesis of 2-methyl-1-(2-oxo-aryl)-1,2,3,4-tetrahydroisoquinolines and their biological applications have been scarcely documented in literature.[9] Hence, derivatization of cotarnine could pave the way to simplified noscapine analogues as potential anticancer agents.[8] Importantly, quaternized tetrahydroisoquinolines were synthesized from cotarnine iminium methyl sulfate (prepared in multiple steps from 3,4,5-trihydroxybenzoic acid), and their biological activities were investigated (Table 1).[10] Thus, various biological activities have been found such as filamin a-binding anti-inflammatory analgesic,[10a] inhibition of tau phosphorylation,[10b] and inhibition of growth of cancer cells.[10c]

Green chemistry is being used in fine chemical and pharmaceutical industries to reduce waste, reduce costs and develop environmentally benign processes.[11] Chemical synthesis from laboratory to industrial level differs with respect to reaction scale, reproducibility of methodology, product purity, atom economy, cost and *E*-factor of the process.[12,13] Herein we disclose the reaction of cotarnine and aryl ketones in green solvents to synthesize gram quantities of 1,2,3,4-tetrahydroisoquinolines in excellent yields. The reaction is simple and requires no metal catalyst. The reaction takes place under base-free and metal-free conditions at room temperature in short reaction times. The products were isolated by simple filtration without work-up. The methodology was developed on the basis of structural similarity of cotarnine derivatives with noscapinoids, which represent an emerging class of microtubule-modulating anticancer agents. [14]

Table 1. Synthesis of some representative examples of cotarnine on gram scale as described in procedure B (Previous work)[14]

2a	R=Phenyl,	80% (3a)
2b	R= 4-Chloro-Phenyl	85% (3b)
2c	R= 3-Nitro-Phenyl	75% (3c)
2d	R= 2-Amino-Phenyl	90% (3d)
2e	R= 2-Hydroxy-5-Bromo-Phenyl	88% (3e)
2f	R= 2-Acetyl Thiofuran	80% (3f)
2g	R= 3-Acetyl Pyridine	90% (3g)
2h	R= 2-Acetyl Furan	80% (3h)
2i	R= 2-Acetyl napthalene	90% (3i)

Conditions: The conditions are consistent with those described in Step B of this *Organic Syntheses* procedure with similar scale. The yields are based on the isolated material obtained after gravity filtration.[14]

References

1. PG Department of Chemistry, Berhampur University, Ganjam, Odisha-760007, India. Institute Curie, Centre de Recherche, Pavillon Trouillet-Rossignol, 4e étage 26 rue d'Ulm, F-75248 PARIS Cedex 05, FRANCE, University of Hydrabad. *E mail: routlaxmi@gmail.com. ldr.chem@buodisha.edu.in.* This work was supported by SERB/EMR/2016/006898) DST, India; Science & Technology department Govt. of Odisha/ 27562800512107/20/1919. Planning and Coordination Department, Government of Odisha, India (no. L.N.

/716/P/2016), University Grants Commission (UGC) start-up grant [F-4-5(58)/2014(BSR/FRP)].

2. (a) Bentley, K. W. *Nat. Prod. Rep.* **2006**, *23,* 444–463. **(b)** Scott, J. D.; Williams, R. M. *Chem. Rev.* **2002**, *102*, 1669–1730.

3. (a) Bennani, Y. L.; Gu, W.; Canales, A.; Díaz, F. J.; Eustace, B. K.; Hoover, R. R.; Barbero, J. J.; Nezami, A.; Wang, T. *J. Med. Chem.* **2012**, *55*, 1920–1925. (b) Manchukonda, N. K.; Naik, P. K.; Santoshi, S.; Lopus, M.; Joseph, S.; Sridhar, B.; Kantevari, S. *PLOS ONE*, **2013**, *8*, e77970.

4. Ye, K.; Ke, Y.; Keshava, N.; Shanks, J.; Kapp, J. A.; Tekmal, R. R.; Petros, J.; Joshi, H. C. *Proc. Natl. Acad. Sci. USA, Cell Biology* **1998**, *95*, 1601–1606.

5. Anderson, J. T.; Ting, A. E.; Boozer, S.; Brunden, K. R.; Crumrine, C.; Danzig, J.; Dent, T.; Faga, L.; Harrington, J. J.; Hodnick, W. F.; Murphy, S. M.; Pawlowski, G.; Perry, R.; Raber, A.; Rundlett, S. E.; Krongrad A. S.; Wang, J., Bennani, Y. L. *J. Med. Chem.* **2005**, *48*, 7096–7098.

6. Sonneville, A. *Allerg. Immunol.* **1988,** *20*, 365–368.

7. Hahn, F.; Teschendrof, H. J.; Kretzsmar, R.; Gossou, U.; Glanzmann, C.; Filipowski, P.; Somorjai, K. *Arzeneim-Forsch*, **1970**, *20*, 1490.

8. (a) Courme, C.; Gillon, S.; Gresh, N.; Vidal, M.; Garbay, C.; Florent, J.-C.; Bertounesque, E. *Eur. J. Med. Chem.* **2010**, *45*, 244–255. (b) Bergonzini, G.; Schindler, C. S.; Wallentin, C.-J.; Jacobsen, E. N.; Stephenson, C. R. J. *Chem. Sci.* **2014**, *5*, 112–116.

9. (a) Liebermann, C.; Kropf, F. *Chem. Ber.* **1904**, *37*, 211–216. (b) Hope, E.; Robinson, R. *J. Chem. Soc., Trans.* **1913,** *103*, 361–377. (c) Sud, A.; Sureshkumar, D.; Klussmann, M. *Chem. Commun.* **2009**, 3169-3171. (d) Yang, Q.; Zhang, L.; Ye, C.; Luo, S.; Wu, L.-Z.; Tung, C-H. *Angew. Chem. Int. Ed. Eng.* **2017**, 56, 3694–3698.

10. (a) Burns, B. L.; Wang, H.-Y.; Lin, N.-H.; Blasko, A.; Filamin a-binding anti-inflammatory analgesic. Patent WO 2010051476 A1,PCT/US2009/062823, 2009. (b) Wang, H.-Y.; Burns, B. L. A method of inhibiting tau phosphorylation. Patent WO 2014011917 A2. Patent PCT/US2013/050126, **2014**.

11. (a) Li, C.-J.; Trost, B. M. *Proc. Nat. Acad. Sci.* **2008**, *105*, 13197–13202. (b) Bryan, M. C.; Dillon, B.; Hamann, L. G.; Hughes, G. J.; Kopach, M. E.; Peterson, E. A.; Pourashraf, M.; Raheem, I.; Richardson, P.; Richter, D.; Sneddon, H. F. *J. Med. Chem.* **2013**, *56*, 6007–6021.

12. (a) Kolb, H. C.; Finn, M. G.; Sharpless, K. B. *Angew. Chem. Int. Ed. Eng.* **2001**, *40*, 2004–2021. (b) Hill, C. L. *Nature* **1999**, *401*, 436–437. (c) Rout, L.; Sen, T. K.; Punniyamurthy, T. *Angew. Chem. Int. Ed. Eng.* **2007**, *46*, 5583–5586.

13. (a) Walter, S. M.; Kniep, F.; Rout, L.; Schmidtchen, F. P.; Herdtweck, E.; Huber, S. M.; *J. Am. Chem. Soc.* **2012**, *134*, 8507–8512; (b) Rout, L.; Harned, A. M.; *Chem. Eur. J.* **2009**, *15*, 12926–12928. (c) Punniyamurthy, T.; Rout, L. *Coord. Chem. Rev.* **2008**, *252*, 134–154.

14. (a) Rout, L.; Parida, B.B.; Choudhury, S. K.; Florent, J.-C.; Johanne, L.; Phaomei, G.; Bertounesque. E.; *Eur JOC* **2017**, *35*, 5275–5292 ; (b) Rout, L.; Parida, B. B.; Florent, J-C.; Johhanne, L.; Choudhury, S, K.; Phaomei, G.; *Chem. Eur. J.* **2016**, *22*, 14812–14815.

Appendix
Chemical Abstracts Nomenclature (Registry Number)

(*S*, *R*)-Noscapine; (128-62-1)
Nitric Acid; (7697-37-2)
Acetophenone; (98-86-2)
Potassium Hydroxide; (1310-58-3)
1,4-Diiodobenzene; (624-38-4)
Cotarnine: 5, 6, 7, 8-Tetrahydro-4-methoxy-6-methyl-1, 3-dioxolo[4, 5-*g*]isoquinolin-5-ol; (82-54-2)
2-(5,6,7,8-Tetrahydro-4-methoxy-6-methyl-[1,3]dioxolo[4,5-g]isoquinolin-5-yl)-1-phenylethanone; (2220-06-6)

Laxmidhar Rout was born in small village Kismat Krushnapur, Basudevpur, Bhadrak in 1976. He graduated in 1997 from A.B. College Basudevpur, Bhadrak. After earning a M. Sc. from Utkal University in 2003 with Prof S. Jena, he completed his Ph. D. with Prof. T. Punniyamurthy in IIT Guwahati in synthetic organic chemist in 2008. He worked as a postdoctoral associate with Prof. A. M. Harned in Minnesota and Prof. C-G. Zhao in University of Texas. He was awarded a Humboldt (AvH) fellowship in Germany in 2011–2013 with Prof. S. M. Huber in Technical University Munich and in the Institute of Curie Paris with Dr. E. Bertounesque. Since 2014, he has been working as an Assistant Professor in Berhampur University Odisha, India. His research interest is organic synthesis and drug discovery.

Bibhuti Bhusan Parida received his M. Sc. degree from Utkal University-Bhubaneswar with advanced organic chemistry as his specialization. He qualified via national level entrance examinations CSIR-JRF and GATE to pursue a Ph. D. in synthetic organic chemistry at Indian Institute of Chemical Technology (CSIR-IICT)-Hyderabad under the supervision of Dr. S. Chandrasekhar (Director, IICT). Subsequently, he has over 7-years of post-Ph.D. research experiences including postdoctoral research at Wayne State University (Detroit, Michigan-USA), Institute Curie (Paris-France), teaching at IIIT-Bhubaneswar. Currently, he is working as an Assistant Professor at Berhampur University, Odisha.

Emmanuel Bertounesque received his Ph.D. in synthetic organic chemistry from the University of Paris VII (ITODYS) in 1987, under the supervision of Professor Jacques Emile Dubois. He did his post-doctoral research in total synthesis of natural products at the University of Pennsylvania with Professor Amos B. Smith, III. He joined ESPCI Paris as a tenured researcher in the Simon Group and subsequently moved to Institut Curie - CNRS UMR 176 as a medicinal chemist in oncology drug discovery, in collaborations with Pierre Fabre Laboratories and Servier. He is currently involved in new therapeutic developments against Shiga toxin-producing Escherichia coli infection in the Chemical Biology of Membranes and Therapeutic Delivery Unit (UMR3666/U1143) of Dr. Ludger Johannes at Institut Curie.

Akhila Kumar Sahoo received his masters from Utkal University Odisha in 1994. He completed Ph.D. with Prof G. Pandey in NCL Pune, after which he started working as a postdoctoral fellow with Prof. Hans-Joachin Gais in RWTH Germany. He worked with Prof. T. Hyiama at Kyoto University on a JSPS Fellowship. After service as a postdoc with Prof. K. Osuka in Kyoto University he joined the University of Hydrabad as an Assistant Professor in 2007. He has been working as a full Professor from 2016 in the field of organic synthesis.

Organic
Syntheses

Dr. Ganngam Phaomei received his M.Sc. degree from Jamia Millia Islamia, New Delhi, India and Ph.D. from Manipur University, Imphal, India with Prof. W.R. Singh. He has research interests in nanotechnology, rare earth luminescence nanomaterials and functionalized nanomaterial. During his Ph. D, research he worked jointly with Dr. R. S. Ningthoujam, Scientist F, BARC, Mumbai, India and as a Research Associate in DST, India research project under Prof. N. Rajmuhon Singh at Manipur University,. He is currently working on nano-drug delivery.

Lee Mains was born in Newcastle, England in 1992. He received his BSc (Hons) in Biopharmaceutical Science from University of Sunderland in 2016. He is currently a doctoral student under the direction of Prof. John L. Wood at Baylor University with research interests in natural product total synthesis.

Lauren Markham was born in Arlington, Texas in 1998. She graduated from Keller High School in Keller, Texas in 2016. She currently attends Baylor University in pursuit of a B.S. in Chemistry to be completed in 2019. As an undergraduate research student, Lauren has worked on synthesizing 8-bromoguanosine analogs under the direction of Dr. Jesse W. Jones, and she currently is a member of Dr. John L. Wood's organic synthesis research group.

Organic
Syntheses

Austin Medley is a native-born Texan, who moved to Canada at a young age. Currently, he works in the research group of John L. Wood as an undergraduate organic synthetic researcher at Baylor University.

Carbonyl-Olefin Metathesis for the Synthesis of Cyclic Olefins

Marc R. Becker, Katie A. Rykaczewski, Jacob R. Ludwig, and Corinna S. Schindler*[1]

Department of Chemistry, University of Michigan, Willard Henry Dow Laboratory, 930 North University Avenue, Ann Arbor, Michigan 48109, USA

Checked by Yang Cao and Kevin Campos

Procedure (Note 1)

A. *2,6-Dimethyl-1-phenyl-2-vinylhept-5-en-1-one (2)*. (Note 2) A flame-dried 500-mL three-necked, round-bottomed flask is fitted with a Teflon-coated magnetic stir bar (egg-shaped, 3.5 cm length x 1.2 cm diameter) and equipped with two rubber septa, one of which is pierced with a needle connected to a nitrogen source. The three-necked flask is charged with activated zinc powder (14.09 g, 215 mmol, 2.5 equiv) (Note 3), lithium chloride (5.11 g, 121 mmol, 1.4 equiv) and anhydrous THF (230 mL). A flame-dried 125-mL pressure-equalizing addition funnel containing a solution of geranyl bromide (21.4 mL, 108 mmol, 1.25 equiv) (Note 4) in

Org. Synth. **2018**, *95*, 472-485
DOI: 10.15227/orgsyn.095.0472

Published on the Web 11/16/2018
© 2018 Organic Syntheses, Inc.

anhydrous THF (20 mL) is attached to the reaction flask and sealed with a rubber septum. The flask is immersed in a 23 °C water bath and the geranyl bromide solution added dropwise over 10 min to the vigorously stirring suspension (Figure 1A). After complete addition, the grey, heterogeneous mixture is stirred for 1.5 h at 23 °C under an atmosphere of nitrogen, at which time stirring is stopped and excess zinc allowed to settle. A flame-dried 500-mL round-bottomed flask is charged with a Teflon-coated magnetic stir bar (egg-shaped, 3.5 cm length x 1.2 cm diameter), benzoyl chloride (**1**) (10.0 mL, 86 mmol, 1.0 equiv.) and anhydrous THF (80 mL) and fitted with a rubber septum with nitrogen inlet and thermometer-probe.

Figure 1. A) Preparation of allyl zinc reagent (left); B) Addition of allyl zinc reagent to benzoyl chloride via cannula (right) (photo provided by submitter)

The benzoyl chloride containing flask is immersed in an ice/water bath and the supernatant zinc reagent solution is added via a cannula at a rate that kept the internal temperature below 5-6 °C (Figure 1B). After addition is complete, the flask is maintained in the ice/water bath and the slightly cloudy solution is stirred for 30 min under an atmosphere of nitrogen until full conversion of starting material occurs, as judged by TLC (Note 5). An aqueous saturated NH$_4$Cl solution (50 mL) and deionized water (50 mL) are added sequentially. After the ice/water bath is removed and the mixture is stirred for 10 min, the biphasic mixture is transferred to a 1-L separatory funnel and the reaction flask rinsed with EtOAc (3 x 10 mL), which is added to the separatory funnel. The aqueous layer is separated, extracted with

EtOAc (3 x 100 mL) and the combined organic layers concentrated under reduced pressure with a rotary evaporator (100 mmHg, bath temperature 40 °C). The oily residue is taken up in EtOAc (200 mL), transferred to a 500-mL separatory funnel and washed with deionized water (2 x 50 mL), an aqueous 1 M NaOH solution (2 x 50 mL) and brine (2 x 50 mL) (Note 6). The organic layer is dried over Na_2SO_4 (75 g), filtered and concentrated under reduced pressure with a rotary evaporator (100 mmHg, bath temperature 40 °C) and then with a vacuum pump (0.5-1 mmHg) (Note 7). The crude residue is transferred into a 100-mL round-bottomed flask (Note 8) and distilled using a short-path distillation apparatus under vacuum (0.5-1 mmHg) (Figure 2) (Note 9). A forerun (2–3 mL) is collected and discarded, and the fraction distilling at 110 °C is collected to afford aryl ketone **2** (5:1 mixture of olefin isomers) as a pale-yellow oil (18.44 g, 89%) (Notes 10 and 11). Purity of the oil was determined to be 90% by quantitative NMR using dimethyl terephthalate as internal standard (Note 12).

Figure 2. Distillation apparatus for the distillation of aryl ketone 2 (photo provided by submitter)

B. *(5-Methyl-5-vinylcyclopent-1-en-1-yl)benzene (3)*. A flame-dried 500-mL round-bottomed flask is fitted with a Teflon-coated magnetic stir bar (egg-shaped, 3.5 cm length x 1.2 cm diameter) and charged with iron(III) chloride (502 mg, 3.1 mmol, 0.05 equiv) (Note 13), which is immediately suspended in anhydrous CH_2Cl_2 (230 mL) (Notes 14 and 15).

Figure 3. Color change before (left) and after (right) addition of aryl ketone 2 to FeCl₃ suspension (photo provided by submitter)

The flask is equipped with a flame-dried 125-mL pressure-equalizing addition funnel containing a solution of aryl ketone **2** (15.00 g, 61.9 mmol, 1.0 equiv) (Note 16) in anhydrous CH₂Cl₂ (20 mL) and the apparatus sealed with a rubber septum with nitrogen inlet. The aryl ketone solution is added to the suspension of FeCl₃ over 2 min and the resulting dark-brown solution stirred at 23 °C for 2.5 h (Figure 3), which resulted in full consumption of starting material as determined by TLC analysis (Note 17). The reaction mixture is passed through a short silica plug (180-200 g silica gel equilibrated in CH₂Cl₂ in a glass column with 8.5 cm diameter) positioned over a 2-L round-bottomed flask, and the plug is rinsed with CH₂Cl₂ (500 mL) applying a brief positive pressure to move the solution through the silica gel (Figure 4).

The eluent is concentrated under reduced pressure with a rotary evaporator (100 mmHg, bath temperature 40 °C) and then with a vacuum pump (0.5-1 mmHg). The residual orange oil is transferred to a 50-mL round-bottomed flask and distilled using a short-path distillation apparatus under vacuum (0.5-1 mmHg) (Note 18). The fraction distilling at 87–91 °C is collected to afford cyclopentene **3** as a colorless oil (10.45 g, 93%) (Notes 19 and 20). Purity of the oil was determined to be 97% by quantitative NMR using dimethyl terephthalate as an internal standard (Note 21).

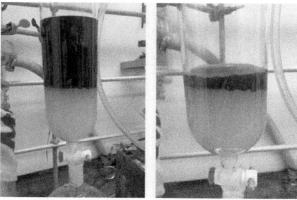

Figure 4. Silica plug before (left) and after (right) filtration and rinsing (photo provided by submitter)

Notes

1. Prior to performing each reaction, a thorough hazard analysis and risk assessment should be carried out with regard to each chemical substance and experimental operation on the scale planned and in the context of the laboratory where the procedures will be carried out. Guidelines for carrying out risk assessments and for analyzing the hazards associated with chemicals can be found in references such as Chapter 4 of "Prudent Practices in the Laboratory" (The National Academies Press, Washington, D.C., 2011; the full text can be accessed free of charge at https://www.nap.edu/catalog/12654/prudent-practices-in-the-laboratory-handling-and-management-of-chemical. See also "Identifying and Evaluating Hazards in Research Laboratories" (American Chemical Society, 2015) which is available via the associated website "Hazard Assessment in Research Laboratories" at https://www.acs.org/content/acs/en/about/governance/committees/chemicalsafety/hazard-assessment.html. In the case of this procedure, the risk assessment should include (but not necessarily be limited to) an evaluation of the potential hazards associated with geranyl bromide, tetrahydrofuran (THF), hexane, ethyl acetate (EtOAc), sodium sulfate (Na$_2$SO$_4$), zinc (Zn), lithium chloride (LiCl), benzoyl chloride, ammonium chloride (NH$_4$Cl), sodium hydroxide (NaOH), sodium

chloride (NaCl), iron(III) chloride (FeCl₃), silica and dichloromethane (CH₂Cl₂), as well as the proper procedures for vacuum distillation.

2. The procedure for the synthesis of aryl ketone **2** was adapted with small modifications from a report by Sämann *et al.*[2]

3. Zinc powder (99.3%) was obtained from Fischer Chemical and activated prior to use. For the reported scale, a 250-mL Erlenmeyer flask, fitted with a Teflon-coated magnetic stir bar (5 cm length x 1 cm diameter), was charged with zinc powder (~20 g) and 1 M HCl (aq., 75 mL). The suspension was stirred vigorously for 10 min, then filtered with a Büchner funnel and subsequently washed with 1 M HCl (aq., 2 x 50 mL), water (1 x 50 mL), ethanol (2 x 50 mL) and diethyl ether (2 x 50 mL) and the activated zinc powder dried under reduced pressure with a nitrogen atmosphere (Figure 5). Lithium chloride (99%, for molecular biology) was obtained from Acros Organics and dried prior to use at 200 °C under reduced pressure with a vacuum pump. Benzoyl chloride (99%) was obtained from Sigma Aldrich and distilled prior to use. Tetrahydrofuran was obtained from Fisher (Optima) and dried by being passed through a column of activated alumina under argon (using a JC-Meyer Solvent Systems).

Figure 5. Drying activated zinc

4. Geranyl bromide (96%) was obtained from Alfa Aesar and distilled prior to use, or purchased from Aldrich (94.6wt% by QNMR) and used as received. The submitters observed that distilled geranyl bromide (colorless oil) turned orange within a few days indicating decomposition. Alternatively, the submitters found that geranyl bromide can be prepared from geraniol following a procedure by Baer et al.[3], which, when performed on a 30 g scale, afforded geranyl bromide in 83-86% yield and 97wt% purity after vacuum distillation.

5. The reaction progress was followed by TLC analysis on silica gel with 98:2 hexane/EtOAc as eluent and visualization with UV (254 nm) and ceric ammonium molybdate (CAM) stain. The starting material benzoyl chloride **1** has $R_f = 0.59$ (no CAM activity) and the ketone product **2** has $R_f = 0.41$ (blue with CAM).

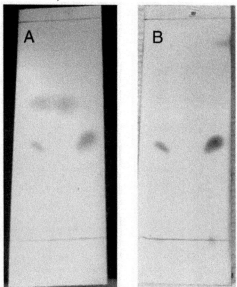

Figure 6. TLC analysis at the end of the reaction and visualized with A) UV light, and B) ceric ammonium molybdate solution

6. A basic wash was necessary to remove trace amounts of benzoic acid to avoid precipitation of evaporated benzoic acid in the distillation apparatus during distillation. A basic wash without prior removal of THF proved to be ineffective.

7. An aliquot (29.8 mg) of the crude material was taken and the purity assessed as 93wt% by quantitative ^{1}H NMR using dimethyl terephthalate as internal standard.

8. The flask used for the distillation needs to be free from any Lewis acid impurities to avoid undesired side reactions at the high temperatures employed during the distillation. Therefore, a wash with strong acid (e.g. concentrated hydrochloric acid) is suggested.

9. It is suggested to apply vacuum carefully while stirring vigorously to prevent bumping. When the desired product started to distill, the distillation head was mantled in aluminum foil to keep the head temperature constant.

10. The checkers reported that product **2** was obtained as a 5:1 mixture of inseparable olefin isomers. The checkers reported that use of a lower distillation temperature (110 °C versus 125 °C for the submitters) led to a reduced appearance of the minor isomer. Compound **2** is characterized as follows: ^{1}H NMR (500 MHz, CDCl$_3$) δ : 1.40 (s, 3H), 1.47 (s, 3H), 1.65 (s, 3H), 1.78 (d, J = 8.4 Hz, 2H), 1.96 (d, J = 8.2 Hz, 2H), 5.05 (s, 1H), 5.18–5.32 (m, 2H), 6.19 (dd, J = 17.6, 10.7 Hz, 1H), 7.39 (t, J = 7.7 Hz, 2H), 7.47 (t, J = 7.4 Hz, 1H), 7.86 (d, J = 7.8 Hz, 2H); ^{13}C NMR (125 MHz, CDCl$_3$) δ : 17.4, 22.9, 23.0, 25.6, 39.0, 53.6, 114.7, 124.0, 127.9, 129.0, 131.5, 131.9, 137.8, 143.2, 204.7. IR (film): 2968, 2914, 1677, 1630, 1445, 1411, 1375, 1222, 1175, 1001, 964, 916, 828, 792, 717, 693, 664 cm^{-1}. HRMS: calcd for C$_{17}$H$_{23}$O$^+$ [M + H]$^+$: 243.1743; Found: 243.1756.

11. A second run at identical scale provided 18.27 g (90wt%, 79% yield) of the same colorless oil.

12. The purity was assessed as 90wt% by quantitative ^{1}H NMR using dimethyl terephthalate as internal standard (the combined integral of the signals at 5.02 and 4.65 ppm was used in reference to the signal at 3.94 ppm of the internal standard). It was observed that the isomer level increases significantly during high temperature vacuum distillation.

13. FeCl$_3$ (98%) was obtained from Strem Chemicals and used as received.

14. Anhydrous CH$_2$Cl$_2$ was obtained from Fisher (Stabilized/Certified ACS) and dried by being passed through a column of activated alumina under argon (using a JC-Meyer Solvent Systems).

15. FeCl$_3$ was weighed out onto a metal spatula (1.5 cm width) and washed into the round-bottomed flask using CH$_2$Cl$_2$.

16. The isomeric ratio of aryl ketone **2** does not have an impact on the outcome of the carbonyl-olefin metathesis reaction due to isomerization, which occurs under the metathesis conditions; therefore, the isomeric

purity is not considered in the calculation of the starting material or in the reaction's yield.

17. The reaction progress was followed by TLC analysis on silica gel with 19:1 hexane/EtOAc as eluent and visualization with UV (254 nm). The starting material aryl ketone **2** has $R_f = 0.38$ and the product cyclopentene **3** has $R_f = 0.62$. Note: A side product with a similar R_f value as the starting material was observed by TLC during the course of the reaction. To determine full consumption of starting material, a 0.1 mL aliquot was taken from the reaction mixture, passed through a pipette silica plug (~3 cm length) and eluted with 10 mL CH_2Cl_2. The eluent was concentrated in vacuo and analyzed by ^1H NMR.

18. The same distillation setup as illustrated in Figure 2 was used. The submitters obtained slightly higher yields, when the flask used to collect the product was immersed in a water/ice bath.

19. The product **3** was characterized as follows: ^1H NMR (500 MHz, CDCl$_3$) δ : 1.33 (s, 3H), 1.87–1.93 (m, 1H), 2.00–2.05 (m, 1H), 2.36–2.49 (m, 2H), 5.05–5.12 (m, 2H), 6.05 (d, J = 2.6 Hz, 1H), 6.09 (dd, J = 17.5, 10.6 Hz, 1H), 7.22 (dd, J = 8.3, 6.3 Hz, 1H), 7.25-7.32 (m, 2H), 7.43 (d, J = 7.2 Hz, 2H); ^{13}C NMR (125 MHz, CDCl$_3$) δ : 23.4, 29.6, 41.9, 52.2, 111.3, 126.7, 127.1, 127.9, 128.8, 137.0, 146.0, 148.9; IR (film): 3081, 3053, 2929, 2844, 1634, 1598, 1491, 1444, 1370, 1297, 1131, 1097, 1075, 1033, 1000, 959, 908, 828, 758, 695, 670 cm^{-1}. HRMS: m/z calcd. for $C_{14}H_{17}^+$ [M+H]: 185.1325; Found: 185.1331.

20. A second reaction on identical scale provided 10.46 g (92%) with 96wt% purity.

21. The purity was assessed as 97wt% by quantitative ^1H NMR using dimethyl terephthalate as internal standard (the average integral of the signals at 7.42, 7.27 and 7.22 ppm was used in reference to the signal at 8.12 ppm of the internal standard).

Working with Hazardous Chemicals

The procedures in *Organic Syntheses* are intended for use only by persons with proper training in experimental organic chemistry. All hazardous materials should be handled using the standard procedures for work with chemicals described in references such as "Prudent Practices in the Laboratory" (The National Academies Press, Washington, D.C., 2011;

the full text can be accessed free of charge at
http://www.nap.edu/catalog.php?record_id=12654). All chemical waste
should be disposed of in accordance with local regulations. For general
guidelines for the management of chemical waste, see Chapter 8 of Prudent
Practices.

In some articles in *Organic Syntheses*, chemical-specific hazards are
highlighted in red "Caution Notes" within a procedure. It is important to
recognize that the absence of a caution note does not imply that no
significant hazards are associated with the chemicals involved in that
procedure. Prior to performing a reaction, a thorough risk assessment
should be carried out that includes a review of the potential hazards
associated with each chemical and experimental operation on the scale that
is planned for the procedure. Guidelines for carrying out a risk assessment
and for analyzing the hazards associated with chemicals can be found in
Chapter 4 of Prudent Practices.

The procedures described in *Organic Syntheses* are provided as
published and are conducted at one's own risk. *Organic Syntheses, Inc.*, its
Editors, and its Board of Directors do not warrant or guarantee the safety of
individuals using these procedures and hereby disclaim any liability for any
injuries or damages claimed to have resulted from or related in any way to
the procedures herein.

Discussion

The carbonyl-olefin metathesis reaction is characterized by the exchange
of double-bonded atoms in a carbonyl and an olefin to form a new carbonyl
and a new olefin. The most common protocol for this reaction involves
using precious metal alkylidenes as reagents that provide ring-closing
carbonyl olefin metathesis products with a corresponding amount of a
catalytically inactive metal-oxo by-product.[4,5] Our lab recently developed a
carbonyl-olefin metathesis reaction relying on $FeCl_3$ as an environmentally
benign Lewis acid catalyst that allows for the synthesis of cyclic olefins
under mild conditions.[6-8] The reported method is operationally facile and
employs catalyst loadings as low as 5 mol%. Notably, the reaction occurs at
ambient temperature, utilizes a cheap, abundant metal salt as a catalyst, and
produces an easily removable, organic compound as the sole by-product.
Table 1 demonstrates a small selection of the broad substrate scope for the

carbonyl-olefin metathesis reaction. Generally, aromatic ketones with electron-donating or withdrawing groups are converted to the corresponding cyclopentene products in good to excellent yields. Furthermore, the reaction protocol allows for efficient access to cyclohexenes and structurally complex motifs such as tricycles and spirocycles.

Table 1. Scope of the iron(III)-catalyzed carbonyl-olefin metathesis reaction

*Conditions: ketone (1 equiv.), FeCl₃ (5 mol%), dichloroethane (0.1-0.01M), rt, 1-24 h.

In this work, we report a carbonyl-olefin metathesis protocol that provides an operationally simple and easily scalable synthesis of cyclic olefin **3**. We demonstrate that the required starting material for the carbonyl-olefin metathesis can be prepared from readily accessible and cheap reagents allowing for rapid access to large quantities of compound. The carbonyl-olefin metathesis as a key transformation was carried out on 15 g-scale using $FeCl_3$ as a cheap and environmentally benign Lewis acid catalyst. Notably, the reported protocol gives the desired product in excellent yield and purity. This manuscript demonstrates the potential of carbonyl-olefin metathesis as an economical and sustainable approach for the synthesis of cyclic olefins.

References

1. Department of Chemistry, University of Michigan, Willard Henry Dow Laboratory, 930 North University Avenue, Ann Arbor, Michigan 48109, USA. E-mail: corinnas@umich.edu. This publication is based on work supported by the NIH/National Institute of General Medical Sciences (R01-GM118644), the David and Lucile Packard Foundation and Alfred P. Sloan Foundation (fellowships to C.S.S.). J.R.L. thanks the National Science Foundation for a predoctoral fellowship.
2. Sämann, C.; Knochel, P. *Synthesis* **2013**, *45*, 1870–1876.
3. Baer, P.; Rabe, P.; Citron, C. A.; de Oliveira Mann, C. C.; Kaufmann, N.; Groll, M.; Dickschat, J. S. *ChemBioChem* **2014**, *15*, 213–216.
4. Fu, G.C.; Grubbs, R.H. *J. Am. Chem. Soc.* **1993**, *115*, 3800–3801.
5. Hong, B.; Li, H.; Wu, J.; Zhang, J.; Lei, X. *Angew. Chem. Int. Ed.* **2015**, *54*, 1011–1015.
6. Ludwig, J.R.; Zimmerman, P.M.; Gianino, J.B.; Schindler, C.S. *Nature* **2016**, *533*, 374–379.
7. McAtee, C.C.; Riehl, P.S; Schindler, C.S. *J. Am. Chem. Soc.* **2017**, *139*, 2960–2963.
8. Ludwig, J.R.; Phan, S.; McAtee, C.C.; Zimmerman, P.M.; Devery III, J.J.; Schindler, C.S. *J. Am. Chem. Soc.* **2017**, *139*, 10832–10842.

Appendix
Chemical Abstracts Nomenclature (Registry Number)

Geranyl bromide: 2,6-Octadiene, 1-bromo-3,7-dimethyl-, (2*E*); (6138-90-5)
Zinc; (7440-66-6)
Lithium chloride; (7447-41-8)
Benzoyl chloride; (98-88-4)
Iron(III) chloride: Iron chloride; (7705-08-0)

Marc R. Becker was born in Germany and received his B.S. and M.S. degree in chemistry at the University of Muenster, Germany. In 2016, he started his graduate studies at the University of Michigan, where he is currently pursuing his Ph.D. under the supervision of Prof. Corinna S. Schindler. His research interests are method development and their application in natural product synthesis.

Katie A. Rykaczewski was born in 1996 in Alaska. She is currently enrolled in Seattle University's undergraduate program working towards a B.S. in Biochemistry. She joined the Langenhan group in April 2016 and studies the synthesis of *N*-linked glycoproteins. She was awarded an NSF REU fellowship to work at the University of Michigan under the direction of Prof. Corinna S. Schindler during the summer of 2017.

Jacob R. Ludwig received his B.Sc. in chemistry in 2014 from Michigan State University, where he performed research in the laboratory of Jetze Tepe. After graduation, he joined the Schindler lab at the University of Michigan to pursue his Ph.D. degree.

Corinna S. Schindler received her diploma in chemistry from the Technical University of Munich. After a research stay with K.C. Nicolaou at the Scripps Research Institute, she joined the group of Erick Carreira at ETH Zürich for her graduate studies. She then returned to the US to conduct postdoctoral studies with Eric Jacobsen at Harvard before starting her independent career at the University of Michigan in 2013.

Yang Cao received his Ph.D. in medicinal chemistry from University of Toledo. He joined the SPRI in 2007, and is currently working in the Project Chemistry Department of the Process Research, MRL, Rahway, Merck & Co., Inc.

Hexafluoro-2-propanol-promoted Intramolecular Friedel-Crafts Acylation

Rakesh H. Vekariya, Matthew C. Horton, and Jeffrey Aubé[*1]

Division of Chemical Biology and Medicinal Chemistry, UNC Eshelman School of Pharmacy, 125 Mason Farm Road, CB 7363, University of North Carolina at Chapel Hill, Chapel Hill, NC, 27599-7363, USA

Checked and modified by Feng Peng and Kevin Campos

Procedure (Note 1)

A. *4-(3,4-Dimethoxyphenyl)butanoyl chloride* (**2**). An oven-dried, 250-mL, three-necked, 14/20 round-bottomed flask is equipped with an egg-shaped magnetic stir bar (2 cm), a pressure-equalizing addition funnel (25 mL), a rubber septum, and an argon inlet (Figure 1). The flask is charged with 4-(3,4-dimethoxyphenyl)butanoic acid (**1**) (7.25 g, 32.3 mmol, 1.0 equiv) (Note 2), anhydrous DCM (40 mL) (Note 3) and DMF (50. μL, 0.65 mmol, 0.02 equiv) (Note 4). Oxalyl chloride (5.5 mL, 64 mmol, 2.0 equiv) (Note 5) is added via syringe to the addition funnel, and the stopcock is opened such that the oxalyl chloride is added over 4 min, resulting in effervescence. The reaction mixture is stirred at 23 °C for 30 min from the start of the addition of oxalyl chloride (Note 6). The argon inlet is replaced with a rubber septum, the addition funnel is removed, and the flask placed on a rotary evaporator. Concentration under reduced pressure (35 °C, 30 mm Hg) afforded 8.35 g of crude acid chloride (**2**) as a yellow oil (Note 7 and 8), which is used in the next step without purification.

Figure 1. Left to right: reaction assembly for Step A, reaction appearance for Step A (photos provided by submitters)

B. *6,7-Dimethoxy-3,4-dihydronaphthalen-1(2H)-one (3)*.   An oven-dried, 100-mL, three-necked, 14/20 round-bottomed flask is equipped with an egg-shaped magnetic stir bar (2 cm), a water condenser, a thermocouple, and an argon inlet. Hexafluoroisopropanol (HFIP) (17 mL, 162 mmol, 5.0 equiv) (Note 9) is added. To this three-necked flask is charged a solution of acid chloride (**2**) in 3 mL of dichloroethane (Note 10) via plastic syringe at a rate that maintains the internal temperature below 35 °C. The reaction is allowed to cool and stirred at 23 °C for 2 h (Note 11). The stir bar is removed and rinsed with HFIP (1.0 mL) and the mixture is concentrated using a rotary evaporator (45 °C, 30 mmHg) to afford a dark brown oil (Figure 2), which is further dried for 15 min at 0.5 mmHg. The crude product is dissolved in DCM (50 mL) (Note 12), transferred to a 250-mL

Figure 2. Reaction appearance following Step B

separatory funnel, and washed with saturated, aqueous sodium bicarbonate (2 × 50 mL) and brine (50 mL). The combined aqueous layers are extracted with DCM (75 mL) and the combined organic layers dried over 31 g of Na$_2$SO$_4$ (20 min) and gravity-filtered through a 185 mm Whatman qualitative circle into a 500 mL, single-necked, 24/40 round-bottomed flask. The flask and the Na$_2$SO$_4$ are washed with additional DCM (3 × 20 mL). Celite (13.6 g) (Note 13) is added to the flask and the mixture is concentrated (40 °C, 160 mmHg). In a 150-mL, coarse-fritted Büchner funnel with a 24/40 lower vacuum assembly with an attached 250-mL round-bottomed flask, 28 g of sand is layered over the frit followed by a slurry of 25 g of silica gel (Note 14) and hexanes (50 mL). Additional hexanes (50 mL) is used to rinse leftover silica gel into the funnel. After allowing the slurry to settle (2 min), the dry-loaded product is added. At this stage, the 250-mL flask is switched with a 500-mL flask and 150 mL of hexanes followed by 100 mL of 9:1 hexanes:EtOAc (Note 15) are added and pulled through with house vacuum (Note 16) such that the solvent level does not fall below the top of the Celite. The 500-mL flask is switched with a 1-L round-bottomed flask and 200 mL of 4:1 hexanes:EtOAc followed by 500 mL of 7:3 hexanes:EtOAc are run through the silica in the same fashion as before.

The filtrate from the 500-mL flask is discarded and the filtrate from the 1-L round bottom flask is concentrated by rotary evaporator (40 °C, 30 mmHg). The resulting white solid is scraped out of this flask and transferred to a 250-mL round-bottomed flask. The 1-L flask is rinsed with EtOAc (3 × 30 mL) into the 250-mL flask and the solution concentrated (40 °C, 80 mmHg) resulting in 6.4 g of pink solid (Note 17). The flask is allowed to stand overnight (17 h), at which time isopropanol (22 mL) (Note 18) is added to the flask. The mixture is heated and swirled every 25 sec (Note 19) to dissolve the solid. Once the solvent vapor condensate reaches the opening of the flask (2 min), the flask is removed from heat and covered with a laboratory wipe secured with a rubber band. After cooling to 23 °C and sitting overnight (24 h), the supernatant is decanted, and the crystals washed with –20 °C isopropanol (6 mL) (Note 20) and dried overnight under vacuum (0.5 mm Hg) to afford the product as light pink crystals (5.23 g, 78%) (Notes 21, 22, and 23) (Figure 3).

Figure 3. Left to right: Filtration setup after filtration, appearance of product before recrystallization, and appearance of product after recrystallization (photos provided by Submitters)

Notes

1. Prior to performing each reaction, a thorough hazard analysis and risk assessment should be carried out with regard to each chemical substance and experimental operation on the scale planned and in the

context of the laboratory where the procedures will be carried out. Guidelines for carrying out risk assessments and for analyzing the hazards associated with chemicals can be found in references such as Chapter 4 of "Prudent Practices in the Laboratory" (The National Academies Press, Washington, D.C., 2011; the full text can be accessed free of charge at https://www.nap.edu/catalog/12654/prudent-practices-in-the-laboratory-handling-and-management-of-chemical). See also "Identifying and Evaluating Hazards in Research Laboratories" (American Chemical Society, 2015) which is available via the associated website "Hazard Assessment in Research Laboratories" at https://www.acs.org/content/acs/en/about/governance/committees /chemicalsafety/hazard-assessment.html. In the case of this procedure, the risk assessment should include (but not necessarily be limited to) an evaluation of the potential hazards associated with 4-(3,4-dimethoxyphenyl)butanoic acid, dichloromethane, oxalyl chloride, dimethylformamide, HFIP, celite, silica gel, hexanes, EtOAc, and isopropanol.

2. 4-(3,4-Dimethoxyphenyl)butanoic acid (**1**) (99%) was purchased from Sigma–Aldrich and used as received.

3. Anhydrous DCM was purchased from Alfa Aesar (99.7+ %, packaged under argon in resealable ChemSeal bottles, stabilized with amylene) and used as received.

4. *N,N*-Dimethylformamide (99.8%) was purchased from Sigma–Aldrich and used as received.

5. Oxalyl chloride (98%) was purchased from Beantown Chemical and is used as received. CAUTION: Oxalyl chloride is irritating, toxic, and prone to release gas when used in a chemical reaction.

6. TLC analysis is performed with silica gel plates (10 × 20 cm, glass backed, purchased from Miles Scientific) with EtOAc–hexanes (1:1) and visualized using a 254 nm UV lamp. The acid chloride (**2**) is converted to the corresponding methyl ester for analysis purposes by dissolving a small aliquot in methanol prior to TLC. 4-(3,4-Dimethoxyphenyl)butanoic acid (**1**) R_f = 0.29, methyl 4-(3,4-dimethoxyphenyl)butanoate (from acid chloride) R_f = 0.57. In the latter case, the acid chloride hydrolyzes on the TLC plate to yield acid.

7. A second reaction on identical scale provided 8.67 g of the same product. The excess mass over the theoretical yield is attributed to leftover DCM not removed by rotary evaporation.

8. ^{1}H NMR (500 MHz, CDCl$_3$) δ: 2.04 (m, 2H), 2.66 (t, J = 7.5 Hz, 2H), 2.91 (t, J = 7.2 Hz, 2H), 3.89 (s, 3H), 3.90 (s, 3H), 6.70 (d, J = 2.0 Hz, 1H), 6.73 (br d, J = 7.9 Hz, 1H), 6.83 (d, J = 8.0 Hz, 1H); ^{13}C NMR (125 MHz, CDCl$_3$) δ: 26.7, 33.9, 46.2, 55.8, 55.9, 111.4, 111.7, 120.3, 132.9, 147.6, 149.1, 173.7.

9. Hexafluoroisopropanol was purchased from Oakwood Products, Inc. (>99%) and used as received (bp = 59 °C).

10. Dichloroethane was purchased from Sigma–Aldrich (99%) and used as received.

11. TLC analysis was performed with silica gel plates (10 × 20 cm, glass backed, purchased from Miles Scientific) with EtOAc–hexanes (1:1) and visualized with a 254 nm UV lamp. Ketone (**3**) R_f = 0.43.

12. DCM was purchased as a 19-L drum from BDH and used as received.

13. Celite (545) was purchased from Sigma–Aldrich and used as received.

14. Silica gel was purchased from SiliCycle (P60, 230–400 mesh) and used as received.

15. Hexanes and EtOAc were purchased in 19-L drums from BDH and used as received.

16. The vacuum measured 260 mmHg; care was taken to ensure solvent level remained above the Celite.

17. The solid is initially white after concentration, but during the time it takes to transfer the solid and measure its mass, the solid turns pink. The melting point of this solid was 98–100 °C.

18. Isopropanol was purchased in a 19-L drum from BDH and used as received.

19. Appropriate PPE was worn (insulated glove) when handling hot glassware. The solution was heated either by a heat gun or by an aluminum block set to 160 °C. Care was taken to avoid spilling isopropanol.

20. Isopropanol was placed in a –20 °C freezer in a covered Erlenmeyer flask for at least 1 h to chill before being used in this step.

21. A second run performed on the same scale yielded 5.15 g.

22. Physical properties and spectroscopic analysis of **3**: mp 99–101 °C (lit.[2] mp 98–100 °C). IR (powder) 2029, 1660, 1597, 1505, 1255, 1220 cm^{-1}; ^{1}H NMR (500 MHz, CDCl$_3$) δ: 2.14 (m, 2H), 2.62 (t, J = 6.5 Hz, 2H), 2.91 (t, J = 6.1 Hz, 2H), 3.93 (s, 3H), 3.95 (s, 3H), 6.69 (s, 1H), 7.54 (s, 1H); ^{13}C NMR (125 MHz, CDCl$_3$) δ: 23.5, 29.3, 38.4, 55.8, 55.9, 108.4, 110.1, 125.7, 139.2, 147.8, 153.4, 197.1; HRMS (ESI) m/z calcd. for C$_{12}$H$_{15}$O$_3$ [M+H]$^+$ 207.1021, found 207.1017.

23. Purity was measured at 99% by quantitative NMR using or trimethoxybenzene as the standard. The compound is bench stable in open air.

Working with Hazardous Chemicals

The procedures in *Organic Syntheses* are intended for use only by persons with proper training in experimental organic chemistry. All hazardous materials should be handled using the standard procedures for work with chemicals described in references such as "Prudent Practices in the Laboratory" (The National Academies Press, Washington, D.C., 2011; the full text can be accessed free of charge at http://www.nap.edu/catalog.php?record_id=12654). All chemical waste should be disposed of in accordance with local regulations. For general guidelines for the management of chemical waste, see Chapter 8 of Prudent Practices.

In some articles in *Organic Syntheses*, chemical-specific hazards are highlighted in red "Caution Notes" within a procedure. It is important to recognize that the absence of a caution note does not imply that no significant hazards are associated with the chemicals involved in that procedure. Prior to performing a reaction, a thorough risk assessment should be carried out that includes a review of the potential hazards associated with each chemical and experimental operation on the scale that is planned for the procedure. Guidelines for carrying out a risk assessment and for analyzing the hazards associated with chemicals can be found in Chapter 4 of Prudent Practices.

The procedures described in *Organic Syntheses* are provided as published and are conducted at one's own risk. *Organic Syntheses, Inc.*, its Editors, and its Board of Directors do not warrant or guarantee the safety of individuals using these procedures and hereby disclaim any liability for any injuries or damages claimed to have resulted from or related in any way to the procedures herein.

Discussion

The Friedel–Crafts acylation is a storied and often-used procedure for preparing aromatic ketones; accordingly, a very extensive bibliography and review literature is available for this reaction.[3] Classically, the reaction is promoted by acids such as AlCl$_3$, FeCl$_3$, SnCl$_4$ or H$_2$SO$_4$, generally requiring a stoichiometric amount of catalyst for full conversion due to complex formation between ketone products and Lewis acid catalysts, resulting in product inhibition. In such cases, the reactions entail an aqueous workup and produce metal-containing acidic waste streams. More recent methods use sub-stoichiometric, including heterogenous, catalysts.[3f,4] An important early example was Kobayashi's conditions of Hf(OTf)$_4$ in LiClO$_4$–nitromethane;[5] indeed, numerous examples employ ionic liquids and other unconventional media.[6] Another general approach is to modify the substrate, with a great deal of effort devoted to the study of highly electrophilic acylating agents as reaction partners.[7]

Here, we provide a detailed procedure for an intramolecular Friedel–Crafts acylation reaction that is notable for its simplicity: the substrate is merely dissolved in HFIP solvent at room temperature or below.[8] Following reaction, the workup consists of an aqueous wash to remove residual acid followed by removal of solvent under reduced pressure and purification of the product by appropriate means. This differs from the traditional reaction insofar as no aqueous metal waste streams are generated. Other workers have also reported the use of HFIP under similar conditions for Friedel–Crafts alkylation reactions.[9]

The mechanism of this variant of the Friedel–Crafts reaction is not known, but preliminary experiments have ruled out the possible intermediacy of an HFIP ester derived from the acyl chloride.[8] HFIP chemistry is often dominated by its strong hydrogen bonding potential[10] and any reasonable mechanism likely involves HFIP H-bonding to the acyl chloride. This could lead to formation of an acylium ion in a process reminiscent of the textbook mechanism for AlCl$_3$-promoted Friedel–Crafts reaction but direct addition of the arene is not out of the question. It has not been possible to distinguish between these possibilities, but it is worth mentioning that we have not been able to identify any reaction intermediates using *in situ* infrared spectroscopy.

Electron-rich arenes and heteroarenes worked well under these conditions (Table 1). In general, six- and seven-membered cyclic ketones were obtained in good yields but five-membered cyclic ketones proved more challenging (entries 18-20). Substrates without electron-donating groups on them resulted in lower yields (entries 8 and 12), and these reaction conditions do not succeed on electron-poor substrates. In cases where multiples isomers are possible, only one is formed (entries 1, 11, 12, and 17). Despite these constraints, this variation of the Friedel–Crafts reaction provides easy and efficient access to a good range of attractive carbo- and heterocyclic ketones, and for many of these substrates, will be a method of choice.

Table 1. Substrate scope

entry	acid chloride	product	yield (%)
1			97^a 78^b
2			86
3			73
4			81
5			99
6			90

Table 1. (cont.)

entry	acid chloride	product	yield (%)
7			70
8			77
9			72
10			77
11			93
12			56
13			61
14			72
15			68

Table 1. (cont.)

entry	acid chloride	product	yield (%)
16			72
17			81
18			14
19			99
20			12

[a]Reaction was performed on 0.30 mmol scale with a full column and no recrystallization. [b]Yield from this work.

References

1. Rakesh H. Vekariya, Department of Medicinal Chemistry, University of Kansas, Lawrence, KS, 66045, USA; Matthew C. Horton, Department of Chemistry, University of North Carolina at Chapel Hill, Chapel Hill, NC, 27599-7363, USA; Jeffrey Aubé, Division of Chemical Biology and Medicinal Chemistry, UNC Eshelman School of Pharmacy, 125 Mason Farm Road, CB 7363, University of North Carolina at Chapel Hill, Chapel Hill, NC, 27599-7363, USA. Email: jaube@unc.edu. JA and MH thank the UNC Eshelman School of Pharmacy for financial support, Alexander Li for early experimental assistance, and the checkers for a useful modification of Step B.

2. Hashem, M. M.; Berlin, K. D.; Chesnut, R. W.; Durham, N. N. *J. Med. Chem.* **1976**, *19*, 229–239.

3. (a) Friedel, A.; Crafts, J.; Ador, E. *Ber. Dtsch. Chem. Ges.* **1877**, *10*, 1854–1858; (b) Friedel, C.; Crafts, J. M. *Compt. Rend.* **1877**, *84*, 1450–1454; (c) Gore, P. H. *Chem. Rev.* **1955**, *55*, 229–281; (d) Olah, G. A. *Friedel-crafts chemistry*; 1st ed.; Wiley-Interscience: New York, 1973; (e) Heaney, H. In *Comprehensive Organic Synthesis*; Trost, B. M., Fleming, I, Ed.; Pergamon: Oxford, 1991; Vol. 2, p 753–768; (f) Sartori, G.; Maggi, R. *Advances in Friedel-Crafts Acylation Reactions: Catalytic and Green Processes*; CRC Press: Boca Raton, 2010.

4. (a) Pearson, D. E.; Buehler, C. A. *Synthesis* **1972**, *1972*, 533–542; (b) Sartori, G.; Maggi, R. *Chem. Rev.* **2011**, *111*, PR181–PR214.

5. Hachiya, I.; Moriwaki, M.; Kobayashi, S. *Tetrahedron Lett.* **1995**, *36*, 409-412.

6. (a) Earle, M. J.; Hakala, U.; Hardacre, C.; Karkkainen, J.; McAuley, B. J.; Rooney, D. W.; Seddon, K. R.; Thompson, J. M.; Wahala, K. *Chem. Commun.* **2005**, 903–905; (b) Tran, P. H.; Do, N. B. L.; Le, T. N. *Tetrahedron Lett.* **2014**, *55*, 205–208; (c) Dupont, J.; de Souza, R. F.; Suarez, P. A. Z. *Chem. Rev.* **2002**, *102*, 3667–3692.

7. (a) Hwang, J. P.; Surya Prakash, G. K.; Olah, G. A. *Tetrahedron* **2000**, *56*, 7199–7203; (b) Fillion, E.; Fishlock, D.; Wilsily, A.; Goll, J. M. *J. Org. Chem.* **2005**, *70*, 1316–1327; (c) Yin, W.; Ma, Y.; Xu, J.; Zhao, Y. *J. Org. Chem.* **2006**, *71*, 4312–4315; (d) Kangani, C. O.; Day, B. W. *Org. Lett.* **2008**, *10*, 2645–2648; (e) Nishimoto, Y.; Babu, S. A.; Yasuda, M.; Baba, A. *J. Org. Chem.* **2008**, *73*, 9465–9468; (f) Andrews, B.; Bullock, K.; Condon, S.; Corona, J.; Davis, R.; Grimes, J.; Hazelwood, A.; Tabet, E. *Synth. Commun.* **2009**, *39*, 2664–2673; (g) Xu, Y.; McLaughlin, M.; Chen, C.-y.; Reamer, R. A.; Dormer, P. G.; Davies, I. W. *J. Org. Chem.* **2009**, *74*, 5100v5103; (h) Wilkinson, M. C. *Org. Lett.* **2011**, *13*, 2232-2235; (i) Liu, Y.; Meng, G.; Liu, R.; Szostak, M. *Chem. Commun.* **2016**, *52*, 6841–6844.

8. Motiwala, H. F.; Vekariya, R. H.; Aubé, J. *Org. Lett.* **2015**, *17*, 5484–5487.

9. (a) Champagne, P. A.; Benhassine, Y.; Desroches, J.; Paquin, J.-F. *Angew. Chem., Int. Ed.* **2014**, *126*, 14055–14059; (b) Cativiela, C.; García, J. I.; Mayoral, J. A.; Salvatella, L. *Can. J. Chem.* **1994**, *72*, 308–311; (c) Trillo, P.; Baeza, A.; Nájera, C. *J. Org. Chem.* **2012**, *77*, 7344–7354; (d) Li, G.-X.; Qu, J. *Chem. Commun.* **2010**, *46*, 2653–2655; (e) Weisner, N.; Khaledi, M. G. *Green Chem.* **2016**, *18*, 681–685; (f) Arai, T.; Yokoyama, N. *Angew. Chem., Int. Ed.* **2008**, *47*, 4989–4992; (g) Willot, M.; Chen, J.; Zhu, J. *Synlett* **2009**,

Organic
Syntheses

577–580; (h) Li, C.; Guo, F.; Xu, K.; Zhang, S.; Hu, Y.; Zha, Z.; Wang, Z. *Org. Lett.* **2014**, *16*, 3192–3195.

10. Khaksar, S. *J. Fluorine Chem.* **2015**, *172*, 51–61.

Appendix
Chemical Abstracts Nomenclature (Registry Number)

4-(3,4-Dimethoxyphenyl)butanoic acid; (13575-74-1)
4-(3,4-Dimethoxyphenyl)butanoyl chloride; (348143-75-9)
6,7-Dimethoxy-3,4-dihydronaphthalen-1(2*H*)-one; (13575-75-2)
Oxalyl chloride, (79-37-8)
Hexafluoroisopropanol: 1,1,1,3,3,3-Hexafluoro-2-propanol; (920-66-1)

Rakesh Vekariya was born in Gujarat, India. He received a B.S. degree in Pharmacy in 2008 from JSS College of Pharmacy, India. He received M.S. degree in medicinal chemistry at Virginia Commonwealth University working in the research group of Professor Richard A. Glennon. He received his Ph.D. from the University of Kansas, working on synthetic method development and medicinal chemistry projects in the laboratory of Professor Jeffrey Aubé.

Matthew C. Horton grew up in Sugar Land, TX and received a B.S. degree in Chemistry from Louisiana State University in Baton Rouge, LA. He joined the lab of Professor Jeffrey Aubé at UNC in 2015 and is now a doctoral student. He has been working on the total synthesis of a virulence factor.

Organic
Syntheses

Jeffrey Aubé attended the University of Miami, where he did undergraduate research with Professor Robert Gawley and earned a B.S. degree in 1980. He received his Ph.D. in chemistry in 1984 from Duke University, working with Professor Steven Baldwin, and was an NIH postdoctoral fellow at Yale University with Professor Samuel Danishefsky. In 1986, he moved to the University of Kansas, where he worked until his retirement from that institution in 2015. He is currently an Eshelman Distinguished Professor at the University of North Carolina at Chapel Hill.

Feng Peng joined the Process Research Department of Merck & Co., Inc. in 2012. His research focuses on using state-of-art organic chemistry to address critical problems in drug development. He received his B. S. degree from Beijing Normal University. He obtained his M.S. under the supervision of Professor Dennis Hall at University of Alberta with a research focus on Boron Chemistry. Feng then moved to New York City, where he obtained Ph.D. in the area of total synthesis (maoecrystal V) with Professor Samuel Danishefsky at Columbia University.

Discussion Addendum for:
Preparation of the COP Catalysts: [(S)-COP-OAc]₂, [(S)-COP-Cl]₂, and (S)-COP-hfacac

Jeffrey S. Cannon[#] and Larry E. Overman[1*&]

[#]Department of Chemistry, Occidental College, 1600 Campus Rd. M-5, Los Angeles, California, 90041-3314, United States; [&]Department of Chemistry, University of California, 1102 Natural Sciences II, Irvine, California 92697-2025, United States

Anderson, C. E.; Kirsch, S. F.; Overman, L. E.; Richards, C. J.; Watson, M. P. Org. Synth. 2007, 84, 148–155.

Alkenes harboring allylic heteroatoms are a ubiquitous functional arrangement in organic molecules.[2,3,4] Their enantioselective synthesis under mild conditions is of broad utility for the preparation of many biologically active and complex organic structures. One opportunity for such enantioselective syntheses is the use of nucleophilic heteroatom species in allylic substitution reactions of prochiral alkenes containing allylic leaving groups. Palladium(II) catalysts are ideal reagents for these transformations because of their chemoselectivity as π-acids,[5] with catalyst complexes having planar chirality proven to be particularly effective for achieving antarafacial nucleopalladation of C–C double bonds.[6] Our group has utilized a family of chiral cobalt oxazoline palladacycle (COP) catalysts (Figure 1), originally discovered by Richards,[6a] for the activation of alkenes for enantioselective nucleopalladation/deoxypalladation rearrangement and allylic substitution reactions.[7,8] In 2007, we reported scalable syntheses of the three COP complexes depicted in Figure 1,[9] as they had proven to be excellent catalysts for the transformation of prochiral allylic trichloroacetimidates to either branched allylic trichloroacetamides or branched allylic esters in good yields and enantioselectivities (Scheme 1).[10] Although some attempts have been made to replicate the effectiveness of the COP catalysts with less elaborate ligand scaffolds, the COP complexes remain the most selective catalysts for these transformations.[6c,11] High enantioselectivities, essentially complete selectivity for formation of the branched allylic product, and mild reaction conditions that tolerate most functional groups—notably those that are base labile—are typical of COP-catalyzed allylic substitution reactions.[10,12] These features provide the COP family of catalysts with complementary reactivity to many catalytic enantioselective allylic functionalizations that proceed via π-allyl intermediates.[13]

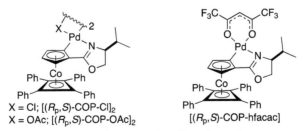

Figure 1. The COP family of catalysts

At the time of our report of the syntheses of the COP complexes, only the enantioselective trichloroacetimidate rearrangement and allylic substitution with carboxylic acid nucleophiles had been described (Scheme 1). In the following years, a number of new COP-catalyzed transformations and synthetic applications of the allylic products have been reported. These advances will be the focus of this Addendum.[14]

Scheme 1. Applications of COP catalysts in allylic trichloroacetimidate rearrangement and allylic ester synthesis

(I) Rearrangements

Additional transformations involving formal [3,3]-sigmatropic rearrangements of 1,3-dihetero-1,5-hexadienes catalyzed by COP catalysts have been reported. It has been found that the complex [COP-Cl]₂, having bridging chloride ligands, is generally the optimal catalyst for these applications.

The enantioselective synthesis of allylic thiocarbamates by the palladium-catalyzed rearrangement of carbamothioates has been described by our and the Clayden research groups (Scheme 2).[15,16] Typical of this family of palladium-catalyzed rearrangements, the substrate scope was fairly broad and included unprotected alcohols and secondary carbamates. Clayden extended this chemistry by utilizing a subsequent lithiation/Smiles-type rearrangement to enable the synthesis of fully-substituted sulfur stereocenters. This tactic overcame the inability of the COP complexes to catalyze rearrangements of allylic precursors having two substituents at the site of C–S bond formation. The stereoselectivity of this two-step sequence was generally high, with the configuration of the newly formed stereocenter being determined by the enantioselectivity of the COP-catalyzed rearrangement.[15]

Scheme 2. Enantioselective rearrangements of allylic carbamothioates

The enantioselective introduction of nitrogen-containing stereocenters by COP-catalyzed allylic rearrangements has been extended to allylic alcohol precursors other than trichloroacetimidates. Batey recognized the opportunity to utilize 2-allyloxypyridines as imidate surrogates in palladium-catalyzed rearrangements using COP complexes (Scheme 3).[17] These transformations provide *N*-allylpyridones in good yields and enantioselectivities using [COP-Cl]₂. It was found that generation of a cationic complex by removal of the chloride ligands with silver trifluoroacetate significantly enhanced the rate of the rearrangement without reduction in enantioselectivity. Batey also demonstrated that other allyloxy-substituted nitrogen heterocycles, including quinolines and benzothiazoles, could be transformed in a similar fashion to *N*-allylated products in good yields and enantioselectivities. In addition, Batey reported the COP-catalyzed rearrangement of iminodiazaphospholidines to provide chiral allylic phosphoramides, which upon reaction with 1 M HCl gave the corresponding allylic tosylamide.[18] In the presence of silver trifluoroacetate and [COP-Cl]₂, this rearrangement proceeded in moderate to good yields and modest to high enantioselectivities. As a method for enantioselective synthesis of chiral amines, the COP-catalyzed allylic trichloroacetimidate rearrangement has a broader scope and provides products readily converted to the corresponding primary amine.

Scheme 3. Enantioselective rearrangements to generate C-N bonds:
allyloxypyridines and iminodiazapholidines

The typically broad scope, mild conditions, and high enantioselectivity of COP-catalyzed allylic trichloroacetimidate rearrangements has led to this reaction being employed as a key step in the enantioselective construction of various nitrogen-containing molecules (Scheme 4). A common tactic is to strategically combine this reaction with ruthenium-catalyzed alkene cross-metathesis or ring-forming metathesis. For example, Han followed the COP-catalyzed enantioselective rearrangement of an allylic alcohol with an aminomercuration/demercuration reaction to construct 2,6-disubstituted piperidines.[19] Cross metathesis was then employed to append a 2-octanone side chain to yield, after alkene hydrogenation, the CNS-active piperidine alkaloid (+)-iso-6-cassine. Combining COP-catalyzed allylic trichloroacetimidate rearrangements with ring-closing metathesis (RCM) to construct nitrogen heterocycles was reported by Aldrich and Richards.[20,21] After palladium-catalyzed rearrangement, both strategies appended an allyl or homoallyl group to the resulting nitrogen. Ruthenium-catalyzed RCM of the resulting dienes provided the targeted heterocycles. Aldrich utilized this strategy to construct a piperidinone inhibitor of biotin synthesis. Richards was able to synthesize a number of functionalized enantioenriched piperidines, pyrrolidines, quinolizidines, and indolizidines, including the natural products anisomycin and coniine. Sutherland also employed a palladium-catalyzed rearrangement/RCM strategy on dienol starting materials to construct amine-substituted carbocycles that were then elaborated to polyhydroxylated aminocarbocycles.[22]

Han:

Scheme 4. Applications of [COP-Cl]$_2$-catalyzed rearrangements in synthesis

(II) Allylic Substitution Reactions

One other important use of the COP family of catalysts is to promote enantioselective S$_N$2′-type reactions of prochiral allylic imidates. Our original report of the intermolecular allylic esterification motivated us and others to pursue the utility of other nucleophiles in the enantioselective synthesis of branched allylic substitution products.[9c,d] In general, successful nucleophiles needed to be relatively acidic (pK$_a$ <12) in order to participate in the S$_N$2′ reactions. This requirement is particularly important for intermolecular reactivity and presumably results from the need to eventually protonate the imidate leaving group. Furthermore, intermolecular substitutions catalyzed by the COP family of catalysts require the use of the Z-allylic imidate in order to suppress competing [3,3]-

sigmatropic rearrangement by increasing the steric demand of the intramolecular iminopalladation step.

Phenols were quickly found to be particularly useful nucleophiles for the generation of allyl aryl ethers (Scheme 5).[23] S_N2' substitution onto prochiral Z-allylic imidates provided the corresponding allyl aryl ethers in good yields and consistently high enantioselectivities. As in the synthesis of allylic esters, the palladacyclic complex [COP-OAc]₂ was ideal for this transformation of the Z-allylic imidates. In the course of these studies, a new complex, [COP-NHCOCCl₃]₂, was discovered in which the bridging acetate ligands were replaced by bridging trichloroacetamidates.[24] This complex was the first catalyst capable of promoting the intermolecular S_N2' substitutions of E-allylic imidates. For reasons still not well understood, [COP-NHCOCCl₃]₂ is a comparatively poor catalyst for the allylic trichloroacetimidate rearrangement, allowing useful yields of allyl aryl ethers to be produced without the formation of significant amounts of allyl amide byproducts. Unfortunately, this reactivity could not be translated to the allylic esterification, as [COP-NHCOCCl₃]₂ readily converts to a carboxylate-ligated species in the presence of carboxylic acids.[12] These carboxylate complexes, like [COP-OAc]₂, are also competent catalysts for the competing allylic trichloroacetimidate rearrangement.

1 mol %
[(R_p,S)-COP-OAc]₂
ArOH (3 equiv)
—————————————→
CH₂Cl₂, 40 °C
up to 97% yield; >90% ee

1 mol %
[(R_p,S)-COP-NHCOCCl₃]₂
ArOH (3 equiv)
—————————————→
CH₂Cl₂, 40 °C
45–88% yield; 80–98% ee

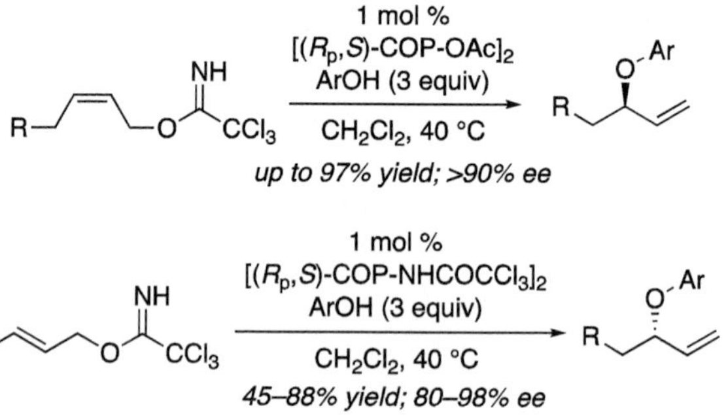

Scheme 5. Enantioselective intermolecular S_N2' reactions with phenol nucleophiles

Catalytic, enantioselective, intramolecular substitution reactions with phenol nucleophiles were also realized with [COP-OAc]₂ (Scheme 6).[25] Vinylchromans and other oxygen heterocycles were synthesized in good yields and enantioselectivities. In particular, E-alkenes were found to be the ideal substrates, as the corresponding Z isomers provided products having significantly reduced enantioselectivities. This intramolecular reactivity allowed two firsts for the COP family of complexes: (1) the stereoselective synthesis of fully-substituted stereocenters in good enantioselectivity, albeit in low yield; and (2) the use of allylic acetates as effective and more atom-economical leaving groups giving allylic substitution products in comparable yields and enantioselectivities. In this latter case, basic potassium fluoride was required to promote the reaction.

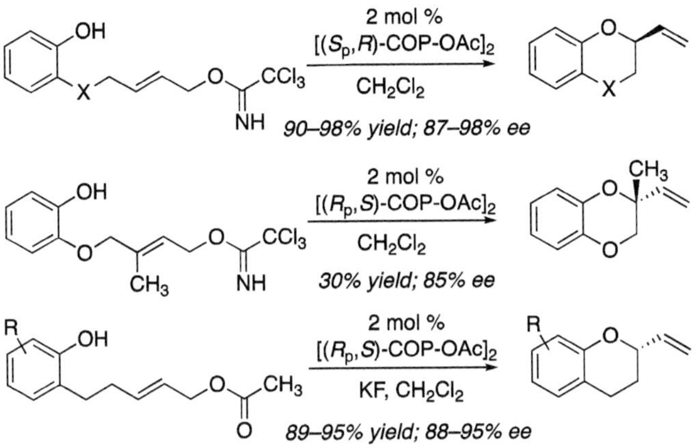

2 mol %
[(S$_p$,R)-COP-OAc]$_2$
CH$_2$Cl$_2$
90–98% yield; 87–98% ee

2 mol %
[(R$_p$,S)-COP-OAc]$_2$
CH$_2$Cl$_2$
30% yield; 85% ee

2 mol %
[(R$_p$,S)-COP-OAc]$_2$
KF, CH$_2$Cl$_2$
89–95% yield; 88–95% ee

Scheme 6. Enantioselective synthesis of 2-vinyl substituted oxygen heterocycles

Jirgensons also reported a COP-catalyzed intramolecular substitution of allylic bis-trichloroacetamidates (Scheme 7).[26] In this case, the *E*-alkene stereoisomer of the bisimidate was required to achieve high yield and diastereoselectivity.

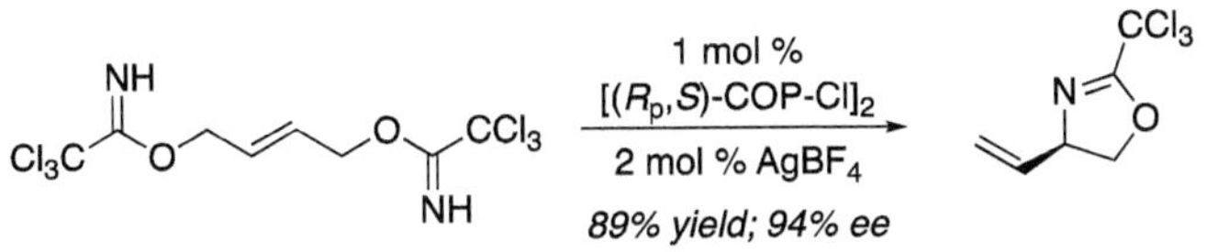

1 mol %
[(R$_p$,S)-COP-Cl]$_2$
2 mol % AgBF$_4$
89% yield; 94% ee

Scheme 7. Synthesis of a vinyloxazoline by a [COP-Cl]$_2$-catalyzed substitution reaction

The allylic substitution reaction has found some use in total synthesis efforts. For instance, Kirsch demonstrated an iterative approach for the synthesis of 1,3-polyols.[27] This strategy utilizes a ring-closing metathesis to generate a 5,6-dihydro-α-pyrone intermediate. Conversion of this lactone to the Z-allylic imidate allowed for a subsequent COP-catalyzed S_N2' reaction to establish the next stereocenter. Kirsch demonstrated that this method could be used to construct all of the possible stereoisomers of 1,3-polyols with high levels of catalyst control, providing programmed access to polyketide scaffolds. Kirsch employed this iterative approach to synthesize polyrhacitides A and B.[28] Kirsch also utilized the [COP-OAc]₂-catalyzed allylic ester synthesis in his syntheses of rugulactone and chloriolide.[27b]

Scheme 8. Iterative strategy for 1,3-polyols utilizing COP-catalyzed S_N2' reactions

Conclusion

Several advances in the utility of the COP family of palladacyclic catalysts have been realized in recent years. These catalysts have proven to be relatively general enantioselective catalysts for both bimolecular and intramolecular allylic substitution reactions involving heteropalladation/deoxypalladation steps. We anticipate that these complexes will continue to find use in similar reactions where mild conditions and high selectivity for forming the branched product are paramount.

References

1. (a) J.S.C.: Department of Chemistry, Occidental College, 1600 Campus Rd. M-5, Los Angeles, California 90041-3314, United States. Email address: jcannon@oxy.edu. (b) L.E.O.: Department of

Chemistry, University of California, 1102 Natural Sciences II, Irvine, California 92697-2025, United States. Email address: leoverma@uci.edu.

2. For a comprehensive review of the synthesis of allylic alcohols, see: Hodgson, D. M.; Humphreys, P. G. In *Science of Synthesis: Houben-Weyl Methods of Molecular Transformations*; Clayden, J. P., Ed.; Thieme: Stuttgart, 2007; Vol. 36, pp 583–665.

3. Lumbroso, A.; Cooke, M. L.; Breit, B. *Angew. Chem., Int. Ed.* **2013**, *52*, 1890–1932.

4. For a comprehensive review of the synthesis of chiral allylic amines see: *Chiral Amine Synthesis* (Ed.: T. C. Nugent) Wiley-VCH, New York, **2008**.

5. McDonald, R. I.; Liu, G.; Stahl, S. S. *Chem. Rev.* **2011**, *111*, 2981–3019.

6. (a) Hollis, T. K.; Overman, L. E. *J. Organomet. Chem.* **1999**, *576*, 290–299. (b) Donde, Y.; Overman, L. E. *J. Am. Chem. Soc.* **1999**, *121*, 2933–2934. (c) Anderson, C. E.; Donde, Y.; Douglas, C. J.; Overman, L. E. *J. Org. Chem.* **2005**, *70*, 648–657.

7. (a) Stevens, A. M.; Richards, C. J. *Organometallics* **1999**, *18*, 1346–1348. (b) For a recent comprehensive review of cobalt sandwich complexes: Kumar, D.; Deb, M.; Singh, J.; Singh, N.; Keshav, K.; Elias, A. J. *Coord. Chem. Rev.* **2016**, *306*, 115–170.

8. (a) Overman, L. E.; Owen, C. E.; Pavan, M. M.; Richards, C. J. *Org. Lett.* **2003**, *5* 1809–1812. (b) Anderson, C. E.; Overman, L. E. *J. Am. Chem. Soc.* **2003**, *125*, 12412–12413.

9. (a) Anderson, C. E.; Kirsch, S. F.; Overman, L. E.; Richards, C. J.; Watson, M. P. *Org. Synth* **2007**, *84*, 148–155. (b) Anderson, C. E.; Overman, L. E.; Richards, C. J.; Watson, M. P.; White, N. *Org. Synth* **2007**, *84*, 139–147.

10. (a) Kirsch, S. F.; Overman, L. E. *J. Am. Chem. Soc.* **2005**, *127*, 2866–2867. (b) Cannon, J. S.; Kirsch, S. F.; Overman, L. E. *J. Am. Chem. Soc.* **2010**, *132*, 15185–15191.

11. Cannon, J. S.; Frederich, J. H.; Overman, L. E. *J. Org. Chem.* **2012**, *77*, 1939–1951.

12. Cannon, J. S.; Kirsch, S. F.; Overman, L. E.; Sneddon, H. F. *J. Am. Chem. Soc.* **2010**, *132*, 15192–15203.

13. (a) Norsikian, S.; Chang, C. W. *Curr. Org. Synth.* **2009**, *6*, 264–289. (b) Onitsuka, K. J. *Synth. Org. Chem. Jpn.* **2009**, *67*, 584–594. (c) Trost, B. M. *J. Org. Chem.* **2004**, *69*, 5813–5837. (d) Trost, B. M.; Crawley, M. L.

Chem. Rev. **2003**, *103*, 2921–2943. (e) Szabo´, K. J. *J. Am. Chem. Soc.* **1996**, *118*, 7818–7826.

14. For a full, recent account of the discovery, use, and proposed mechanisms of action and selectivity, see: Cannon, J. S.; Overman, L. E. *Acc. Chem. Res.* **2016**, *49*, 2220–2231.
15. Overman, L. E.; Roberts, S. W.; Sneddon, H. F. *Org. Lett.* **2008**, *10*, 1485–1488.
16. Mingat, G.; MacLellan, P.; Laars, M.; Clayden, J. *Org. Lett.* **2014**, *16*, 1252–1255.
17. Rodrigues, A.; Lee, E. E.; Batey, R. A. *Org. Lett.* **2010**, *12*, 260–263.
18. Lee, E. E.; Batey, R. A. *J. Am. Chem. Soc.* **2005**, *127*, 14887–14893.
19. Singh, S.; Singh, O. V.; Han, H. *Tetrahedron Lett.* **2007**, *48*, 8270–8273.
20. Shi, C.; Geders, T. W.; Park, S. W.; Wilson, D. J.; Boshoff, H. I.; Abayomi, O.; Barry, C. E., III; Schnappinger, D.; Finzel, B. C.; Aldrich, C. C. *J. Am. Chem. Soc.* **2011**, *133*, 18194–18201.
21. Nomura, H.; Richards, C. J. *Org. Lett.* **2009**, *11*, 2892–2895.
22. (a) Zaed, A. M.; Grafton, M. W.; Ahmad, S.; Sutherland, A. *J. Org. Chem.* **2014**, *79*, 1511–1515. (b) Swift, M. D.; Donaldson, A.; Sutherland, A. *Tetrahedron Lett.* **2009**, *50*, 3241–3244.
23. Kirsch, S. F.; Overman, L. E.; White, N. S. *Org. Lett.* **2007**, *9*, 911–913.
24. Olson, A. C.; Overman, L. E.; Sneddon, H. F.; Ziller, J. W. *Adv. Synth. Catal.* **2009**, *351*, 3186–3192.
25. Cannon, J. S.; Olson, A. C.; Overman, L. E.; Solomon, N. S. *J. Org. Chem.* **2012**, *77*, 1961–1973.
26. Maleckis, A.; Klimovica, K.; Jirgensons, A. *J. Org. Chem.* **2010**, *75*, 7897–7900.
27. (a) Binder, J. T.; Kirsch, S. F. *Chem Commun.* **2007**, 4164–4166. (b) Kirsch, S. F.; Klahn, P.; Menz, H. *Synthesis* **2011**, 3592–3603.
28. Menz, H.; Kirsch, S. F. *Org. Lett.* **2009**, *11*, 5634–5637.

Larry Overman was born in Chicago, Illinois, in 1943 and raised in Hammond, Indiana. He obtained a B.A. degree from Earlham College in 1965 and completed his doctoral dissertation in 1969 with Professor Howard W. Whitlock, Jr. at the University of Wisconsin. After a NIH postdoctoral fellowship with Professor Ronald Breslow at Columbia University, he joined the faculty at the University of California, Irvine in 1971 where he is now Distinguished Professor of Chemistry.

Jeff Cannon obtained his B.A. in Chemistry from Occidental College in Los Angeles. He obtained his Ph.D. from University of California, Irvine working with Prof. Larry Overman. After an NIH postdoctoral fellowship with Prof. Robert Grubbs at Caltech, Jeff returned to Occidental as an assistant professor in 2014. Jeff's current research interests are focused on the development of new stereoselective carbon-carbon bond forming reactions inspired by synthetic needs in natural product synthesis.

Author Index Volume 95